W0080677

Somitogenesis

ADVANCES IN EXPERIMENTAL MEDICINE AND BIOLOGY

Editorial Board:
NATHAN BACK, *State University of New York at Buffalo*
IRUN R. COHEN, *The Weizmann Institute of Science*
ABEL LAJTHA, *N.S. Kline Institute for Psychiatric Research*
JOHN D. LAMBRIS, *University of Pennsylvania*
RODOLFO PAOLETTI, *University of Milan*

A Continuation Order Plan is available for this series. A continuation order will bring delivery of each new volume immediately upon publication. Volumes are billed only upon actual shipment. For further information please contact the publisher.

Somitogenesis

Edited by

Miguel Maroto, PhD
University of Dundee, Dundee, Scotland
Neil V. Whittock, PhD
University of Exeter, Devon, England

Springer Science+Business Media, LLC
Landes Bioscience

Springer Science+Business Media, LLC
Landes Bioscience

Copyright ©2008 Landes Bioscience and Springer Science+Business Media, LLC

All rights reserved.
No part of this book may be reproduced or transmitted in any form or by any means, electronic or mechanical, including photocopy, recording, or any information storage and retrieval system, without permission in writing from the publisher, with the exception of any material supplied specifically for the purpose of being entered and executed on a computer system; for exclusive use by the Purchaser of the work.

Printed in the USA

Springer Science+Business Media, LLC, 233 Spring Street, New York, New York 10013, USA
http://www.springer.com

Please address all inquiries to the publishers:
Landes Bioscience, 1002 West Avenue, Austin, Texas 78701, USA
Phone: 512/ 637 5060; FAX: 512/ 637 6079
http://www.landesbioscience.com

Somitogenesis, edited by Miguel Maroto and Neil V. Whittock, Landes Bioscience / Springer Science+Business Media, LLC dual imprint / Springer series: Advances in Experimental Medicine and Biology

ISBN: 978-0-387-09605-6

While the authors, editors and publisher believe that drug selection and dosage and the specifications and usage of equipment and devices, as set forth in this book, are in accord with current recommendations and practice at the time of publication, they make no warranty, expressed or implied, with respect to material described in this book. In view of the ongoing research, equipment development, changes in governmental regulations and the rapid accumulation of information relating to the biomedical sciences, the reader is urged to carefully review and evaluate the information provided herein.

Library of Congress Cataloging-in-Publication Data

Somitogenesis / edited by Miguel Maroto, Neil V. Whittock.
    p. ; cm. -- (Advances in experimental medicine and biology ; v. 638)
  Includes bibliographical references and index.
  ISBN 978-0-387-09605-6
 1. Somite. I. Maroto, Miguel. II. Whittock, Neil V. III. Series.
  [DNLM: 1. Somites--physiology. 2. Morphogenesis--physiology. W1 AD559 v.638 2008 / WQ 205 S6961 2008]
  QL971.S585 2008
  571.8'636--dc22
                    2008023385

PREFACE

We visualise developmental biology as the study of progressive changes that occur within cells, tissues and organisms themselves during their life span. A good example of a field of developmental biology in which this concept is encapsulated is that of somitogenesis. The somite was identified as the primordial unit underlying the segmented organisation of vertebrates more than two centuries ago. The spectacular discoveries and achievements in molecular biology in the last fifty years have created a gene-based revolution in both the sorts of questions as well as the approaches one can use in developmental biology today. Largely as a result of this, during the 20th and 21st centuries this simple structure, the somite, has been the focus of a deluge of papers addressing multiple aspects of somite formation and patterning both at the cellular and molecular level. One of the main reasons for such interest in the process of somitogenesis stems from the fact that it is such an exquisitely beautiful example of biology working under strict temporal and spatial control in a reiterative manner that is highly conserved across the vertebrate classes.

Our intention is that this book will be of interest to different kinds of scientists, including basic researchers, pathologists, anatomists, teachers and students working in the fields of cell and developmental biology. The nine chapters cover a wide array of topics that endeavour to capture the spirit of this dynamic and ever-expanding discipline by integrating both contemporary research with the classical embryological literature that concentrated on descriptions of morphological changes in embryos and the interactions of cells and tissues during development. In so doing they encompass the main aspects of somitogenesis across four vertebrate classes (frog, fish, mouse and chick) and the hope is that this will enable readers to acquire an appreciation of this developmental process in all its facets. Each of the different animal models offers alternative strategic approaches (including experimental embryology, genetics and cell biology) to tackle the same process and as such each offers an invaluable and unique insight into different aspects of somitogenesis. The topics described in these chapters cover the generation of somitic tissue during gastrulation, the molecular mechanisms by which the unsegmented pre-somitic mesoderm becomes segmented into somites, the generation of polarity within somites and the means by which the somite is directed to differentiate into a number of different cell derivatives. There

are also two chapters devoted to describing the latest developments on relating spontaneous mouse mutations and mutations leading to abnormal vertebral segmentation
in man to the molecular mechanisms already identified as being crucial for somite
formation in the lower vertebrates.

We would like to heartily thank all of the authors that have contributed their time
and effort and whose work has made this book possible. Only they know how difficult
it has been to conclude a book that we started more than four years ago, initially with
another publisher. One year ago the project was cancelled and then re-started again
when the people of Landes accepted for publication. I am especially grateful for
the understanding and patience of Cynthia Conomos and Celeste Carlton who have
worked tirelessly to ensure that this project reached completion. We made it!

Miguel Maroto, PhD
Neil V. Whittock, PhD

ABOUT THE EDITORS...

    MIGUEL MAROTO is a MRC Career Development Fellow and Lecturer at the University of Dundee, UK. He received his PhD in Biochemistry and Molecular Biology from the Department of Biochemistry of the Universidad Autonoma of Madrid, Spain. His research interests include investigating the biochemical basis of different signalling mechanisms implicated in the acquisition of specific cell fates during vertebrate development. In recent years he has been involved in the analysis of the mechanism of the molecular clock in the control of the process of somitogenesis.

ABOUT THE EDITORS...

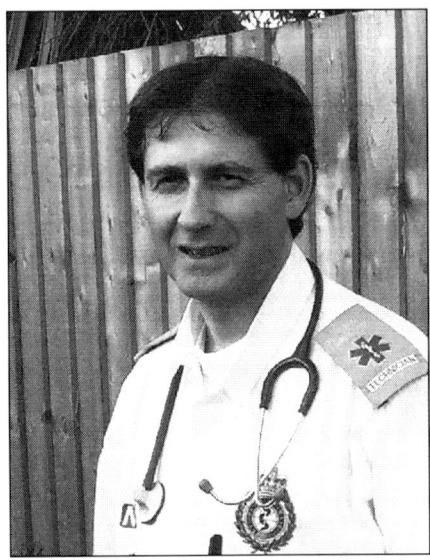

NEIL V. WHITTOCK received his PhD in Human Molecular Genetics whilst working at Guys' and St Thomas' Hospitals in London, UK. His research focussed on developing diagnostic genetic tests for Duchenne muscular dystrophy before moving on to identifying genes involved in bullous skin disorders. He then continued his research as a postdoctoral fellow at the University of Dundee before arriving at the University of Exeter where he spent three years working alongside Dr Peter Turnpenny. The work at Exeter focussed on the identification of genes involved in human genetic disorders that affected the development of the spine and ribs, specifically the spondylocostal dysostoses. He now works as an Ambulance Technician in Devon, UK, and runs his own antique clock restoration business.

PARTICIPANTS

Beate Brand-Saberi
Department of Molecular Embryology
Institute for Anatomy
  and Cell Biology
Freiburg
Germany

Yasumasa Bessho
Graduate School of Biological
  Sciences
Nara Institute of Science
  and Technology
Takayama, Ikoma
Japan

Gavin Chapman
Developmental Biology Program
Victor Chang Cardiac
  Research Institute
University of New South Wales
Sydney, New South Wales
Australia

Bodo Christ
Institut für Anatomie
  und Zellbiologie II
Universität Freiburg
Freiburg
Germany

J. Kim Dale
College of Life Sciences
University of Dundee
Dundee, Scotland
UK

Sally L. Dunwoodie
Developmental Biology Program
Victor Chang Cardiac
  Research Institute
University of New South Wales
Sydney, New South Wales
Australia

Harun Elmasri
Department of Physiological
  Chemistry I Biocenter
University of Wuerzburg
Germany
and
Childrens Hospital
Harvard Medical School
Boston, Massachusetts
USA

Anton J. Gamel
Institute for Anatomy
  and Cell Biology
Department of Molecular Embryology
Freiburg
Germany

Tadahiro Iimura
Stowers Institute for Medical Research
Kansas City, Missouri
USA

Kenro Kusumi
School of Life Sciences
Arizona State University
Tempe, Arizona
USA

Miguel Maroto
College of Life Sciences
University of Dundee
Dundee, Scotland
UK

Megan L. O'Brien
Johns Hopkins Bloomberg School
   of Public Health
Baltimore, Maryland
USA

Stefan Rudloff
Department of Molecular Embryology
Institute for Anatomy
   and Cell Biology
Freiburg
Germany

Yumiko Saga
Division of Mammalian Development
National Institute of Genetics
Mishima
Japan

Martin Scaal
Institut für Anatomie
   und Zellbiologie II
Universität Freiburg
Freiburg
Germany

William Sewell
School of Life Sciences
Arizona State University
Tempe, Arizona
USA

Duncan B. Sparrow
Developmental Biology Program
Victor Chang Cardiac Research
   Institute
Darlinghurst, New South Wales
Australia

Yu Takahashi
Cellular & Molecular Toxicology
   Division
National Institute of Health Sciences
Tokyo
Japan

Peter D. Turnpenny
Peninsula Medical School
Clinical Genetics Department
Royal Devon & Exeter Hospital
Gladstone Road
Exeter
UK

Neil V. Whittock
University of Exeter
Devon, England
UK

Christoph Winkler
Department of Physiological
   Chemistry I Biocenter
University of Wuerzburg
Germany
and
Department of Biological Sciences
National University of Singapore
Singapore

CONTENTS

4. OLD WARES AND NEW: FIVE DECADES OF INVESTIGATION OF SOMITOGENESIS IN *XENOPUS LAEVIS* 73

Duncan B. Sparrow

5. ROLE OF DELTA-LIKE-3 IN MAMMALIAN SOMITOGENESIS AND VERTEBRAL COLUMN FORMATION 95

Gavin Chapman and Sally L. Dunwoodie

6. MESP-FAMILY GENES ARE REQUIRED FOR SEGMENTAL PATTERNING AND SEGMENTAL BORDER FORMATION 113

Yumiko Saga and Yu Takahashi

7. BHLH PROTEINS AND THEIR ROLE IN SOMITOGENESIS 124

Miguel Maroto, Tadahiro Iimura, J. Kim Dale and Yasumasa Bessho

8. MOUSE MUTATIONS DISRUPTING SOMITOGENESIS
AND VERTEBRAL PATTERNING 140

Kenro Kusumi, William Sewell and Megan L. O'Brien

9. DEFECTIVE SOMITOGENESIS AND ABNORMAL VERTEBRAL
SEGMENTATION IN MAN 164

Peter D. Turnpenny

CHAPTER 1

Formation and Differentiation of Avian Somite Derivatives

Bodo Christ and Martin Scaal*

Abstract

During somite maturation, the ventral half of the epithelial somite disintegrates into the mesenchymal sclerotome, whereas the dorsal half forms a transitory epithelial sheet, the dermomyotome, lying in between the sclerotome and the surface ectoderm. The dermomyotome is the source of the majority of the mesodermal tissues in the body, giving rise to cell types as different as muscle, connective tissue, endothelium and cartilage. Thus, the dermomyotome is the most important turntable of mesodermal cell fate choice in the vertebrate embryo. Sclerotome development is characterized by a cranio-caudal polarization, resegmentation and axial identity. Its formation is controlled by signals from the notochord, the neural tube, the lateral plate mesoderm and the myotome. These signals and cross-talk between somite cells lead to the separation of various subdomains, like the central, ventral, dorsal and lateral sclerotome. Here, we discuss the current knowledge on the formation of the dermomyotome and the mechanisms leading to the development of the various dermomyotomal derivatives, with special emphasis on the development of musculature and dermis. We further discuss the molecular control of sclerotomal subdomain formation and cell type specification.

Introduction

During embryonic development, the nascent organism develops from the unicellular zygote into the fully developed body, which is composed of a multitude of different tissues intricately arranged as functionally interacting organs. Vertebrate embryos differ greatly in their early developmental stages, e.g., the gastrulae of amphibia and mammals, but their developmental program converges in the bottleneck of the pharyngula, which represents the general Bauplan of the vertebrate body and is therefore designated as the vertebrate phylotypic stage, before it diverges again to build the diverse anatomical concepts realized within the chordate phylum.[1,2]

The vertebrate phylotypic stage is characerized, among other features, by the segmental organization of the paraxial mesoderm which is arranged as metameric balls of epithelial spheres aligned along both sides of the axial organs, the notochord and the neural tube (Fig. 1). These segmental portions of paraxial mesoderm, the somites, give rise to the majority of mesodermal organs of the body wall. The term "somite" was introduced by Balfour[3] to characterize these segmental units of the paraxial mesoderm which had been previously described as protovertebrae.[4] The first detailed description of the avian somite was given by Williams.[5]

The somites bud off from the presomitic mesoderm (segmental plate) in cranio-caudal direction.[6,7] Prior to morphologically overt segmentation, the cranial part of the presomitic mesoderm undergoes a mesenchymal-to-epithelial transition (MET), with the exception of the cells located

*Corresponding Authors: Bodo Christ, Martin Scaal—Institut für Anatomie und Zellbiologie II Universität Freiburg, Albertstr. 17, 79104 Freiburg, Germany.
Email: bodo.christ@anat.uni-freiburg.de, martin.scaal@anat.uni-freiburg.de

Somitogenesis, edited by Miguel Maroto and Neil V. Whittock. ©2008 Landes Bioscience and Springer Science+Business Media.

inside which retain their mesenchymal organization to form the cells of the somitic core, the somitocoele cells. The other cells rearrange to surround these core cells as an epithelial ball. The somites constitute a metameric pattern within the embryonic body wall that determines the segmental arrangement of the vertebral column, ribs, muscles, tendons, ligaments, dorsal root ganglia, peripheral nerves and blood vessels.[8, 9] The metamerism of these structures is the prerequisite for the ability of the vertebrate body to perform bending and rotating movements.

Each somite gives rise to various cell lineages such as myocytes, chondrocytes, fibrocytes and other cell types which differentiate within different subdomains of the somite.[10,11] If the early epithelial somite is subjected to a dorso-ventral rotation[12] or to an exchange of the dorsal and ventral moieties[13,14] or medial and lateral moieties[15] the somite develops normally corresponding to the adjacent nonrotated somites indicating that the axes of these somites are still labile and the cells are still multipotent. These results clearly show that environmental signals control the formation of the dermomyotome and sclerotome as well as their subdomains and the differentiation program of their cells.[6,16] The determination of the cranio-caudal sclerotome polarization, on the other hand, occurs already in the presomitic mesoderm and is mediated by the Delta-Notch signaling pathway. Lateral to the paraxial mesoderm lies the intermediate and lateral plate mesoderm and dorsally the surface ectoderm is forming the primitive integument. The ontogenetic conservation of this embryonic Bauplan is crucial, because the specification and coordinate development of the great variety of derivatives originating from the somites is regulated by a complex interplay of signals secreted by neighboring structures. The term "sclerotome" was created by Hatschek[17] who worked as an anatomist at the Charles University of Prague. Hatschek corrected the assumption by His[18] that the cells of the sclerotome only contribute to the formation of the aortic wall.

In the first part of this review, we will focus on the dermomyotome, which represents the dorsal portion of the developing somite and we will give an overview on the current knowledge about the formation and subsequent differentiation of this pivotal embryonic structure into dermis, muscle, endothelium and cartilage. In the second part, we will address the formation of the sclerotome from the ventral half of the epithelial somite and from the mesenchymal somitocoele cells. We will describe the development of distinct sclerotomal subdomains and their subsequent differentiation.

Dermomyotome

Morphogenesis of the Dermomyotome

Somites develop progressively in caudal to cranial direction, so that the caudalmost somites represent the youngest stages and the oldest stages are gradually aligned more cranially. Accordingly, Christ and Ordahl[6] established a dynamic staging system denominating the youngest somite as somite number I and the cranially abutting somites in consecutive roman numbers.

Up to somites III to IV, somites are organized as epithelial spheres enclosing a lumen filled with somitocoele cells. The outer surface of these spheres is formed by the basal pole of the somitic epithelium and is covered with a basement membrane, while the apices of the somite cells line the somitocoele. Shortly after the deepithelialization of the sclerotome, from stage V to VII, the remaining dorsal somitic epithelium stretches to form a slightly oblique epithelial sheet that roofs the sclerotomal mesenchyme dorsolaterally, the dermomyotome (Fig. 2).

In dorsolateral view, the dermomyotome adopts a more and more rectangular, sheet-like form, thus abutting closely to the dermomyotomes of the adjacent segments from which it is separated by a thin intersegmental artery. All four margins of the dermomyotomal sheet bend in slightly to form four lip-like structures that bulge towards the underlying sclerotome. As will be detailed later, these dermomyotomal lips are important morphogenetic centers for dermomyotomal growth and myotome formation. The central portion of the dermomyotome in between the marginal lips has been classically referred to as dermatome, as it contributes to the dermis of the back. However, because the origin of the dermis is more complex and not entirely established and the central dermomyotomal cells also gives rise to connective tissue, muscle and endothelium of the back,[6]

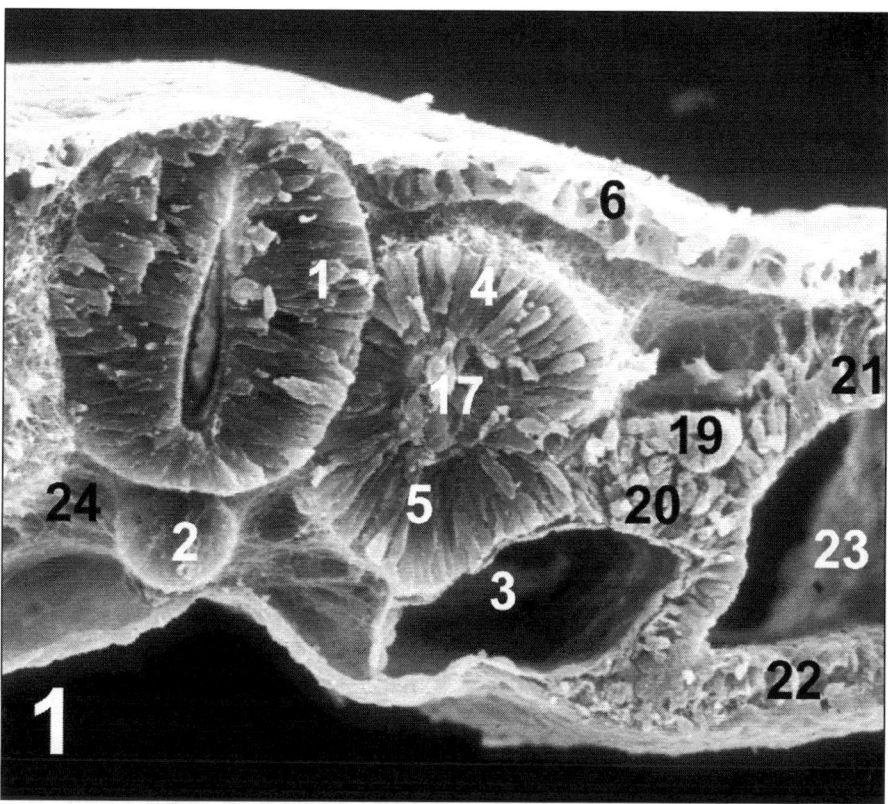

Figure 1. Scanning electron microscopic view on a transverse fracture of a 2-day chick embryo showing an epithelial somite. (Courtesy of Dr. Heinz Jürgen Jacob, Bochum) Explanation of figure numbers of all figures in this chapter. 1) Neural tube, 2) Notochord, 3) Aorta, 4) Dorsal somite half, 5) Ventral somite half, 6) Surface ectoderm, 7) Dermomyotome, 8) Myotome, 9) Central sclerotome, 9a) Cranial half of the sclerotome, 9b) Caudal half of the sclerotome, 10) Ventral sclerotome, 11) Lateral sclerotome, 12) Dorsal sclerotome,13) Spinal nerve, 14) v. Ebner's fissure 15) Meningotome, 16) Syndetome, 17) Somitocoele cells/Arthrotome, 18) Dermis, 19) Wolffian duct, 20) Intermediate mesoderm, 21) Somatopleure, 22) Splanchnopleure, 23) Coelomic cavity, 24) Paranotochordal extracellular matrix, 25) Meninx primitiva, 26) Intersomitic blood vessels, 27) Limb muscle, 28) Intercostal/Abdominal muscle, 29) Dermis, 30) Dorsal root ganglion.

this term should be omitted in favour of central dermomyotome (CD). The epithelial nature of the dermomyotome depends on beta-catenin activity, which is induced by ectodermal Wnt 6 and mediated by the bHLH-transcription factor paraxis.[16]

Later, the CD disintegrates into a loose mesenchyme, the cells of which migrate dorsally as dermal and subcutaneous precursor cells and ventral into the myotome as connective tissue precursors integrating into the epaxial musculature. In contrast, the dermomyotomal lips stay epithelial to enable ongoing myotomal growth. The DML is maintained by the mesoderm-intrinsic epithelialization factor Wnt 11, the VLL by Wnt 6 signaling from the overlying ectoderm which is downregulated over the medial and central dermomyotome by the inhibiting influence of Wnt 11.[20] Finally, the DML and VLL also disintegrate at around embryonic day 7 after the primary pattern of the trunk musculature has been laid down.[21,22]

Initiation and Growth of the Dermomyotome

The dermomyotome is formed from the dorsal half of the epithelial somite, whereas the ventral half-somite gives rise to the sclerotome. Dermomyotomal fate is thus determined by dorsalizing signals from adjacent tissues. In vivo and in vitro experiments in the chick embryo have established that the dorsalizing signals are Wnts secreted by the dorsal neural tube and the dorsal surface ectoderm (reviewed in ref. 16) By long-range diffusion, the dorsal Wnt signals reach the dermomyotomal cells, where they are available for receptor binding due to the desulfation of cell-surface heparan sulfate proteoglycans by the sulfatase Qsulf1,[23] which is not active in ventral somitic cells. Moreover, dorsalizing Wnt signaling are suppressed in ventral somitic cells by the Shh-dependent Wnt-inhibitor Sfrp2,[24] while the ventralizing Shh signal is blocked in the dorsal somite by the membrane-bound glycoprotein Gas1.[25] Within the dorsal compartment, medial and lateral dermomyotomal cells are specified differently. Medial dermomyotome formation depends on Wnt 1 and Wnt 3a from the dorsal neural tube,[13,26-31] whereas the lateral dermomyotome requires contact-mediated signaling by Wnt6 and possibly other Wnt proteins from the surface ectoderm.[27,32,28,31] The specification of dorsomedial and dorsolateral fate seems to be competitive, as in mice lacking both dorsomedial signals, Wnt 1 and Wnt 3a, the medial compartment of the dermomyotome is not formed, whereas the lateral domain, marked by *Sim 1*, is extended medially.[30] Furthermore, Wnt3a seems to be a positive regulator of the proliferation actvity of dermomyotomal cells, which acts upstream of *Pax3* and *Pax7*. Overexpression of *Wnt3a* leads to a mediolateral expansion of the dermomyotomal epithelium due to increased proliferation.[33]

However, the mode of dermomyotomal growth is still under debate. Experiments using quail-chick transplantations and BrdU-labelling of the dermomyotomal lips identified the medial and lateral lip of the dermomyotome as blastema-like centers of mediolateral dermomyotomal growth,[21,22] while DiI-labelling experiments suggested a growth contribution of the entire dermomyotome.[34] The epaxial vs. hypaxial identity of the dermomyotomal cells, which is marked by *En 1* and *Sim 1* expression, respectively, is maintained during the subsequent cell migration processes in the mature somite.[35]

Muscle

Skeletal muscle represents the bulk of the total body mass in vertebrates. In contrast to the smooth musculature of the inner organs, which is of splanchnopleural origin, all skeletal muscles of the vertebrate body originate from the dermomyotome, with the exception of the branchiomeric and ocular muscles which originate from the unsegmented cranial and prechordal mesoderm. Muscle is the phylogenetically oldest derivative of the somites, as a rigid skeleton and a tough dermis were not of primary importance in early aquatic vertebrates (reviewed in.[11]) Still, in modern anamniotes, the myotome represents the largest and earliest formed somitic domain,[37,36] whereas in amniotes, the myotome develops relatively late. Here, we will give an overview on the current knowledge on the initiation of myogenesis and early steps of muscular differentiation in somites.

Initiation of Myogenesis

During gastrulation, the paraxial mesoderm arises from cells of the cranial primitive streak and Hensen's node. Cells determined for myogenic differentiation have been identified from the onset of gastrulation.[38] Cell labelling experiments in mouse suggest that, within the primitive streak, a population of self-renewing stem cells gives rise to spatially coherent paraxial muscle precursor cells, which migrate as cohorts to their destination in the forming paraxial mesoderm.[39] Thus, cells originating from the rostralmost primitive streak contribute largely to epaxial muscle, whereas cells from slightly more caudal regions predominantly give rise to hypaxial muscle.[40-42] However, even prior to gastrulation, chick epiblast cells express the muscle regulatory factor (MRF) MyoD[43] and undergo myogenesis when cultured in protein-free medium,[44] suggesting that myogenesis might be the default differentiation program of pregastrulation epiblast cells. In spite of the spatial coherence of clonal myogenic precursor cells during normal development, classical rotation experiments in chick have shown that all cells of the segmental plate and early somites are competent to differentiate into muscle when positioned appropriately.[12, 13,45] This led to the

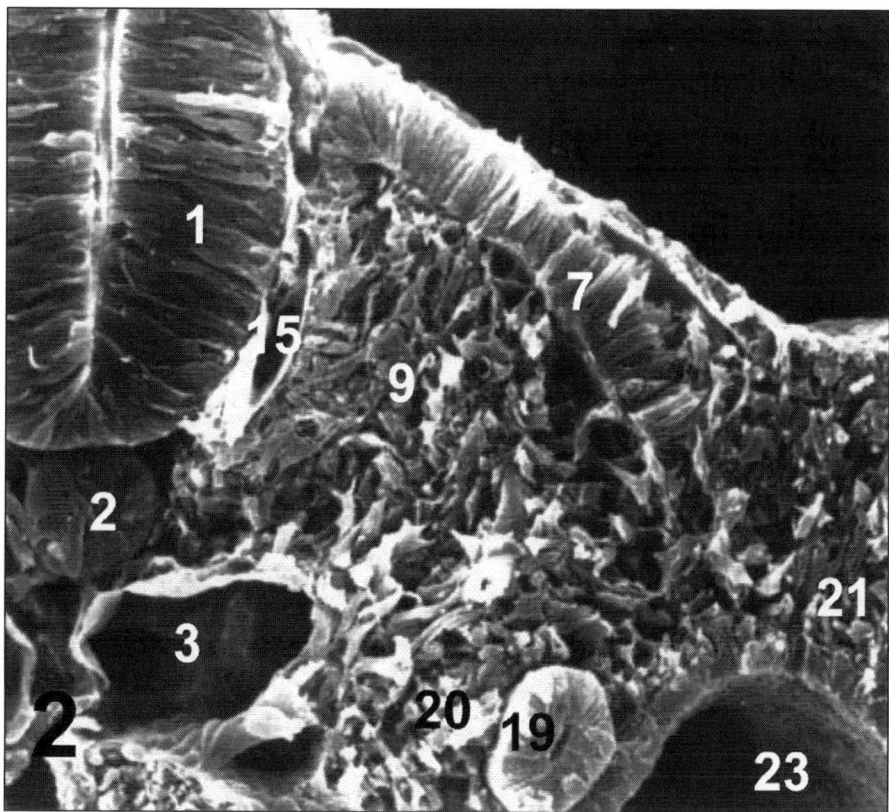

Figure 2. Scanning electron microscopic view on a transverse fracture of a 3-day chick embryo showing the somite derivatives: dermomyotome and sclerotome. (Courtesy of Dr. Heinz Jürgen Jacob, Bochum). See Figure 1 for legends.

assumption that the naive paraxial mesoderm is specified for the myogenic lineage by inductive influences from neighboring tissues.

Ablation experiments identified three structures which are required for myogenesis: The neural tube,[13,46-50] the notochord-floor plate-complex[48,51] and the dorsal ectoderm.[8,50] In molecular terms, it appeared that a delicately balanced level of both, Shh from the notochord and Wnts from the dorsal neural tube and the ectoderm, acts to elicit myogenesis, while high levels of Shh or Wnt alone lead to sclerotomal or nonmyogenic dermomyotomal differentiation, respectively.[52,53] Shh seems to activate dermomyotomal *Myf5* expression via zinc-finger transcription factors of the Gli family, which can bind to a proximal *Myf5* epaxial enhancer[54,55] The view that Shh is involved in the induction of myogenesis has been challenged by the finding that mice lacking Shh show clear, albeit low, expression of the earliest MRF, *Myf5*.[56] However, in the absence of Shh, somitic cells stop to proliferate and eventually die, suggesting that Shh is more likely to be a trophic or maintenance factor that keeps the myogenic program upright, rather than an inducer of myogenesis.[57-59] Likewise, removal of the Wnt-producing dorsal neural tube does not impede the inititation of muscle marker gene expression, but rather the subsequent myogenic differentiation program.[60] In a recent paper, Marcelle and coworkers provide further evidence that myogenesis is a mesoderm-autonomous process, as the segmental plate mesoderm expresses *Myf5* and *MyoD* independent of any neighbouring tissue in vivo and in vitro.[61] They propose that, in the segmental plate, *MyoD* expression is induced by Wnt5b secreted within the segmental plate itself, while *Myf5* does not seem to depend on any inductive signal. Thus, there is accumulating evidence that the myogenic program is intrinsic to the paraxial mesoderm, while its localized realization depends on permissive signals from the embryonic environment.

Once the myogenetic program is initiated, how is it realized? The first molecular manifestation of muscle differentiation is the expression of MRFs. In mouse, the first known MRF to be expressed in myogenic cells is *Myf5* followed by *MyoD* (Fig. 3), whereas in chick the chronology is vice versa. Knockout experiments in the mouse have shown that Myf5 and MyoD have redundant functions, but mice lacking both MRFs lack all skeletal muscles, including the myoblast precursor cells in the paraxial mesoderm.[62] Murine Myf5 is able to induce expression of *MyoD* and is in turn induced by Pax-3, rendering Pax-3 a key regulator of somitic myogenesis in mouse, which can induce the myogenic program even in the absence of surrounding tissues.[63,64] A similar role of Pax-3 in avian myogenesis, or on the putative myogenic inducer *Wnt5b*,[61] remains elusive. In addition to the positive regulation by Pax-3 and Myf5, myogenesis is negatively regulated by BMPs.[65] The onset of muscle differentiation in the medial epithelial somites[66,672] is likely to be due to the release from inhibitory BMP signaling by synthesis of the BMP-antagonist Noggin in the cranial segmental plate and the dorsomedial somite.[68,61]

Early Myotome Development

The first morphological manifestation of myogenesis in the avian embryo is the fomation of primitive muscle fibers subjacent to the dermomyotome. These primitive myotubes, which are mononucleate, postmitotic from the onset of their extension and span the craniocaudal extent of one somitic segment, represent the myotome (reviewed in.[6]) It is generally accepted that all myotomal cells derive exclusively from the dermomyotome, without any myogenic contribution from the sclerotomal mesenchyme.[69] This requires localized deepithelialization of myogenic precursor cells from the dermomyotomal epithelium and subsequent translocation of the myogenic precursor cells into the myotomal domain. The mechanisms of this myotomal cell recruitment have been the topic of an ongoing debate over many years. Early investigators assumed that the myotome develops from the entire dermomyotome.[18,705] Later, some authors argued that the myotome forms only from the DML,[71,72] while others believed that the myotome is of sclerotomal origin.[73] More recently, using DiI-cell labelling techniques, the laboratories of Ordahl and Kalcheim came to conflicting results concerning localization and morphogenetic behavior of myotomal precursor cells.[74-82,21,22,16] Briefly, Ordahl and coworkers found that the early myotome is exclusively formed by cells from the dorsomedial (DML) and ventromedial (VLL) lip of the dermomyotome, which enter the myotomal domain by direct ventral translocation after deepithelialization and subsequently elongate bidirectionally towards the cranial and caudal margins of the segment. In this view, the DML and VLL are blastema-like growth zones enabling the mediolateral growth of both, the myotome and the dermomyotomal epithelium, in an incremental mode of growth. In contrast, Kalcheim and colleagues claimed that the earliest myotomal precursor cells first migrate cranially to the craniomedial corner of the dermomyotome. From there, they elongate as "pioneer cells" unidirectionally in caudal direction until they reach the caudal margin of the segment. Thus, they form a scaffold for later immigrating myoblasts over the entire mediolateral extent of the dermomyotome. According to this model, there is no direct contribution of the DML or VLL to myotomal growth, but ongoing intercalatory addition of myotubes, originating from precursor cells in the cranial and caudal dermomyotome, to existing fibres, resulting in mediolateral myotomal growth. For the later myotome, both models agree on the contribution of cranial and caudal dermomyotomal lips to myotomal growth as was previously suggested by Christ et al.[69] Recently, Marcelle and coworkers have presented a model based on GFP labelling of myogenic precursor cells via electroporation in ovo,[83] which provides an unequivocal model for the mechanics of avian myotome formation and represents a synthesis of aspects of both earlier hypotheses.[84] They demonstrated that in a first step, myotomal cells are only provided by the DML by direct ingression and bidirectional extension. In a second step, firstly the caudal dermomyotomal border, secondly the cranial dermomyotomal border and lastly the VLL, also start to release myotomal precursor cells. The cells originating from the DML and VLL contribute exclusively to the epaxial

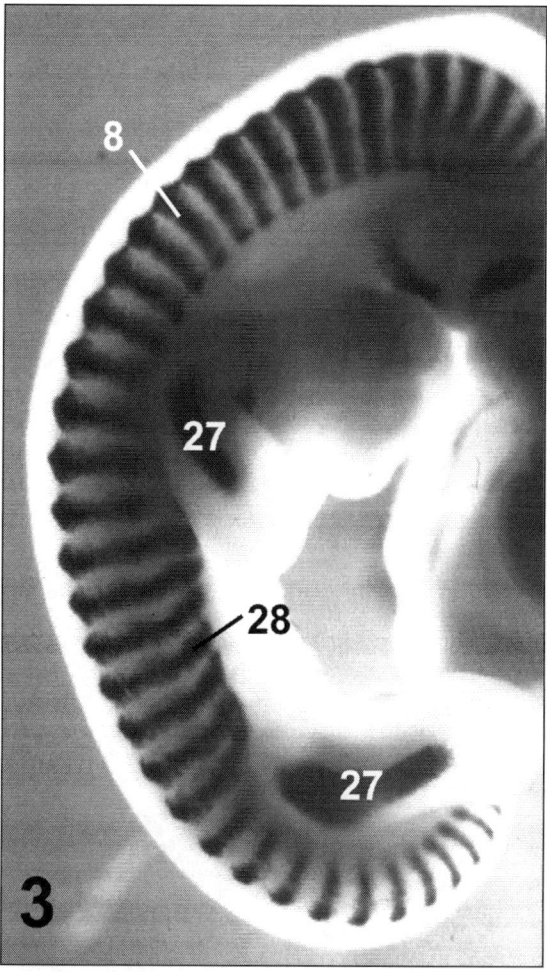

Figure 3. In situ hybridization for MyoD in a 4-day chick embryo. MyoD expression can be seen in the segmentally arranged myotome, the limbs, the tongue and the branchial arches. (Courtesy of Dr. Ketan Patel, London).See Figure 1 for legends.

and hypaxial domain, respectively, whereas the cells from the cranial and caudal margins populate both mediolateral compartments. Thus, all four margins of the dermomyotome provide a distinct contribution to the formation and growth of the myotome, in a process combining incremental growth from the DML and VLL and intercalatory growth from the cranial and caudal dermo-myotomal borders. Prior to translocation into the myotome, mitotically active dermomyotomal cells change the plane of cell division from planar to apico-basal.[85,86] The basal daughter cells of these asymmetric divisions give rise to dermal precursor cells, the apical daughter cells express N-cadherin and give rise to the myotomal lineage.[87,86]

Late Myotome Development

Approximately two days after the onset of myotome formation, ingression of myogenic cells from the dermomyotome ceases as the dermomyotome has started to dissociate. Thenceforward, growth of the myotome-derived musculature depends on the continuous contribution of myogenic cells from a proliferative pool of precursor cells, which are thought to express the transmembrane tyrosine kinase receptor FGFR4 (FREK).[88,89,78] The late recruited myotomal cells and satellite precursor cells, have been shown to originate from the mature dermomyotome.[90] However, the mechanistic and molecular processes during later myotomal growth are still largely unknown. During later stages of myogenesis, the mononucleate, unisegmental myotomal muscle fibres become multinucleate myotubes, probably by fusion with satellite cells. In the epaxial domain, the myotubes give rise to the intrinsic muscles of the back and, in the superficial strata, fuse longitudinally to bridge the segment borders. A similar process is thought to form the abdominal and intercostal muscles in the hypaxial domain (see below).

Hypaxial Muscle Formation

Within the hypaxial compartment of the dermomyotome, the myogenic lineage develops in two fundamentally different modes, according to the axial level of the somites: At cervical and trunk levels, the VLL organizes the formation of a hypaxial myotome similar to the situation at the epaxial DML (see above). The lateral growth of both, the dermomyotome and the hypaxial myotome, are coordinated such that the blastema-like VLL extends ventrolaterally into the prospective lateral and ventral body wall, forming in its wake the ventrolateral myotome, which will give rise to the abdominal and intercostal muscles. The molecular regulation of this process is still obscure, the commitment of VLL cells to the nonmigratory myogenic lineage seems to depend on *Paraxis*.[91]

In contrast, at limb level, the cells of the lateral dermomyotome do not form a VLL, but deepithelialize and emigrate into the nascent limb buds to form appendicular, shoulder and hip muscles (Fig. 4).[92,93] Prior to the experimental proof by Christ and coworkers, which was based on quail-chick chimeras, observations by Fischel,[94] Murray[95] and Grim[96] already argued for a somitic origin of the wing musculature, but were challenged by conclusions by other workers who claimed a somatopleural origin of wing muscles.[97-99]

The lateral plate mesoderm in the limb field secretes the glycoprotein SF/HGF, which activates the *c-met* receptor in the hypaxial dermomyotomes of adjacent somites and thus elicits localized depithelialization of myogenic precursor cells at limb level.[100] Mice deficient for *SF/HGF* or the *c-met* receptor lack limb muscles because muscle precursor cells do not delaminate from the hypaxial dermomyotome.[101,102] The same phenotyp is seen in mutants lacking *Pax-3*, which is essential for *c-met* transcription in the dermomyotome.[103,104] Following the delamination, the onset of the emigration of precursor cells into the limb mesenchyme depends on the expression of the homeo-domain containing transcription factor Lbx1. In *Lbx1* mutants, the deepithelialized precursor cells stay close to the somite of origin and are unable to enter the dorsal premuscular mass in the limb. Intriguingly, the emigration to the ventral premuscular mass seems to be unaffected.[104] During their migratory route, the myogenic cells express *Msx1*,[105-107] *Pax3*[108] and the Wnt-antagonist *Sfrp2*[109] and are guided on their trajectory by SF/HGF secreted by the stationary limb mesenchyme.[110,111] Once in their target location, the precursor cells stop to proliferate in response to FGFR4 signaling[112] and activate the myogenic programme via *MyoD* and *Myf5*,[113] the latter upon induction by ectodermal Wnt6.[114,115] Prior to Wnt-induced myogenesis, the Wnt antagonist *Sfrp2* is downregulated in the myoblasts.[109] The subsequent differentiation of the myoblasts into myofibers depends on multiple factors including Myogenin, Mef2 and MRF4.[116]

Similar to the situation in the limb field, in the hypaxial domain of the cervical somites 2 to 6, a population of migratory cells detach from the VLL and emigrate cranially to form the tongue muscle.[117-119]

Dermis

The dermis, which is also known as corium, is a layer of fibrous connective tissue between the superficial epidermis and the underlying subcutis of the amniote integument. The taut yet elastic structure of the skin and the formation of skin-appendages like hairs, feathers and scales, depends on the proper differentiation of dermal tissue and its interactions with the overlying epidermis. Of all somite derivatives, the development of the dermis is least understood. Here, we will summarize

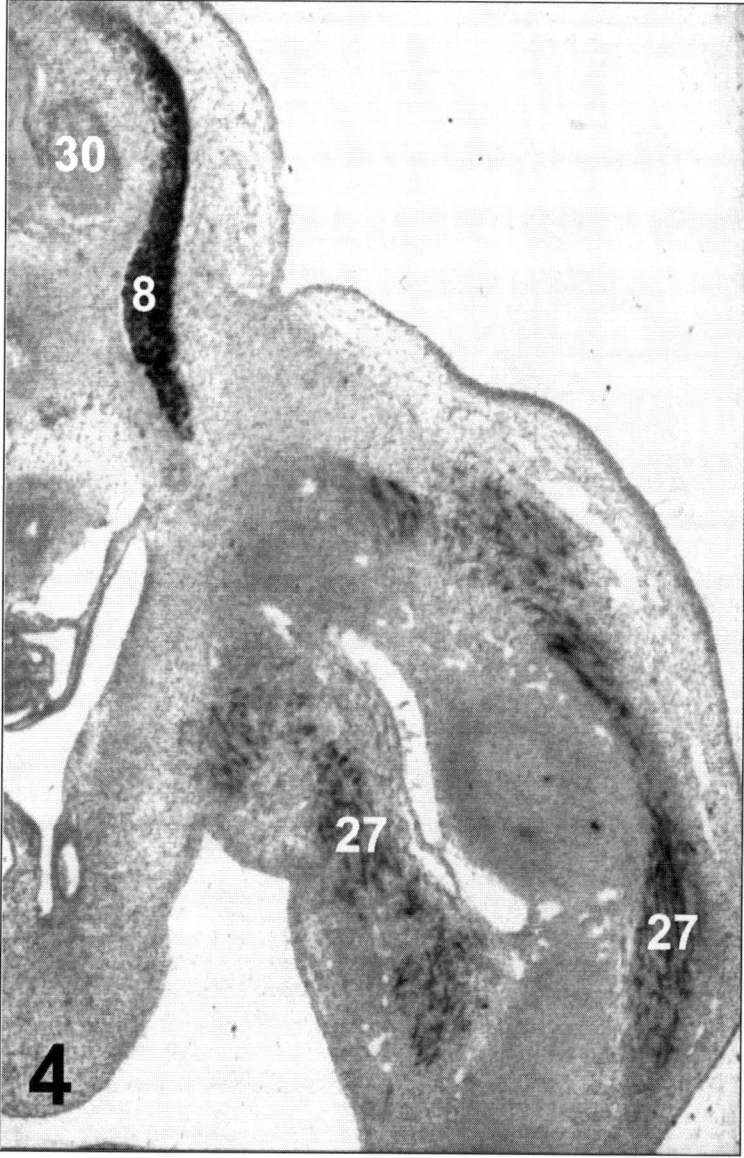

Figure 4. Transverse section of a 4-day chick embryo at limb level. Muscle anlagen in the epaxial and hypaxial domains are stained with an anti-desmin antibody. See Figure 1 for legends.

the current knowledge on the embryonic origin of dermal precursor cells in the dermomyotome and in other structures and give an overview on dermal histogenesis.

Origin of Dermal Precursor Cells

In the 19th century, embryologists realized that skin tissue is of dual developmental origin, the dermis deriving from the mesoderm as opposed to the ectodermal epidermis. Based on the analysis of histological sections, early investigators like Reichert,[120] Remak[4] and Kölliker[121] concluded that the dermis of the ventral and lateral body wall derives from the somatopleure. However, they shared the erroneous view that also the dorsal dermis is formed from cells of the somatopleure that migrate dorsomedially as coherent sheets or "Hautplatten" (skin plates) until they fuse in the dorsal midline. In 1875, Goette[122] found that the "Bildungsgewebe" or mesenchyme, including the dermal precursor cells, arises from the somites. Minot[123] adopted this concept and detailed that dermis forms from the "outer wall of the mesothelium" of the mature "primitive segments", i.e., from the dermomyotome in modern terminology. These findings, although valid in retrospective, remained entirely hypothetical as they were solely based on descriptive embryology. With the identification of the somites as source of mesenchyme, the notion of the somatopleural origin of the ventrolateral dermis was given up. In contrast, it was now assumed that the dermal mesenchyme formed by the somitic "dermatome" spreads out underneath the totality of the embryonic epidermis to give rise to the dermis of all body regions. This remained textbook knowledge, until, with the advent of experimental embryology, Murray[124,95] demonstrated by grafts of lateral plates to host chorioallantoic membranes that lateral and ventral dermis are indeed formed by the somatopleure, thus re-establishing the primary hypothesis by Remak and colleagues.

Henceforward, the idea that dorsal dermis arises from the somites and ventrolateral dermis arises from the somatopleure, was well established. Nevertheless, the experimental proof that this view is correct was pending until the early 1970ies, when Le Douarin established the heterospecific grafting technique between quail and chick to follow the developmental fate of migrating precursor cells.[125,126] Studies using this technique confirmed clearly that cells giving rise to the dorsal and dorsolateral dermis of the trunk, originate in the dermomyotome (Fig. 8), whereas the ventral and ventrolateral trunk dermis originates from precursor cells in the somatopleure.[127,128] Furthermore, quail-chick chimerization identified cranial neural crest cells as precursors of the dermis of the head and cranial neck.[129] Although Noden[130] argued that a subset of cephalic dermis was formed from the head paraxial mesoderm, Couly et al[131] demonstrated that the dermis of the entire head region, including the subcutis, derives exclusively from the neural crest.

With the dermomyotome being identified as the cradle of dorsal dermis, the question was as to the existence of a specific dermogenic compartment, or dermatome in a strict sense, in the somite. Olivera-Martinez[132] and coworkers replaced lateral and medial halves of presomitic mesoderm with homotopic quail transplants and looked for the presence of quail nuclei in the dermis. They found that dermal derivatives formed only from the medial presomitic mesoderm and thus from the medial halves of the epithelial somites, whereas grafts of the lateral presomitic mesoderm did not yield dermal cells. The subcutis underlying the dermis, which is also of somitic origin, behaved likewise. These results have lately been challenged by the group of Kalcheim.[34] After labelling of cells in medial or lateral locations in the epithelial somites with DiI, these workers identified labelled dermal cells originating from the entire mediolateral extent of the somite, including the lateral half. They found the relative mediolateral position of the somitic precursors to be conserved both in the dermomyotome and in the resulting dermal mesenchyme. Thus, at present, the early fate restriction of somitic dermis precursor cells remains controversial.

Dermal Induction and Specification

While it is still under debate whether the lateral dermomyotome contains dermal precursor cells, the contribution of the medial and central dermomyotome to dorsal dermis is generally accepted from morphological observations. There is accumulating evidence that signals from the neural tube are responsible to specify dermal fate. It is well known that removal of the neural tube leads to cell death and degeneration in the medial dermomyotome. Under these conditions, the dorsal dermis as well as epaxial muscles, both deriving from the medial dermomyotome, are defective or absent.[127,46,58] However, implantation of dorsal neural tube, or of Wnt-1 expressing cells that replace the neural tube, can specifically rescue proper dermis formation, but not the formation of other dermomyotomal derivatives like epaxial muscle.[133] This argues for a specific inductive and/or trophic action of Wnt-1 from the dorsal neural tube on prospective dermal cells in the dermomyotome. An additional ectodermal signal, which seems very likely due to the subectodermal situation of the dermomyotome, has not yet been identified. In the mouse, beta-catenin activity, which is

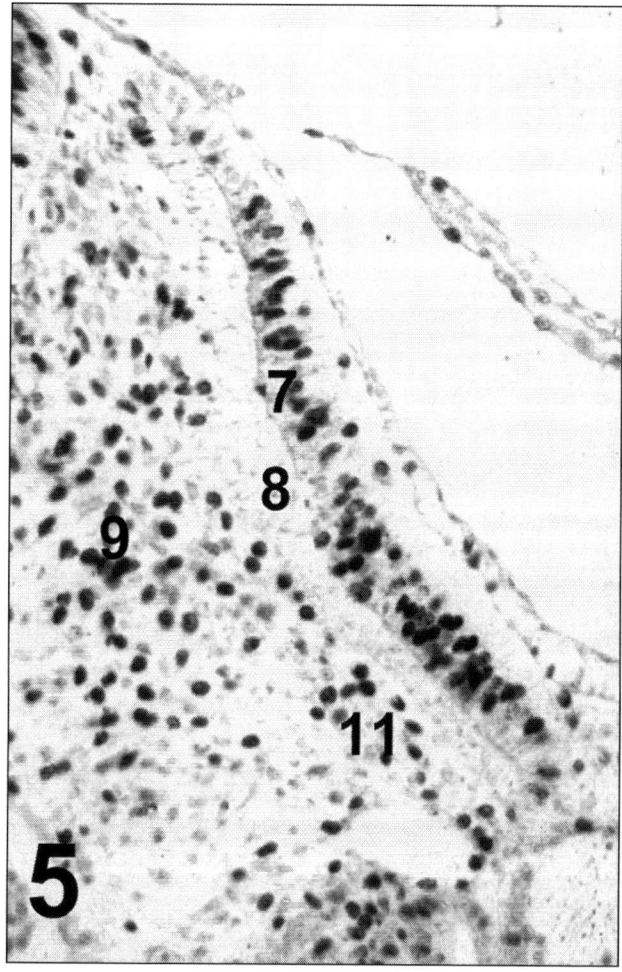

Figure 5. Proliferation study with the BrdU method of a 3-day embryo. See Figure 1 for legends.

downstream of Wnt-signaling, activates the dermal lineage and disrupts the muscle lineage.[134] This indicates that the unknown ectodermal signal might also be a member of the Wnt family.

Epithelio-Mesenchymal Transition of Dermal Precursor Cells

Starting at E3, cells in the CD, that is in between the marginal lips of the dermomyotome, start to deepithelialize and to emigrate towards the overlying ectoderm.[135] This population, which has formerly been designated as dermatome, is generally thought to give rise to dermis. Brill et al[136] have found that shielding of the neural tube at E2.5, when somites have already formed, specifically blocks this deepithelialization process without effects on muscle or cartilage. Moreover, deepithelialization and subsequent formation of subectodermal mesenchyme can be rescued by neurotrophin-3 (NT-3), a growth factor produced by the neural tube. Thus, signals from the neural tube seem to be responsible not only for the specification, but also for the deepithelialization of dermal precursor cells in the somite. Clonal analysis in ovo argues for common precursor cells of both, dermal and muscle fate, in the dermomyotome.[87]

Dermal Precursor Cell Migration

There is accumulating evidence that in addition to the classical "dermatome" in the CD, the dorsomedial lip of the dermomyotome (DML) gives rise to dermal precursor cells, too. The DML is an infolding of the epithelial dermomyotomal sheet in close vicinity to the dorsal neural tube, which is well known to give rise to myotomal cells during somite maturation (reviewed in refs. 6,137) According to a recent model,[21] the DML represents a stem cell population which does not only provide myotomal precursor cells, but also a growing pool of dermomyotomal cells which enable mediolateral growth of the dermomyotomal epithelium. In addition, a third subset of DML-cells might be specified to migrate to the dorsomedial subectodermal space overlying the dorsal neural tube. In contrast to the dermal precursors in the CD, these cells migrate extensively to reach their target location. By quail-chick-transplantations (Ruijin Huang, unpublished data) and by labelling of DML cells with GFP (M.S., unpublished data), offspring of DML cells could be identified in the dorsomedial dermis. Interestingly, the dorsomedial and dorsolateral dermal precursor population differ not only with respect to their migratory behaviour, but also regarding gene expression. Studies in the mouse have identified the homeobox containing gene *Msx-1* as a marker of migratory dermal precursors from the DML[138] *Msx-1* is also expressed in other migratory populations like limb muscle progenitor cells, where it is thought to act as a repressor of differentiation[105-107.] Lateral to the DML, cells in the murine CD and in the respective dermogenic mesenchyme express the EGF-repeat-containing gene *MAEG*.[139] The function of MAEG in dermal morphogenesis is unknown. In the chick, a comparable role of Msx-1 still needs to be determined. In an effort to distinguish the mediodorsal predermal cells in the DML from the dorsolateral cells in the CD, Olivera-Martinez et al[140] showed that a known marker of the DML, *Wnt-11*,[60] continues to be expressed in the dermal precursor cell which migrate dorsomedially. Dermal precursor cells in the CD, instead, do not express *Wnt-11* but *En-1*. These latter cells, however, turn off *En-1* expression after deepithelialization. Interestingly, Olivera-Martinez et al[140] found that only *Wnt-11* expression in the DML depends on Wnt-1 signaling as previously shown,[60] whereas *En-1* expression in the CD depends on an unknown signal from the overlying ectoderm. As both, *Wnt-11* and *En-1* expressing dermomyotomal cells, give rise not only to dermal precursor cells, but also to other lineages like muscle, they are not suitable as dermomyotomal markers of the dermal lineage, all the more as it is controversial if the lateral limit of *En-1* expression corresponds to the lateral border of the dermal precursor pool.[34]

Little is known about the movements of dermal precursor cells after they have left the dermomyotomal epithelium. Dermal cells from the CD express the gene for the matrix-metalloprotease MMP-2 after they have detached from the epithelium, which might facilitate their movements in the subectodermal fibrous substrate.[141] Dermal precursor cells from the DML seem to receive guiding cues from the axial organs, as they have been shown to migrate medially even after reversal of the mediolateral axis of the dermomyotome.[140]

Dermal Histogenesis

In the literature, the whole of the subectodermal mesenchyme is frequently denominated dermis. This is misleading, as the subectodermal mesenchyme contains not only dermal precursor cells, but also cells of prospective subcutis, endothelium and, in dorsomedial locations, the vertebral arches.[11,142] Dermis is a highly differentiated stratum of the integument which is characterized by a specific collagenous matrix, high cell density and multiple distinct subpopulations of fibroblasts.[143]

In chick, until E2.5, dermomyotomes and lateral plates are not in direct contact with the surface ectoderm, but separated by subectodermal extracellular matrix containing a fibrous lattice of mainly collagens and elastins. From E.3 onwards, mesodermal cells invade the subectodermal space as discussed above and colonize it to form a loose mesenchyme.[135] Starting at E.5, a subset of the mesenchyme which is closest to the ectoderm starts to condense and becomes morphologically distinguishable from the underlying subcutaneous mesenchyme by its high cell density from E.6.[144] In featherless regions of the avian integument, mesenchymal condensation starts several days later.[135]

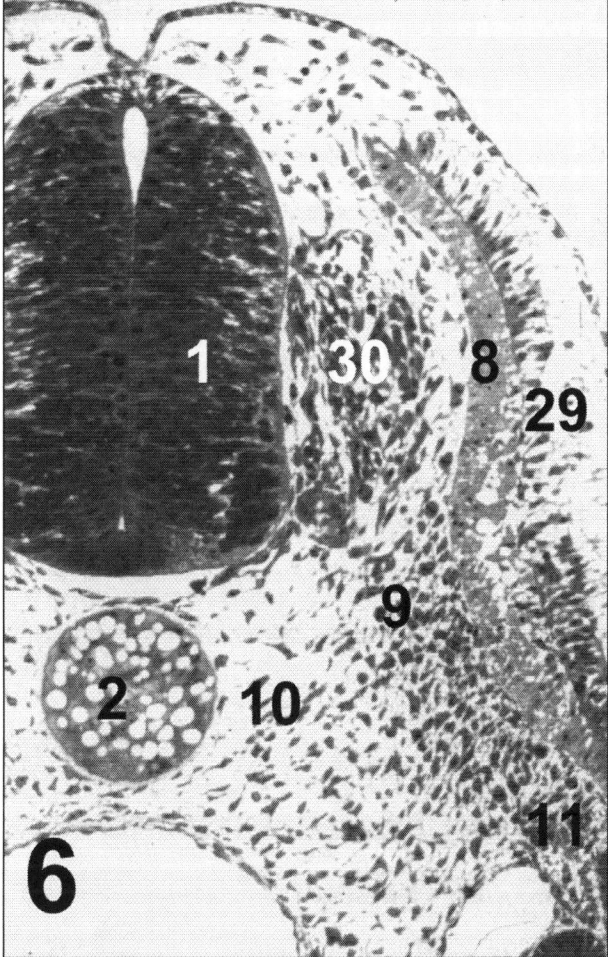

Figure 6. Transverse semithin section of a 3-day chick embryo showing sclerotomal subdomains, epaxial myotome and EMT of dermis precursors. See Figure 1 for legends.

The molecular basis of dermal differentiation is only beginning to be understood. Differentiating dermis is characterized by the expression of the bHLH transcription factor Dermo-1.[145,146] Taking advantage of this marker, Scaal et al[147] demonstrated that BMP2, which is expressed in the ectoderm at the onset of dermal differentiation, induces *cDermo-1* expression and subsequent dermal condensation, which leads eventually to the development of feather tracts. Within the feather tracts, dermal cells accumulate in local spots which represent the future feather buds. Complex regulatory mechanisms involving multiple steps of reciprocal dermo-epidermal signaling lead to the spacing of the feather anlagen in a hexagonal array and the subsequent outgrowth of feathers.[135,148-151]

Dermomyotome-Derived Endothelium

The endothelial lining of blood vessels is formed by migratory precursor cells which originate from two strictly different mesodermal cell pools: The endothelium of vessels in the neural tube, body wall, limbs, kidney and roof of the aorta derive from the paraxial mesoderm, the endothelium of vessels in the visceral organs and the floor of the aorta derive from the splanchnopleura.

In contrast to the paraxial mesoderm, the splanchnopleura also produces hematopoietic cells.[152] Quail-chick-transplantations revealed that the segmental plate and the entire epithelial somite contains angioblasts which develop into endothelial cells. However, according to their location in the somite, angioblasts preferentially populate certain body regions. The ventral half-somite gives rise to the endothelium of ventrolateral blood vessels. Within the dermomyotome, angioblasts from the dorsomedial quadrant migrate predominantly into the dorsal dermis, angioblasts from the dorsolateral quadrant populate the ventrolateral body wall and the limbs.[153] Thus, the angioblastic migratory routes correspond largely with those of other lineages of the same somitic origin, like muscle and connective tissue.[154] Clonal analysis in chick has shown that angioblasts and myoblasts originate, at least in part, from the same precursor cells in the dermomyotome.[155] This indicates that the three distinct cell types migrating from the lateral dermomyotome to the limb bud, blood vascular endothelial cells, myogenic cells and lymphatic cells, are specified by environmental cues of as yet unknown nature.[156]

Dermomyotome-Derived Cartilage

Experiments based on quail-chick-chimaeras have established that the entire chondrogenic lineage giving rise to the postcranial skeleton is of dual origin: The sclerotome gives rise to the axial skeleton and the ribs and the somatopleura gives rise to the limb skeleton including the shoulder and pelvic girdle. However, recent experiments by Huang and coworkers[157] demonstrated that, in birds, there is one skeletal element originating fom the dermomyotome: the scapular blade, which is the caudal portion of the scapula overlying the ribs. In contrast, the cranial elements of the scapula, caput and collum, are of somatopleural origin. Specifically, the dermomyotomes of somites 17 to 24 provide chondrogenic cells and contribute to craniocaudally sequential sectors in the scapular blade which correspond to the segmental level of origin of the precursor cells in the somites. The scapula precursor cells are located in the hypaxial compartment of the dermomyotome and are specified upon BMP4 signaling from the lateral plate mesoderm.[158] Remarkably, prior to differentiation, the scapular chondrogenic precursor cells express *Pax1*, which is otherwise restricted to the sclerotome. These results are in line with the finding of Tajbakhsh et al[159] that dermomyotomal cells prevented to form muscle can adopt nonmuscle fates including cartilage.

Sclerotome

Origin and Initiation of the Sclerotome

The avian epithelial somite is surrounded by a basement membrane which is connected with the adjacent structures by extracellular matrix components. These structures include medially the neural tube and notochord, dorsally the surface ectoderm, ventrally the endoderm and aorta. Laterally, the epithelial somite is connected with the lateral plate mesoderm by a continuous cell layer, the intermediate mesoderm, with the Wolffian duct on its dorsal side.

The first and most striking morphological characteristic of sclerotome formation in the avian embryo is an epithelio-mesenchymal transition (EMT) of the ventro-medial half of the somite. Matrix metalloprotease-2 (MMP-2) plays an essential role in EMT,[141] which is accompanied by a reduction of apical junctions between somitic cells and an expansion of the extracellular space by hyaluronate.[160] The cells acquire numerous filopodia and leave the epithelium, keeping their original polarity during emigration.[161] The EMT does not depend on signals emanating from the neural tube or the notochord. Christ et al[8] and Hirano et al[162] have shown that somites, experimentally separated from both the neural tube and notochord, show a dissociation of the ventro-medial somite wall, thus forming a mesenchymal sclerotome in which, however, the cell number is considerably reduced. Differentiation of the sclerotome does not depend on the preceding epithelial cell organization since in *paraxis* knockout mice, in which somite epithelialization does not occur, sclerotome markers are still expressed.[163,164] It is not quite clear if, in the mouse, prospective sclerotomal cells do also pass through an epithelial configuration, as somitic cells in avian and human embryos do[165]; (G. Steding, Göttingen, personal communication). Ectopically implanted Wnt-producing cells seem to inhibit the formation of the mesenchymal sclerotome.[31] The EMT in the ventro-medial

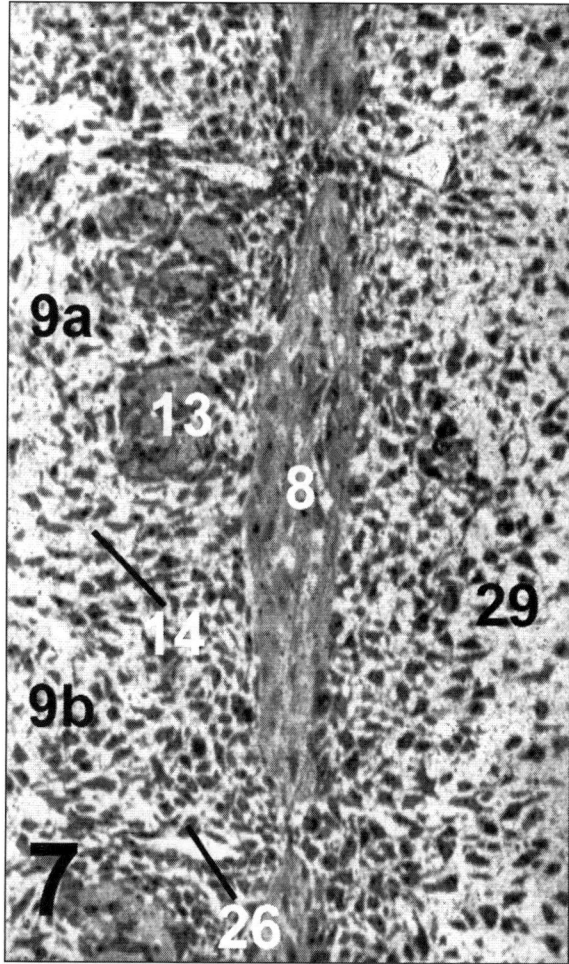

Figure 7. Coronal semithin section of a 4-day chick embryo showing the derivatives of one somite. See Figure 1 for legends.

somitic wall of the avian embryo is acompanied by a down-regulation of N-Cadherin which leads to a decrease in cell-adhesion and an increase in cell motility.[166-168]

The sclerotome is defined not only by its mesenchymal organization, but also by the expression of sclerotome specific marker genes, such as *Pax1* and *Pax9*.[169-171] These genes form one group within the family of nine vertebrate *Pax* genes, which are characterized by the paired box that encodes a DNA-binding domain.[172,173] The 128 amino acid long DNA-binding paired domains of *Pax1* and *Pax9* are almost identical.[171]

Pax1 has been found to be already expressed in the ventro-medial cells of the still epithelial somite.[174-178] In the avian embryo, its expression precedes the EMT, whereas *Pax9* mRNA is not expressed until the sclerotomal mesenchyme has actually formed.[178] Downstream signal response genes of Sonic hedgehog, *Gli2* and *Gli3*, mediate sclerotome induction.[179,180]

It has been shown that the initiation and maintenance of the sclerotome depends on both sonic hedgehog (Shh) and noggin, a BMP4 antagonist, which both are expressed in the notochord at the time when the sclerotome is forming.[32,178,181-184] Shh knock-out mice lack the entire vertebral column.[56] However, *Pax1* expression and sclerotome formation take place in these mice although the sclerotomes are smaller and *Pax1* expression is drastically reduced, indicating that Shh does not initiate but maintain the sclerotome. McMahon et al[184] have shown that Noggin is sufficient to induce low levels of *Pax1* expression and that in homozygous *noggin* mutants *Pax1* expression is delayed. It can therefore be assumed that both signals, Shh and noggin, cooperate during sclerotome formation.[185] After experimental removal of the notochord prior to neural tube patterning, the acquisition of *Pax1* positivity in the ventral part of the somite is lost.[175]

The size of the sclerotome is a result of the balance between dorsal and ventral signals. Dorsal signals promote the development of the dorsally located dermomyotome and suppress sclerotome formation.[30,133] These signals are produced by the dorsal neural tube and the surface ectoderm and are mediated by the Wnt family of signaling molecules and their receptors.[186,187] *Wnt1* and *Wnt3a* are expressed in the dorsal neural tube and *Wnt6* in the ectoderm.[29,30,187-189] Wnt 1, 3a and 4 producing cells ectopically grafted between axial structures and somites, induce an expansion of the dorsal somite compartment, the dermomyotome, at the expense of the ventral compartment, the sclerotome, which is reduced in size and shows a down-regulation of *Pax1*.[31] The ventral signals including Shh and Noggin promote sclerotome formation.[24,25,181,190] The competitive interaction between dorsal Wnts and ventral Shh has also been shown by Lee et al.[24] Wnts are also involved in cell adhesion. The Wnt signal transduction pathway leads to an accumulation of beta-catenin and also stabilizes the binding of free beta-catenin to the cadherins.[191] Schmidt et al[192] have recently shown that Wnt6 plays an important role in the initiation and maintenance of epithelial structure in the paraxial mesoderm.

Moreover, the size of the sclerotome may also depend on a signaling network along the medio-lateral axis, where Bmp4 produced by the intermediate and lateral plate mesoderm[65] becomes antagonized and modulated by medial signals such as Shh, noggin and follistatin produced by the notochord.[184,193,194]

Proliferation of Sclerotomal Cells

When formation of the sclerotome starts, almost all somitic cells are able to replicate. Only a subset of somitic cells in the dorsomedial aspect of the somite is already postmitotic.[75] These cells have been proposed to bend underneath the medial portion of the nascent dermomyotome to form the early myotome. The boundary between these postmitotic cells and the ventrally adjacent cells of the ventral somitic wall represents medially the boundary between the dorsal and ventral compartments of the somite.

All sclerotomal cells are able to undergo mitosis. Using both anti-PCNA staining and BrdU-labelling, Wilting et al[195] were able to show, that the cells of the caudal sclerotomal halves have a higher proliferative activity than those in the cranial halves. These differences in the rate of proliferation could be the cause for the cell condensation in the caudal sclerotome halves and for the formation of the morphologically visible segment polarity of the somites. Within the caudal half of the sclerotome, a gradient in proliferation activity in the central sclerotome is found in a lateral-to-medial direction. Later, a higher proliferation activity can be seen in the ventral sclerotome in connection with the formation of the vertebral bodies and intervertebral discs, where the latter function as proliferation zones providing cells for the adjacent vertebral bodies.

The expression of Pax1 seems to coincide with high proliferative activity.[196] Proliferation of sclerotomal cells has been found to be controlled by different signaling mechanisms. The cells of the neurotome require signals from the axons and neural crest cells to survive and to differentiate into cells of the perineurium and endoneurium. Huang et al[197] have shown that Fgf8 which is normally secreted by a subset of myotomal cells, is involved in the control of proliferation of sclerotomal cells. Proliferation might also be promoted by Fgf-8 secreted by the notochord.[198] Shh produced by the notochord and the floor plate of the neural tube is also important for the proliferation of sclerotomal cells.[181,182,199] Another gene expressed in the sclerotome is Twist, which is a basic helix-loop-helix (bHLH) transcription factor that inhibits muscle and cartilage differentiation.[200-204] Hornik et al[204]

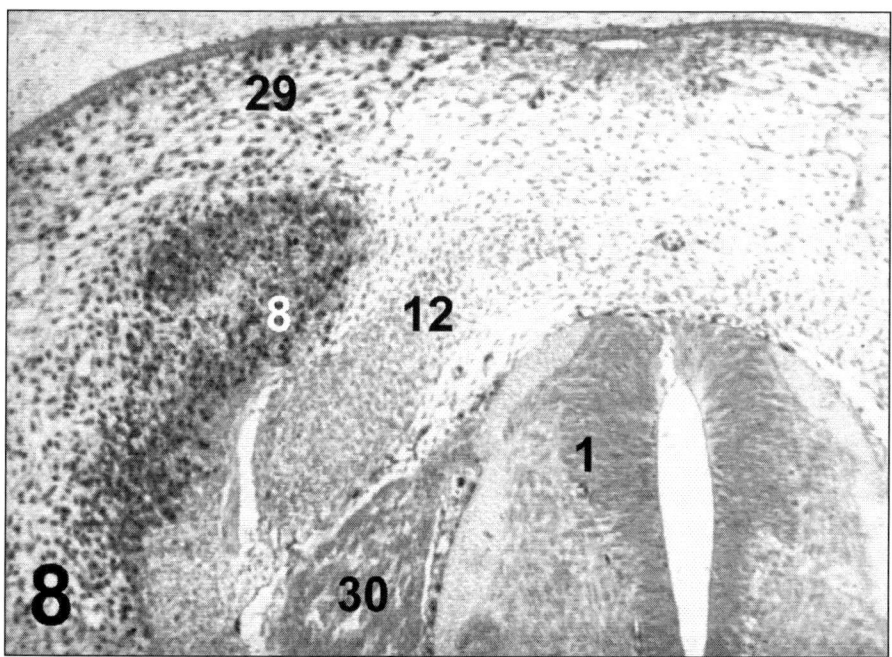

Figure 8. Chick embryo after interspecific (quail to chick) grafting of the dermomyotome; Reincubation period: 4 days. Note that muscle and dermis have originated from the graft. See Figure 1 for legends.

concluded from their experiments with avian embryos that Twist could play a role in keeping sclerotomal cells in a proliferative state.

Pax1 cooperates with mesenchymal forkhead-1 (*Mfh1*), a member of the winged helix/forkhead family of proteins, that is expressed in the sclerotome under the control of Shh. The expression domains of *Mfh1* extend more dorsally and laterally than those of *Pax1*. The functions of *Mfh1* include the maintenance of proliferation activity of sclerotomal cells.[205]

Apoptosis of Sclerotomal Cells

Sanders[206] has studied the occurrence of apoptotic cells in the sclerotomes of chick embryos in detail. In 2.5 day embryos, the cranial somites (1-18) showed dying cells mainly in the cranial half-sclerotomes. At midlevels (somites 19-26), apoptotic cells were located in the ventral sclerotome. At caudal levels (somites 27-32), the number of dying cells was sharply reduced and those present were primarily located at the caudal side of the intervertebral (v. Ebner's) fissure, where the somitocoele cells are located.[207,208] These results show, that the localization and number of apoptotic cells depend on the differentiation state and the axial level of the sclerotome. Somitocoele cells have been found to be especially sensitive to undergo apoptosis.[8,162] The most important signal to prevent cell death in the sclerotome is Shh emanated from the notochord and the floor plate of the neural tube. After extirpation of the axial structures, apoptotic pathways are activated in the sclerotome cells leading to a down-regulation of bcl-2 protein and an upregulation of caspase-2 activity.[209] This high rate of cell death can be prevented by the implantation of a notochord or the application of Shh protein.[199] This is in line with the observation of an increased cell death in *Shh* mutant mice.[56] Furumoto et al[205] have suggested that *Pax1* and *Mfh1* in concert with *Pax9* and *MF2* might act as mediators of Shh-dependent survival of sclerotomal cells. Apoptosis can be induced by an upregulation of Bmp4[210] which is accompanied by a down-regulation of *Pax1*.[211]

If the cranial half of the sclerotome is not invaded by neural crest cells, the apoptotic activity of mesodermal cells is drastically increased.[212] The signaling mechanisms involved in these interactions have still to be elucidated.

Cranio-Caudal Polarization of the Sclerotome

It has been well known for a long time that the somites become subdivided in a cranial and a caudal half.[4,213,11] The border between these two halves is characterized by elongated and transversally oriented sclerotomal cells which mark the intervertebral (v. Ebner's) fissure.[212] It has been shown that both halves of the sclerotome differ with respect to cell density, produced proteins and their influence on migrating neural crest cells and outgrowing axons. Neural crest cells and axons exclusively invade the cranial half of the sclerotome, whereas the caudal half has a repellent influence on them.[214-217] During the last decades various genes have been identified whose expression domains are restricted to cranial or caudal half-somites. When cranial and caudal halves of chick and quail somites were microsurgically placed adjacent to each other in ovo, it was found that sclerotome cells from unlike half-somites did not mix and kept separated by a boundary.[320] These results indicate that cells from both halves differ in their behavior; e.g., in their adhesive properties. It has been shown that cell-cell communication mediated by the Delta-Notch signaling pathway is involved in the formation of half-somites and the maintenance of compartment borders.[218-224] The initiation of this somitic polarity takes place before somite formation in the presomitic mesoderm by interactions of the Delta/Notch signaling pathway with the basis helix-loop-helix transcription factor Mesp2[224,225] and becomes morphologically visible after sclerotome formation.[226] It is maintained by the T-box transcription factor Tbx18.[227] The morphological realization of the cranio-caudal polarity is dependent on the *paraxis* mediated epithelialization of the somites.[228]

Resegmentation

It was a great step forward in the research of sclerotome development, when Robert Remak in 1850[4] published the first of his three volume work entitled *"Untersuchungen über die Entwicklung der Wirbelthiere: Über die Entwicklung des Hühnchens im Ei"* (Studies on the development of vertebrates. On the development of the chick embryo in the egg). Remak called the somite "protovertebra" and the sclerotome "Wirbelkernmasse", in which he already identified differences between cranial and caudal halves: He suggested the caudal half to form the "vertebral arch" In the perinotochordal mesenchyme from which the vertebral bodies and the intervertebral discs develop, he described a "Neue Gliederung" (new segmentation) by which the vertebral bodies are being formed. Corning[229] supported Remak's view and described an intrasegmental fissure that is separating both halves of the sclerotome in the chick embryo. The existence of this "intervertebral fissure" was also shown in the reptilian embryo by von Ebner.[213]

Our present view of resegmentation in the avian embryo is based on the studies by Bagnall et al,[230,231] Huang et al,[207,208,232] Ewan and Everett,[233] Aoyama and Asomoto[234] and Evans.[235] According to this work, each vertebral body is formed by the caudal half of one sclerotome and the cranial half of the caudally adjacent sclerotome. This is also true for the ribs with the exception of their head, neck and distal-most part. The intervertebral disc originates from the somitocoele-derived cells which are located in the cranial part of the caudal half-sclerotome. The neural arches with their processes and the proximal ribs are mainly formed by cells of the caudal half-sclerotome. As the result of the resegmentation process, the somitocoele cells, which are centrally located within the epithelial somite, are later located intervertebrally in the center of the vertebral motion segment in which two adjacent vertebrae move against each other. Thus, the segmental muscles are functional as they are attached by tendons to two adjacent "resegmented" vertebrae, in this way overspanning the movable intervertebral structures like discs, joints and ligaments (Fig. 7). The origin, specification and behavior of cells forming the vertebral ligaments have not been studied to date.

Axial Identity of the Sclerotome

It is obvious that vertebrae at different levels display distinct morphological features, making it easy to distinguish between a cervical vertebra and one of the lumbar region. An additional

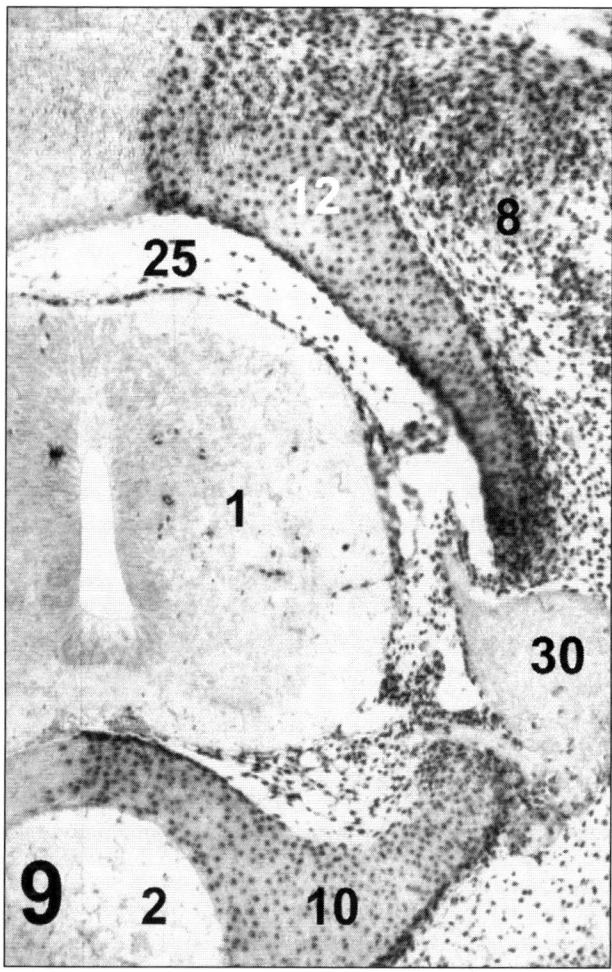

Figure 9. Transverse section of a chimeric embryo after unilateral transplantation of one somite; Reincubation period: 6 days. See Figure 1 for legends.

regional particularity of the thoracic segments is given by the formation of ribs. This segment specificity becomes already determined in the presomitic mesoderm. After heterotopic grafting of presomitic mesoderm from the prospective thoracic region into the cervical or lumbosacral regions, the formation of ectopic ribs could be observed.[236,237] Cervical somites grafted into the thoracic region fail to develop ribs.

The segment-specific identity of the paraxial mesoderm is achieved by the function of members of the *Hox* gene family.[238] In higher vertebrates, four clusters of *Hox* genes can be distinguished (*HoxA, HoxB, HoxC* and *HoxD*), which are located on different chromosomes. The individual genes of the vertebrate *Hox* complexes are activated according to their ordered arrangement in the *Hox* complex. This phenomenon is called "colinearity".[239-241] Transcriptional activation of *Hox* genes according to this colinearity is suggested to be responsible for the development of axial sclerotome identities. Gruss and Kessel[242] and Kessel[243] have introduced the *Hox* code model which is characterized by a segment-specific combination of *Hox* gene expression controlling the identities of axial segments.[244] Alterations of the *Hox*-code change the segment identities resulting in the formation of segment-atypical vertebrae, a phenomenon that is known as homeotic transformations.[245-248] For instance in gain-of-function mutations, where ectopic expression of *Hoxd-4* was achieved, the occipital bone and atlas gained segmental identities characteristic of more caudal cervical vertebrae.[249,250] It has been shown that a *Hox* gene with a more caudal expression boundary dominates *Hox* genes with more cranial boundaries.[251]

Hox genes are sensitive to retinoic acid which can cause homeotic transformations.[245,252,253] Recent studies have shown that Fgf signaling is involved in the coordination of the segmentation process and the spatiotemporal *Hox* gene activation.[254,255] The question of how *Hox* genes control down-stream genes and eventually the behavior of sclerotomal cells in different segments remains to be studied.[256]

Subdomains of the Sclerotome

To better understand the signaling network controlling sclerotome differentiation it is helpful to recall the structures adjacent to the sclerotome. Ventrally, the aorta is moving medially and becomes situated between endoderm and sclerotome. Medially, the neural tube, which is increasingly becoming polarized dorso-ventrally, develops a Shh-producing floor plate and a Wnt- and Bmp4-producing roof plate. Ventral to the neural tube, the notochord with its surrounding ECM is interacting with sclerotomal cells. Dorsally, the sclerotome is bordering on the dorsal somite compartment, the dermomyotome and, later, on the myotome which is being formed underneath the dermomyotome. Laterally, the Wolffian duct, nephrogenic mesenchyme and the dorsal cardinal vein are embedded between the sclerotome and the epithelial lining of the coelomic cavity.

Due to gradients of different signaling molecules produced by the structures adjacent to the sclerotome, various subdomains appear (Figs. 6,13). *Pax1* expression is downregulated in the dorsal and the lateral parts of the sclerotome that develop independent of notochordal signals.[175,257] The differentiation program of these cells is assumed to depend on Bmp-4 produced by the dorsal neural tube, the kidney rudiment and the lateral plate mesoderm.[211,258]

The central part of the sclerotome that is located close to the myotome reveals a high cell density. Since a subpopulation of myotomal cells produce Fgfs, e.g., Fgf8, it has been suggested that myotomal Fgf8 promotes the proliferation of neighboring sclerotomal cells.[197] Recently, a *scleraxis* expressing zone has been described within the central sclerotome close to the myotome.[259] This subdomain has been named the "syndetome" because its cells represent progenitors for the tendons of the epaxial domain.[260]

The ventral part of the sclerotome is formed by the invasion of *Pax1* expressing cells from the ventromedial edge of the central sclerotome. These cells migrate into the perinotochordal space where they proliferate and form the perinotochordal tube from which vertebral bodies and intervertebral discs develop.[261-263]

Close to the lateral surface of the neural tube, a medial subdomain of the sclerotome can be identified that gives rise to the blood vessels and meninges of the spinal cord[263,264] and thus can be named "meningotome".

An additional subdomain of the sclerotome originates from the somitocoele cells which form the mesenchymal core of the epithelial somite. Since these cells mainly contribute to the formation of intervertebral discs and joints of the vertebral column[208] we propose to name this sclerotomal subdomain the "arthrotome".[265]

The cranial half of the developing sclerotome becomes invaded by outgrowing axons and neural crest cells. Cells from this sclerotomal subdomain participate in the formation of dorsal root ganglia and spinal nerves by differentiating into cells of the perineurium and endoneurium. According to Blechschmidt[266] we propose to name this sclerotomal subdomain the "neurotome".

Central Sclerotome

The central part of the sclerotome is characterized by its close neighborhood to the dermomyotome and later, to the myotome and by a morphologically obvious alternation of densely and loosely packed cells in the half-segments.[262,267] The central sclerotome gives rise to the pedicles and ventral parts of the neural arches as well as the proximal ribs. The caudal half of the central sclerotome

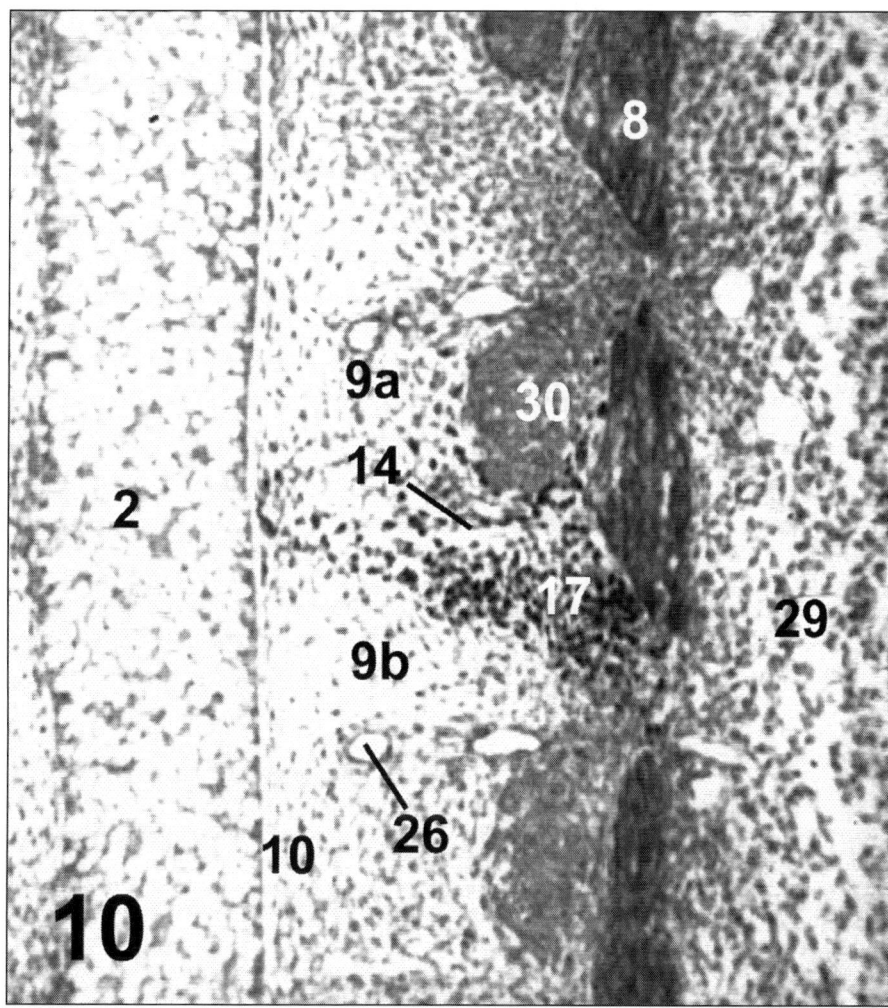

Figure 10. Coronal section of a chimeric embryo after grafting (quail to chick) of somitocoele cells. Reincubation period: 2 days. Quail cells are located in a triangularly shaped area in the caudal sclerotome half. See Figure 1 for legends.

was found to be the main source of the skeletal derivatives.[262] A gene expressed in the caudal half is *Uncx4.1*. This homeobox gene has been suggested to act downstream of the *Notch* signaling pathway since its expression is highly reduced in the somites of *Dll1* and *RBPJK* mutant mice and to less extent in *Notch1* mutant mice.[268,269] *Uncx4.1* is required to maintain the condensation of the caudal half-sclerotome. In homozygous *Uncx4.1* mutant mice, the differential expression of *Pax9* in both halves of the sclerotome is lost due to a reduction of *Pax9* expression in the caudal half, eventually resulting in the absence of the pedicles, proximal ribs and the transverse processes.[270,271] In the avian embryo, *Uncx 4.1* expression in the presomitic mesoderm is independent of signals from the axial structures and presumably induced by the intrinsic Notch/Delta driven oscillator activity that determines craniocaudal somite polarity. In the maturing somite, however, *Uncx 4.1* expression depends on signals from the notochord-floor plate complex.[272]

After removal of the neural tube and notochord in the chick embryo, a total absence of vertebrae and ribs can be seen,[199] indicating that the development even of the distal ribs depends on signals coming from the axial structures. This can be explained by an indirect signaling mechanism as described by Huang et al[197]: Shh emanated from the notochord/floor plate complex induces the expression of *Fgf8* in a subpopulation of myotome cells including the hypaxial domain. Fgf8 has been found to increase the rate of cell division in the adjacent lateral portion of the sclerotome and to promote rib formation.[197]

The development of the distal rib requires Bmp signals emanating from the somatopleural mesoderm to reach the sternum which is formed by lateral plate mesoderm cells.[258,273] Henderson et al[274] postulated *Pax3*-dependent inductive influences from the dermomyotome to be necessary for the early specification of the lateral sclerotome and for rib development; PDGFA and Fgfs from the myotome have been suggested to play a role in this process.[274-276]

Ventral Sclerotome

As described earlier, the sclerotome cells are not homogeneously distributed. At the beginning of sclerotome differentiation, cells are densely packed in the central sclerotome subdomain adjacent to the dermomyotome, whereas the ventral domain in which the vertebral bodies and the intervertebral discs develop is free of cells.[262,263,277] This perinotochordal space only contains extracellular matrix material that connects the sclerotomal cells with the basement membrane of the notochord and functions as a substrate for invading sclerotomal cells.[277] The formation of the extracellular matrix sheath of the notochord and the survival of notochordal cells is dependent on *Sox5* and *Sox6*.[278] The immigrating cells then proliferate and form the perinotochordal tube[279] from which the vertebral bodies and intervertebral discs arise. The invasion of the perinotochordal space by sclerotomal cells and their proliferation depend on signals emanated from the notochord. Shh, noggin and Fgf8 have been identified to be important signaling molecules that induce the expression of *Pax1, Pax9, Twist* and *Mfh1* in sclerotomal cells and promote their proliferation.[175,204,263,280-285] On the other hand, the sclerotomal cells control the morphology and function of the notochord. *Pax1*-deficient mice that lack anlagen of vertebral bodies and intervertebral discs, have an enlarged notochord, the cells of which show an increased proliferation rate.[281] Jun, a major component of the heterodimeric transcription factor AP-1, was found to be required for regulating the survival of notochordal cells and maintaining the vertebral bodies separated.[286]

The vertebral column of *Pax9* deficient mice appears normal indicating that *Pax1* can completely compensate for the absence of *Pax9*, whereas the loss of *Pax1* can be rescued by *Pax9* only to a certain extent. In *Pax1/Pax9* double mutants, however, the derivatives of the ventral sclerotome are totally missing.[287] In addition to *Pax1* and *Pax9*, *Mfh-1* has been shown to be required for the normal development of vertebral bodies and intervertebral discs.[285,288] The chicken homologue of *Mfh1*, designated *cFkh1*, is expressed in the paraxial mesoderm prior to somite formation.[289] As development proceeds, its expression becomes restricted to the developing sclerotome, where it persists until chondrogenesis commences. In *Mfh1/Pax1* double mutants the ventral expansion of the sclerotome and the accumulation of sclerotomal cells around the notochord is significantly impaired.[205] Other homeodomain proteins involved in the control of ventral and lateral sclerotome development are the Meox proteins.[290,291] *Meox1/Meox2* double mutant mice lack the axial skeleton and ribs. The vertebral column is largely replaced by two stripes of fused cartilages, corresponding in position to the neural arches.[290] *Meox* gene function has been suggested to act upstream of *Pax* genes and *Tbx18* that is expressed in the cranial half of the somite,[292] (Dr. B. Mankoo, London, personal communication).

Prior to the advent of molecular approaches, sclerotome formation was found to depend on the presence of notochord and neural tube.[293-303] The interactions between axial structures and paraxial mesoderms can now be explained by the expression of *Pax1, Pax9* and *Mfh1* in response to Shh and noggin signaling from the notochord.[175,178,182,184] *Twist* that is induced by a concerted action of Shh and Fgf8[204] is preferentially expressed in the caudal half of the sclerotome, where it is involved in keeping sclerotomal cells viable and promoting proliferation.[280] *Twist* has also been shown to be involved in the integration of Fgf, Bmp and Shh signaling pathways.[204] Another recently

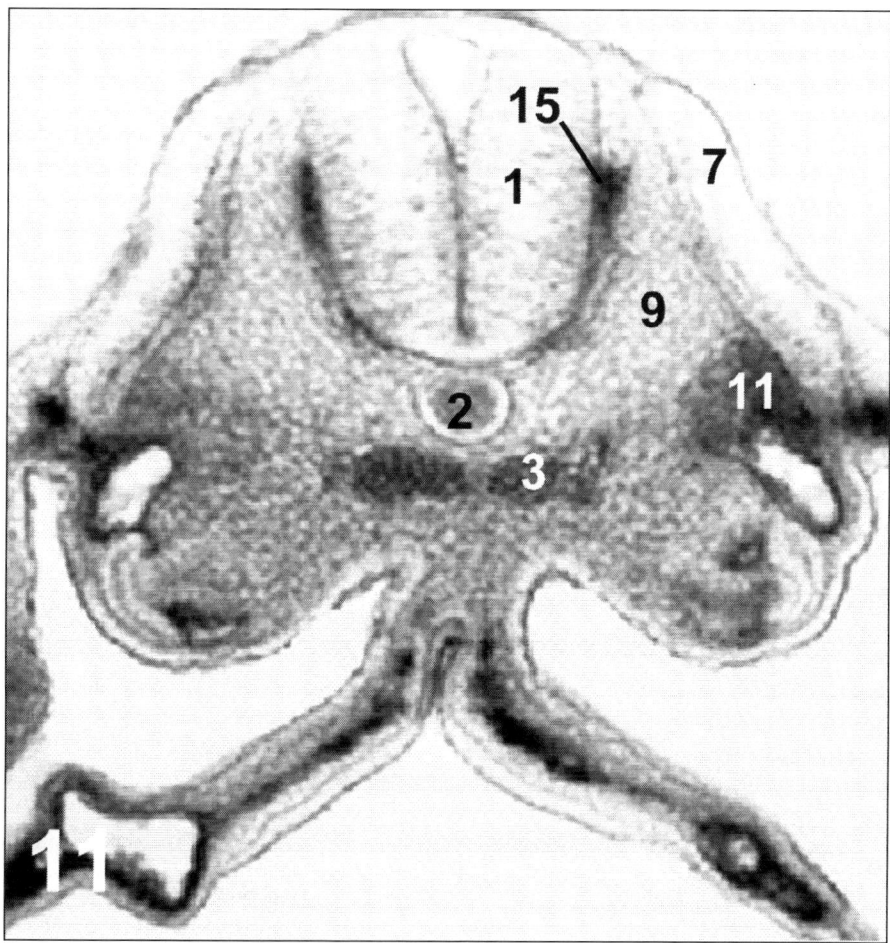

Figure 11. VEGFR-2 expression in a compartmentalized somite. Note the expression in the lateral sclerotome and meningotome. (Courtesy of S. Nimmagadda, Freiburg). See Figure 1 for legends.

discovered function of *Twist* is to suppress the activity of *Runx2* which regulates osteoblast differentiation and bone formation.[203]

In the ventral sclerotome, a metameric pattern of proliferation can be found as soon as the anlagen of the intervertebral discs are recognizable as condensed zones between the anlagen of the vertebral bodies that express chondroitin-6-sulphate proteoglycan indicating the beginning of chondrogenic differentiation.[195] Later, the cells at the borders between vertebral bodies and intervertebral discs continue to proliferate appositionally adding cells to the growing vertebral bodies.[195] Chondrogenic differentiation of sclerotomal cells depends on the expression of the homeobox gene *Bapx1* (also known as *Nkx3-2*).[304,305] Mouse mutants lacking *Bapx1* show a vertebral phenotype strikingly similar to that of mice mutant for both *Pax1* and *Pax9*.[306-308] It has been shown that Meox proteins are required for *Bapx1* expression[309] and that *Bapx1* is a direct target of *Pax1* and *Pax9*.[310] Shh establishes an *Bapx1/Sox9* autoregulatory loop that is maintained by Bmp signals to induce somitic chondrogenesis.[311] Other genes that are required for chondrogenic differentiation and which are expressed under the control of *Pax1* and *Pax9* are *Col2a1*+ encoding collagen II and its transcriptional activator *Sox9*.[312-314]

Dorsal Sclerotome

Compared with the ventral and lateral subdomains, the dorsal sclerotome develops relatively late. Most of its cells first have to migrate dorsally and to invade the space between the surface ectoderm and the growing neural tube before differentiation can commence. The cells of the dorsal sclerotome do not maintain expression of *Pax1*[175] due to a negative effect of Bmp4 that is produced by the roof plate of the neural tube and the surface ectoderm.[211,315] In contrast, these cells express *Msx1* and *Msx2* as soon as they become dorsally positioned but stop to do so as cartilage differentiation begins.[211] The expression of *Msx* genes overlaps with the expression of Bmp4 in the dorsal neural tube and the overlying ectoderm. Grafting of Bmp4 producing cells dorsal to the neural tube results in the increase in number of dorsal mesenchymal cells and in a higher and prolonged expression of *Msx* genes, eventually leading to hypertrophied spinous processes.[257] On the other hand, the ventral structures, notochord and floor plate, inhibit the expression of *Msx* genes and the formation of dorsal parts of neural arches and spinous processes.[257] These results show that the differentiation of the dorsal sclerotome that gives rise to the dorsal part of neural arches and spinous processes is under positive control of Bmp4 produced by the roof plate and negative control of Shh produced by the notochord/floor plate complex. They also explain the observation that *Pax1/Pax9* double mutant mice are able to develop neural arches.[287]

Another gene involved in the control of dorsal sclerotome differentiation is *Mfh1*.[284,285,316] In *Mfh1*-deficient mice, defects of the neural arches are observed resulting in a spina bifida. Since *Mfh1* is expressed in the dorsal sclerotome under the control of Shh signals from the notochord, it can be suggested that at least one step in the formation of dorsal sclerotome derivatives must be dependent on Shh signaling.[205] The *Splotch* mouse in which the *Pax3* gene is mutated, also shows defects of the skeletal dorsal sclerotome derivatives. Since *Pax3* is normally expressed in the dorsal neural tube it, can be assumed that the disturbed closure of the neural tube interferes with an altered induction of *Msx*-gene expression.

Lateral Sclerotome

The early differentiation of lateral presomitic cells, that later contribute to the lateral sclerotome, depends on the presence of their medial counterparts.[317] The lateral sclerotome is characterized by a downregulation of *Pax1*.[175] Its further differentiation seems to depend on Fgfs from the myotome and Bmp4[65,318] from the lateral plate mesoderm and gives rise to distal ribs and contributes to tendons.[132,207,197,270,319] Additionally, cells of the lateral sclerotome express *VEGFR2* under the control of Bmp4 and form endothelial cells of blood vessels.[320-323] The exact origin and development of this sclerotomal subdomain remains to be studied.

Neurotome

The term "neurotome" was proposed by Blechschmidt[266] to characterize that part of the cranial half-sclerotome in which the dorsal root ganglion and the spinal nerve develop. Functions of this sclerotomal subdomain, which arises as a consequence of the cranio-caudal polarization of the somite, include to be invaded by neural crest cells as well as outgrowing axons and to supply the spinal nerves with sheath cells forming the endo- and perineurium.[214,215,217,264,324]

The caudal half of the sclerotome as well as the perinotochordal mesenchyme repel migrating neural crest cells and axons.[214,325] Davies et al[326] have shown that PNA-binding proteins have some role in cranio-caudal barrier function. Additional players in the repellent property of the caudal half-sclerotome have recently come to light in the form of the ephrins. *Ephrin B1* and *B4*, in the chick embryo, are expressed in the caudal sclerotome and repel motor axons.[327] Corresponding receptors for these ephrins, EphB2 and EphB3, are present in the motoneuron axons,[328-330] thus restricting their outgrowth in the cranial half-sclerotomes. After invasion of the neurotome by neural crest cells and axons, the mesodermal cells of the cranial half-sclerotome, which are, in contrast to their caudal counterpart, rather loosely packed, express cytotactin,[331] tenascin[332] and

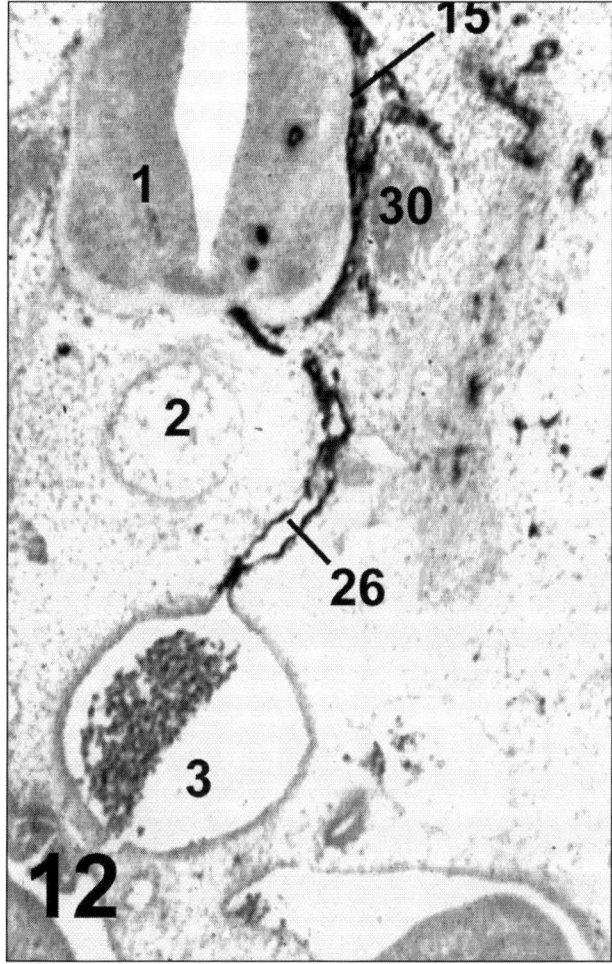

Figure 12. Chick embryo after transplantation of a quail somite. Blood vessel endothelia and angioblasts of quail origin are stained with the QH1 antibody. See Figure 1 for legends.

show butyrylcholinesterase activity.[333] This indicates that there must be an interaction between the neurotome cells, on the one hand and axons as well as neural crest cells, on the other hand. This was confirmed by the observation that neurotome cells undergo apoptosis when neural crest cells and axons are experimentally prevented to invade the sclerotome.[212] Moreover, it has been shown that cells of the sclerotomal neurotome control the size of the dorsal root ganglia.[334,335] After heterotopic grafting of segmental plate mesoderm from cranial to brachial levels and *vice versa,* the sclerotomes developed with a craniocaudal extent corresponding to their level of origin in the donor embryo. As a consequence, dorsal root ganglia that formed in the grafted sclerotomes attained a size appropriate to that of ganglia in the sclerotomes at the original axial levels. If the cranio-caudal polarization of the sclerotome is abolished, as it can be found in *Uncx-1* mutant mice, for instance, the boundaries of the dorsal root ganglia disappear, leading to a fusion of ganglia.[270,271] The same result can be achieved after creation of a paraxial mesoderm consisting only of cranial somite halves.[336,337]

Neural crest cells that migrate on a ventral pathway pass through the sclerotomal neurotome and avoid the perinotochordal space. A number of molecules produced by the notochord or in the cranial half-sclerotomes have been identified to inhibit the migration of the neural crest cells in the vicinity of the notochord: chondroitin-6-sulphate-proteoglycan,[338,339] collagen type XI[362] peanut agglutinin binding molecules,[338,361] versican,[340] collapsin-1[341] and F-spondin.[342] After experimental removal of the notochord, the dorsal root ganglia fuse ventrally to the neural tube.[212]

In addition to their guiding functions and interactions with invaded neural crest cells, the neurotome cells contribute to the formation of the spinal nerves.[264] The epithelium-like cells of the perineurium that envelopes bundles of nerves as wells as the connective tissue cells of the endoneurium surrounding single nerve fibers develop from local mesoderm and thus, in the dorsal body wall, from the sclerotome. This is also the case for the connective tissue inside and outside the dorsal root ganglia.

Meningotome

Quail-chick chimerization experiments have clearly shown that the meninges surrounding the spinal cord are derivatives of the sclerotome (Fig. 9).[264] The primitive meninx becomes subdivided into two layers, the ectomeninx and endomeninx. The ectomeninx gives rise to the dura mater, neurothelium and outer arachnoid layer; the endomeninx to the inner arachnoid and the pial layer.[343-345] The space that is delineated by the ectomeninx and endomeninx is the cerebrospinal fluid space.[346] The main functions of the meninges are to protect the central nervous system and to supply it with blood. It has recently been shown that the first step in the development of the spinal cord meninges is the expression of *VEGFR-2* and *Slug* in a narrow medial-most zone of the sclerotome immediately adjacent to the lateral surface of the neural tube.[322,347] In this zone, which we propose to call the "meningotome", the first blood vessels of the perineural vascular plexus arise.[348] The *VEGFR-2* expression was found to be balanced by promoting Bmp4 signals from the neural tube and inhibiting noggin signals from the notochord[323] indicating that it is the neural tube itself that induces the initiation of meninx formation. One of the signals produced by the neural tube that promotes vascularization of the meningotome, is vascular endothelial growth factor A (VEGFA).[349] This is in line with the observation that meninx formation does not occur after ablation of the neural tube.[303] It can be assumed that, in later stages, the differentiation of the ectomeninx is additionally controlled by signals from the inner surface of the vertebral column anlage. Other signal molecules involved in meninx differentiation remain to be identified.

Syndetome

The term "syndetome" has been proposed by Brent et al[260] to characterize a sclerotomal subdomain from which tendons develop. It is located at the cranial and caudal borders of the sclerotome beneath the myotome and arises from the sclerotome just as the myotome emerges from the dermomyotome. Its cells express *scleraxis*, a specific marker for tendons and ligaments[259] and give rise to epaxial tendons. Brent et al[260,350] have shown that *scleraxis* expression in the tendon-forming part of the sclerotome is induced by Fgf8 signals from the cells located in the center of the myotome. Fgf8 binds to the Fgf receptor FREK which is restricted to cells located at the cranial and caudal borders of the myotome. These cells, in turn, are suggested to activate *scleraxis* in the parts of the sclerotome abutting the *FREK*-expressing myotome via a secondary to date unknown signal. Münsterberg and coworkers have shown that FGF 8 acitvates *scleraxis* in the target cells via the MAP-ERK pathway.[351] The spatial pattern of muscle-forming, tendon-forming and skeleton-forming domains of the somite shows a setting that is later required for the functions of muscle, tendons and vertebrae. It is well known that during later steps of development the superficial part of the segmentally arranged muscle anlagen in the epaxial domain fuse to form segment-overbridging muscles. As a consequence, already existing tendons must lose their connection with the muscle and disappear. The molecular mechanisms controlling this polymerization process remain to be studied. The vertebrae are not only connected by muscles and tendons, but also by ligaments that also originate from the sclerotomes. It is not quite clear how the development of ligaments can be separated from that of tendons.

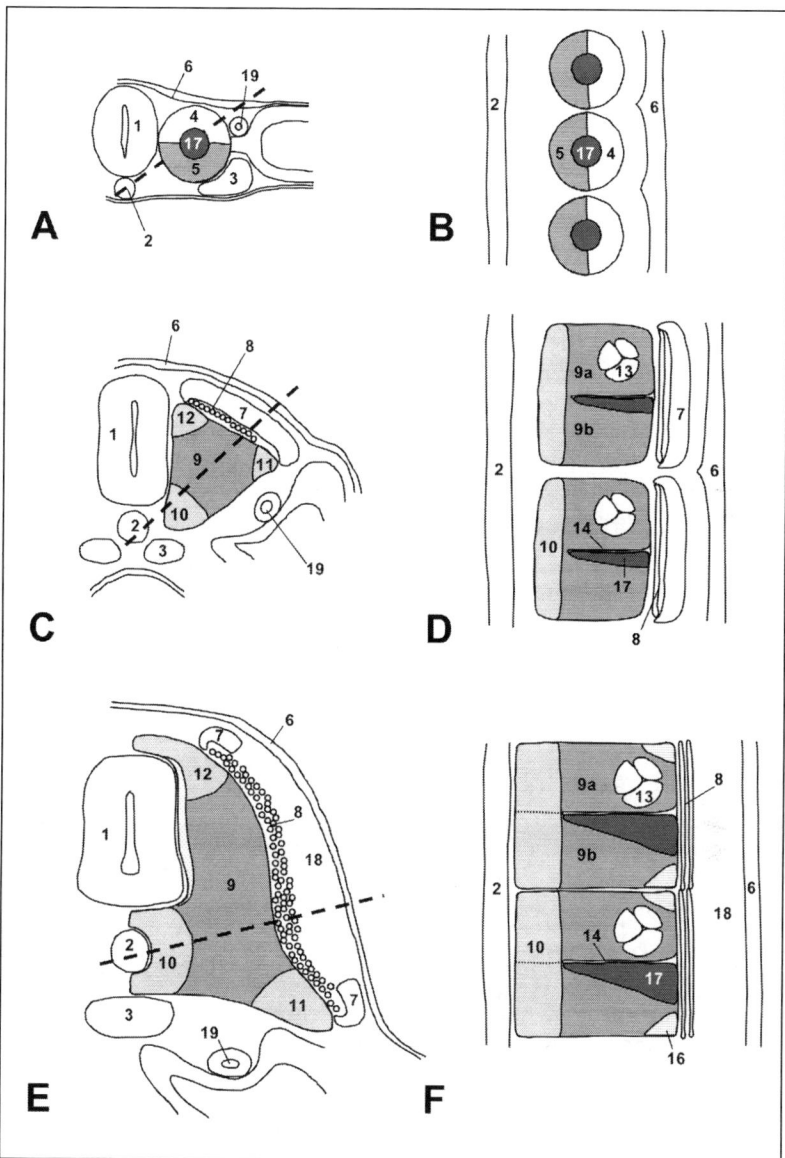

Figure 13. Schematic representation of transverse and longitudinal sections of chick embryos at different developmental stages showing somite derivatives. See Figure 1 for legends.

Arthrotome

An obvious compartment of the somite is formed by the *Pax1* expressing somitocoele cells. Unlike all the other somitic cells, they do not undergo a mesenchymal-to-epithelial transition. They maintain their mesenchymal organization from the presomitic mesoderm up to the sclerotome, where they are integrated into its caudal half close to the intervertebral (v. Ebner's) fissure,[207,208] there forming a triangular area with its tip reaching the notochord (Fig. 10). The somitocoele cells give rise to the vertebral joints, the intervertebral discs and the proximal ribs.[207,208,230,231] A further indication showing the important role of somitocoele cells in vertebral joint formation can be seen after microsurgical removal of these cells and preventing epithelial somite cells from entering the somitocoele: in the operated segments, joints and intervertebral discs do not develop, resulting in a fusion of the articular processes and vertebral bodies.[265] Since the somitocoele cells located centrally both in the epithelial somites and in the vertebral motion segments, where single vertebrae can move against each other, we propose to name the somitocoele-derived sclerotomal subdomain the "arthrotome". Further studies are necessary to identify the molecular mechanisms of arthrotome specification.

Sclerotome-Derived Blood Vessels

The basic structure of blood vessels consists of an inner lining comprising a single layer of flattened endothelial cells supported by a basement membrane and pericytes or the tunica media comprising smooth muscle cells and fibrocytes.[152,352] Precursors of the endothelial cells are angioblasts that are present in various mesodermal compartments. They form cords that undergo tubulogenesis to form vessels with a central lumen. This vascular tube formation has recently been shown to be regulated by Egfl7, an endothelial cell-derived secreted factor, containing an EGF-like domain.[353] The somites have been identified to be important sources of angioblasts that either differentiate within the somite derivatives or invade the neural tube, ventro-lateral body wall, limb buds, mesonephros and the dorsal part of the aorta.[154,320,354] Dorsal and ventral halves of epithelial somites, both, the prospective dermomyotomes and sclerotomes, have been found to have angiogenic potential after being grafted homotopically between quail and chick embryos (Fig. 12).[152,320] This is also true for somitocoele cells which become integrated into the sclerotome.

Endothelial cells and angioblasts can be identified by the expression of *VEGFR-2* (Fig. 13).[321,360] In the paraxial mesoderm of the avian embryo, expression of *VEGFR-2* can be observed in the lateral portion of both the segmental plate mesoderm and the epithelial somite. After compartmentalization of the somite, *VEGFR-2* is expressed in the lateral portion of the sclerotome and later in a medial expression domain that becomes established in that part of the sclerotome that surrounds the neural tube (anlage of the meningotome).[322] *VEGFR-2* expression in the sclerotome is induced and maintained by Bmp4 and down-regulated by noggin. As a result, upregulation of Bmp4 leads to an increase the number of blood vessels whereas application of noggin results in a reduction of blood vessels.[323] The development of the perineural vascular plexus is stimulated by the neural tube, not only by the production of Bmp4 but also by VEGFA secretion.[349]

Sclerotome-derived angioblasts normally do not cross the midline of the embryo. The notochord has been identified to keep endothelial cells from migrating to the controlateral side.[355] It can be assumed that the repulsive effect of the notochord on migrating angioblasts is mediated by the inhibition of *VEGFR-2* expression by noggin.[323]

The sclerotome does not only give rise to endothelial cells, but also to pericytes as well as smooth muscle cells and fibrocytes of the media.[356] However, this is the case only in the dorsal body wall, whereas, in contrast, the media of vessels in the limbs and in the ventro-medial body wall is a derivative of the local somatopleural mesoderm.[153] The recruitment of pericytes and smooth muscle cells which are needed to stabilize the vessel wall, is a result of an interaction of endothelial cells of vascular tubes and surrounding periendothelial mesenchymal cells. Angiopoietin-1 (Ang-1) and Tie receptors are important players in this cross-talk. Ang-1 produced by mesenchymal cells is suggested to activate endothelial cells to produce signals for mesenchymal cell recruitment (e.g., PDGFB for pericytes). When mesenchymal cells contact endothelial cells the latter activate TGFβ to induce smooth muscle cell differentiation (reviewed in ref. 152) Hagedorn et al[357] have discussed the possibility that pericytes and endothelial cells originate from a common progenitor cell. This could mean that sclerotome-derived pericytes become distributed in a more extensive manner. Although it has been shown that the somites give also rise to lymphatic endothelial cells,[358,359] it remains to be studied from which somite compartment these cells arise and how they become specified.

Outlook

It turned out in the last years that the somite is a complex structure that gains its complexity by a concerted execution of both intrinsic informations and instructions given by signaling molecules from adjacent structures including other somite compartments. In this way, a wide range of different cell types become specified at different sites of the dermomyotome and sclerotome providing an exact pattern of derivatives which is needed for coordinated functions. The details of specification, movement and differentiation of somitic cells are far from being understood. For a detailed understanding of dermomyotome and sclerotome development, attention will have to be directed to the mechanisms that connect these various processes. It also remains a challenge to determine how *Hox* genes interact with various downstream genes to generate segment specific structures of muscle, dermis, bone, tendons, ligaments and blood vessels, which are needed for the functions of the locomotive apparatus.

Acknowledgements

We thank the members of our laboratory who have contributed to this work and Ulrike Uhl for typing the manuscript. We also thank the Deutsche Forschungsgemeinschaft for continuous financial support.

Contents of this book chapter have partly been published earlier in the following review articles: Christ, B. Huang, R. and Scaal, M. (2004) Formation and differentiation of the avian sclerotome. Anat Embryol (Berl) 208, 333-50. Scaal, M. and Christ, B. (2004) Formation and differentiation of the avian dermomyotome. Anat Embryol (Berl) 208, 411-24. Publication in this book is with kind permission of Springer Science and Business Media. The original articles are available at www.springerlink.com.

References

1. Baer von KE. Entwickelungsgeschichte der Thiere. Beobachtung und Reflexion. Bornträger, Königsberg 1828.
2. Raff RA. Developmental mechanisms in the evolution of animal form: Origins and evolvability of body plans. In: Early life on earth. Columbia University Press, New York, 1994:489.
3. Balfour FM. Handbuch der vergleichenden Embryologie. 2. Band, Fischer Jena, 1881.
4. Remak R. Untersuchungen über die Entwickelung der Wirbelthiere. Reimer, Berlin, 1850.
5. Williams LW. The somites of the chick. Am J Anat 1910; 11:55-100.
6. Christ B, Ordahl CP. Early stages of chick somite development. Anat Embryol 1995; 191:381-396.
7. Gossler A, Hrabe de Angelis M. Somitogenesis. Curr Top Dev Biol 1998; 38:225-287.
8. Christ B, Jacob HJ, Jacob M. Experimental analysis of somitogenesis in the chick embryo. Z Anat Entwickl-Gesch 1972; 138:82-97.
9. Christ B, Schmidt C, Huang R et al. Segmentation of the vertebrate body. Anat Embryol 1998; 197:1-8.
10. Brand-Saberi B, Wilting J, Ebensperger C et al. The formation of somite compartments in the avian embryo. Int J Dev Biol 1996; 40:411-420.
11. Brand-Saberi B, Christ B. Evolution and development of distinct lineages derived from somites. Curr Top Dev Biol 2000; 48:1-42.
12. Aoyama H, Asamoto K. Determination of somite cells: independence of cell differentiation and morphogenesis. Development 1988; 104:15-28.
13. Christ B, Brand-Saberi B, Grim M et al. Local signalling in dermomyotomal cell type specification. Anat Embryol 1992; 186:505-510.
14. Aoyama H. Development plasticity of the prospective dermatome and the prospective sclerotome region of an avian somite. Dev Growth Differ 1993; 35:507-519.
15. Ordahl CP, Le Douarin NM. Two myogenic lineages within the developing somite. Development 1992; 114:339-353.
16. Brent AE, Tabin CJ. Development regulation of somite derivatives: muscle, cartilage and tendon. Curr Opin Genet Dev 2002; 12:548-557.
17. Hatschek 1980, Cited after Williams LW. The somites of the chick. Am J Anat 1910; 11:55-100.
18. His W. Untersuchungen über die erste Anlage des Wirbelthierleibes. Die erste Entwicklung des Hühnchens im Ei. Vogel Leipzig 1868.
19. Linker C, Lesbros C, Gros J et al. beta-Catenin-dependent Wnt signalling controls the epithelial organisation of somites through the activation of paraxis. Development 2005; 132:3895-3905.

20. Geetha-Loganathan P, Nimmagadda S, Huang R et al. Regulation of ectodermal Wnt6 expression by the neural tube is transduced by dermomyotomal Wnt11: a mechanism of dermomyotomal lip sustainment. Development 2006; 133:2897-2904.

21. Ordahl CP, Berdougo E, Venters SJ et al. The dermomyotome dorsomedial lip drives growth and morphogenesis of both the primary myotome and dermomyotome epithelium. Development 2001; 128:1731-1744.

22. Venters SJ, Ordahl CP. Persistent myogenic capacity of the dermomyotome dorsomedial lip and restriction of myogenic competence. Development 2002; 129:3873.

23. Dhoot GK, Gustafsson MK, Ai X et al. Regulation of Wnt signaling and embryo pattering by an extracellular sulfatase Science 2001; 293:1663-1666.

24. Lee CS, Buttitta LA, May NR et al. SHH-N upregulates Sfrp2 to mediate its competitive interaction with WNT1 and WNT4 in the somitic mesoderm. Development 2000; 127:109-118.

25. Lee CS, Buttitta L, Fan CM. Evidence that the WNT-inducible growth arrest-specific gene 1 encodes an antagonist of sonic hedgehog signaling in the somite. Proc Natl Acad Sci USA 21001; 98:11347-11352.

26. Spence MS, Yip J, Erickson CA. The dorsal neural tube organizes the dermamyotome and induces axial myocytes in the avian embryo. Development 1996; 122:231-241.

27. Dietrich S, Schubert FR, Lumsden A. Control of dorsoventral pattern in the chick paraxial mesoderm. Development 1997; 124:3895-3908.

28. Fan CM, Lee CS, Tessier-Lavigne M. A role for WNT proteins in induction of dermomyotome. Dev Biol 1997; 191:160-165.

29. Capdevila J, Tabin C, Johnson R.L. Control of dorsoventral somite patterning by Wnt-1 and beta-catenin. Dev Biol 1998; 193:182-194.

30. Ikeya M, Takada S. Wnt signaling from the dorsal neural tube is required for the formation of the medial dermomyotome. Development 1998; 125:4969-4976.

31. Wagner J, Schmidt C, Nikowits W Jr et al. Compartmentalization of the somite and myogenesis in chick embryos are influenced by wnt expression. Dev Biol 2000; 228:86-94.

32. Fan CM, Tessier-Lavigne M. Patterning of mammalian somites by surface ectoderm and notochord: evidence for sclerotome induction by a hedgehog homolog. Cell 1994; 79:1175-1186.

33. Galli LM, Willert K, Nusse R et al. A proliferative role for Wnt-3a in chick somites. Dev Biol 2004; 269:489-504.

34. Ben-Yair R, Kahane N, Kalcheim C. Coherent development of dermomyotome and dermis from the entire mediolateral extent of the dorsal somite. Development 2003; 130:4325-4336.

35. Ahmed MU, Cheng L, Dietrich S. Establishment of the epaxial-hypaxial boundary in the avian myotome. Dev Dyn 2006; 235:1884-1894.

36. Scaal M, Wiegreffe C. Somite compartments in amniotes. Anat Embryol 2006; (Berl) Epub 2006 Sep 28.

37. Brennan C, Amacher SL, Currie PD. Somitogenesis. Results Probl Cell Differ 2002; 40:271-297.

38. Krenn V, Gorka P, Wachtler F et al. On the origin of cells determined to form skeletal muscle in avian embryos. Anat Embryol 1988; 179:49-54.

39. Nicolas JF, Mathis L, Bonnerot C et al. Evidence in the mouse for self-renewing stem cells in the formation of a segmented longitudinal structure, the myotome. Development 1996; 122:2933-2946.

40. Selleck MA, Stern CD. Fate mapping and cell lineage analysis of Hensen's node in the chick embryo. Development 1991; 112:615-626.

41. Psychoyos D, Stern CD. fates and migratory routes of primitive streak cells in the chick embryo. Development 1996, 122:1523-1534.

42. Eloy-Trinquet S, Mathis L, Nicolas JF. Retrospective tracing of the developmental lineage of the mouse myotome. Curr Top Dev Biol 2000; 47:33-80.

43. George-Weinstein M, Gerhart J, Reed R. Skeletal myogenesis: the preferred pathway of chick embryo epiblast cells in vitro. Dev Biol 1996; 173:279-291.

44. Holtzer H, Schultheiss T, Dilullo C. Autonomous expression of the differentiation programs of cells in the cardiac and skeletal myogenic lineages. Ann NY Acad Sci 1990; 599:158-169.

45. Fomenou MD, Scaal M, Stockdale FE et al. Cells of all somitic compartments are determined with respect to segmental identity. Dev Dyn 2005; 233:1386-1393.

46. Rong PM, Teillet MA, Ziller C et al. The neural tube/notochord complex is necessary for vertebral but not limb and body wall striated muscle differentiation. Development 1992; 115:657-672.

47. Borman WH, Yorde DE. Barrier inhibition of a temporal neuraxial influence on early chick somitic myogenesis. Dev Dyn 1994; 200:68-78.

48. Buffinger N, Stockdale FE. Myogenic specification in somites: induction by axial structures. Development 1994; 120:1443-1452.

49. Stern HM, Brown AM, Hauschka SD. Myogenesis in paraxial mesoderm: preferential induction by dorsal neural tube and by cells expressing Wnt-1. Development 1995; 121:3675-3686.
50. Cossu G, Kelly R, Tajbakhsh S et al. Activation of different myogenic pathways: myf-5 is induced by the neural tube and MyoD by the dorsal ectoderm in mouse paraxial mesoderm. Development 1996; 122:429-437.
51. Stern HM, Hauschka SD. Neural tube and notochord promote in vitro myogenesis in single somite explants. Dev Biol 1995; 167:87-103.
52. Münsterberg AE, Kitajewski J, Bumcrot DA et al. Combinatorial signaling by Sonic hedgehog and Wnt family members induces myogenic bHLH gene expression in the somite. Genes Dev 1995; 9:2911-2922.
53. Münsterberg AE, Lassar AB. Combinatorial signals from the neural tube, floor plate and notochord induce myogenic bHLH gene expression in the somite. Development 1995; 121:651-660.
54. Gustafsson MK, Pan H, Pinney DF et al. Myf5 is a direct target of long-range Shh signaling and Gli regulation for muscle specification. Genes Dev 2002; 16:114-126.
55. Teboul L, Hadchouel J, Daubas P et al. The early epaxial enhancer is essential for the initial expression of the skeletal muscle determination gene Myf5 but not for subsequent, multiple phases of somitic myogenesis. Development 2002; 129:4571-4580.
56. Chiang C, Litingtung Y, Lee E. Cyclopia and defective axial patterning in mice lacking Sonic hedgehog gene function. Nature 1996; 383:407-413.
57. Duprez D, Fournier-Thibault C, Le Douarin N. Sonic hedgehog induces proliferation of committed skeletal muscle cells in the chick limb. Development 1998; 125:495-505.
58. Teillet M, Watanabe Y, Jeffs P et al. Somic hedgehog is required for survival of both myogenic and chondrogenic somitic lineages. Development 1998; 125:2019-2030.
59. Marcelle C, Ahlgren S, Bronner-Fraser M. In vivo regulation of somite differentiation and proliferation by Sonic Hedgehog. Dev Biol 1999; 214:277-287.
60. Marcelle C, Stark MR, Bronner-Fraser M. Coordinate actions of BMPs, Wnts, Shh and noggin mediate patterning of the dorsal somite. Development 1997; 124:3955-3963.
61. Linker C, Lesbros C, Stark MR et al. Intrinsic signals regulate the initial steps of myogenesis in vertebrates. Development 2003; 130:4797-4807.
62. Rudnicki MA, Schnegelsberg PN, Stead RH et al. MyoD or Myf-5 is required for the formation of skeletal muscle. Cell 1993; 75:1351-1359.
63. Maroto M, Reshef R, Munsterberg AE et al. Ectopic Pax-3 activates MyoD and Myf-5 expression in embryonic mesoderm and neural tissue. Cell 1997; 89:139-148.
64. Tajbakhsh S, Rocancourt D, Cossu G et al. Redefining the genetic hierarchies controlling skeletal myogenesis: Pax-3 and Myf-5 act upstream of MyoD. Cell 1997; 89:127-138.
65. Pourquie O, Fan CM, Coltey M et al. Lateral and axial signals involved in avian somite patterning: a role for BMP4. Cell 1996; 84:461-471.
66. Ott MO, Bober E, Lyons G et al. Early expression of the myogenic regulatory gene, myf-5, in precursor cells of skeletal muscle in the mouse embryo. Development 1991; 111:1097-1107.
67. Pownall ME, Emerson CP Jr. Sequential activation of three myogenic regulatory genes during somite morphogenesis in quail embryos. Dev Biol 1992; 151:67-79.
68. Reshef R, Maroto M, Lassar AB. Regulation of dorsal somitic cell fates: BMPs and Noggin control the timing and pattern of myogenic regulator expression. Genes Dev 1998; 12:290-303.
69. Christ B, Jacob HJ, Jacob M. Regional determination of early embryonic muscle primordium. Experimental studies on quail and chick embryos. Verh Anat Ges 1978; 353-357.
70. Bardeen BI. The development of the musculature in the body wall in the pig. Johns Hopkins Hospital Report 1900; 9:367.
71. Hamilton HL. Lillie's development of the chick. An introduction to embryology. Holt, Rinehart and Winston 1952; New York.
72. Boyd ID. Development of the striated muscle. In: Bourne GH (ed) Structure and Function of Muscle. Academic Press 1960; New York.
73. Mestres P, Hinrichsen K. The histogenesis of somites in the chick. J Embryol Exp Morphol 1976; 36:669.
74. Kahane N, Cinnamon Y, Kalcheim C. The cellular mechanism by which the dermomyotome contributes to the second wave of myotome development. Development 1998; 125:4259-4271.
75. Kahane N, Cinnamon Y, Kalcheim C. The origin and fate of pioneer myotomal cells in the avian embryo. Mech Dev 1998; 74:59-73.
76. Cinnamon Y, Kahane N, Kalcheim C. Characterization of the early development of specific hypaxial muscles from the ventrolateral myotome. Development 1999; 126:4305-4315.
77. Cinnamon Y, Kahane N, Bachelet I et al. The sublip domain—a distinct pathway for myotome precursors that demonstrate rostral-caudal migration. Development 2001; 128:341-351.

78. Kahane N, Cinnamon Y, Bachelet I et al. The third wave of myotome colonization by mitotically competent progenitors: regulating the balance between differentiation and proliferation during muscle development. Development 2001; 128:2187-2198.

79. Kahane N, Cinnamon Y, Kalcheim C. The roles of cell migration and myofiber intercalation in patterning formation of the postmitotic myotome. Development 2002; 129:2675-2687.

80. Denetclaw WF Jr, Christ B, Ordahl C.P. Location and growth of epaxial myotome precursor cells. Development 1997; 124:1601-1610.

81. Denetclaw WF, Ordahl CP. The growth of the dermomyotome and formation of early myotome lineages in thoracolumbar somite of chicken embryos. Development 2000; 127:893-905.

82. Denetclaw WF Jr, Berdougo E, Venters SJ et al. Morphogenetic cell movements in the middle region of the dermomyotome dorsomedial lip asociated with patterning and growth of the primary epaxial myotome. Development 2001; 128:1745-1755.

83. Scaal M, Gros J, Lesbros C et al. In ovo electroporation of avian somites. Dev Dyn 2004; 229:643-650.

84. Gros J, Scaal M, Marcelle C. A two-step mechsnism for myotome formation in chick. Dev Cell 2004; 6:875-882.

85. Venters SJ, Ordahl CP. Asymmetric cell divisions are concentrated in the dermomyotome dorsomedial lip during epaxial primary myotome morphogenesis. Anat Embryol 2005; 209:449-460.

86. Cinnamon Y, Ben-Yair R, Kalcheim C. Differential effects of N-cadherin-mediated adhesion on the development of myotomal waves. Development 2006; 133:1101-1112.

87. Ben-Yair R, Kalcheim C. Lineage analysis of the avian dermomyotome sheet reveals the existence of single cells with both dermal and muscle progenitor fates. Development 2005; 132:689-701.

88. Marcelle C, Eichmann A, Halevy O et al. Distinct developmental expression of a new avian fibroblast growth factor receptor. Development 1994; 120:683-694.

89. Marcelle C, Wolf J, Bronner-Fraser M. The in vivo expression of the FGF receptor FREK mRNA in avian myoblasts suggests a role in muscle growth and differentiation. Dev Biol 1995; 172:100-114.

90. Gros J, Manceau M, Thome V et al. A common somitic origin for embryonic muscle progenitors and satellite cells. Nature 2005; 435:954-958.

91. Wilson-Rawls J, Hurt CR, Parsons SM et al. Differential regulation of epaxial and hypaxial muscle development by paraxis. Development 1999; 126:5217-5229.

92. Christ B, Jacob HJ, Jacob M. Origin of wing musculature. Experimental studies on quail and chick embryos. Experientia 1974; 30:1446-1449.

93. Christ B, Jacob HJ, Jacob M. Experimental analysis of the origin of the wing musculature in avian embryos. Anat Embryol 1977; 150:171-186.

94. Fischel A. Zur Entwicklung der vertebralen Rumpf- und der Extremitätenmuskulatur der Vögel und Säugetiere. Morphol Jahrb 1895; 23:544.

95. Murray PDF. Chorio-allantoic grafts of fragments of the two-day-chick, with special reference to the development of the limbs, intestine and skin. J Exp Biol Med Sc 1928; 5:237-256.

96. Grim M. Differentiation of myoblasts and the relationship between somites and the wing bud of the chick embryo. Z Anat Entwicklungsgesch 1970; 132:260.

97. Glücksmann A. Über die Entwicklung der Amniotenextremitäten und ihre Homologie mit den Flossen. Z Anat Entwicklgesch 1934; 102:498-533.

98. Saunders J.W. Do the somites contribute to the formation of the chick wing? Anat Rec 1948; 100:756.

99. Pinot M. The role of somitic mesoderm in the early morphogenesis of the limbs in the fowl embryo. J Embryol Exp Morph 1970; 23:109.

100. Brand-Saberi B, Muller TS, Wilting J et al. Scatter factor/hepatocyte growth factor (SF/HGF) induces emigration of myogenic cells at interlimb level in vivo. Dev Biol 1996; 179:303-308.

101. Schmidt C, Bladt F, Goedecke S et al. Scatter factor/hepatocyte growth factor is essential for liver development. Nature 1995; 373:699-702.

102. Bladt F, Riethmacher D, Isenmann S et al. Essential role for the c-met receptor in the migration of myogenic precursor cells into the limb bud. Nature 1995; 376:768-771.

103. Epstein JA, Shapiro DN, Cheng J et al. Pax3 modulates expression of the c-Met receptor during limb muscle development. Proc Natl Acad Sci USA 1996; 93:4213-4218.

104. Schäfer K, Braun T. Early specification of limb muscle precursor cells by the homeobox gene Lbx1. Nat Genet 1999; 23:213-216.

105. Song K, Wang Y, Sassoon D. Expression of Hox-7.1 in myoblasts inhibits terminal differentiation and induces cell transformation. Nature 1992; 360:477-481.

106. Davidson D. The function and evolution of Msx genes: pointers and paradoxes. Trends Genet 1995; 11:405-411.

107. Houzelstein D, Auda-Boucher G, Cheraud Y et al. The homeobox gene Msx1 is expressed in a subset of somites and in muscle progenitor cells migrating into the forelimb. Development 1999; 126:2689-2701.
108. Goulding M, Lumsden A, Paquette AJ. Regulation of Pax-3 expression in the dermomyotome and its role in muscle development. Development 1994; 120:957-971.
109. Anakwe K, Robson L, Hadley J. Wnt signaling regulates myogenic differentiation in the developing avian wing. Development 2003; 130:3503-3514.
110. Scaal M, Bonafede A, Dathe V et al. SF/HGF is a mediator between limb patterning and muscle development. Development 1999; 126:4885-4893.
111. Dietrich S, Abou-Rebyeh F, Brohmann H. The role of SF/HGF and c-Met in the development of skeletal muscle. Development 1999; 126:1621-1629.
112. Marics I, Padilla F, Guillemot JF et al. FGFR4 signaling is a neccessary step in limb muscle differentiation. Development 2002; 129:4559-4569.
113. Tajbakhsh S, Buckingham ME. Mouse limb muscle is determined in the absence of the earliest myogenic factor myf-5. Proc Natl Acad Sci USA 1994; 91:747-751.
114. Geetha-Loganathan P, Nimmagadda S, Pröls F et al. Two different pathways during limb myogenesis. Verh Ant Ges 2005; 100:20.
115. Geetha-Loganathan P, Nimmagadda S, Pröls F et al. Ectodermal Wnt-6 promotes Myf5-dependent avian limb myogenesis. Dev Biol 2005; 288:221-233.
116. Buckingham M, Bajard L, Chang T et al. The formation of skeletal muscle: from somite to limb. J Anat 2003; 202:59-68.
117. Schemainda H. Experimentelle Untersuchungen zur Entwicklung der Zungenmuskulatur beim Hühnerembryo. Verh Anat Ges 1981; 75:501.
118. Huang R, Zhi Q, Izpisua-Belmonte JC et al. Origin and development of the avian tongue muscles. Anat Embryol 1999; 200:137-152.
119. Huang R, Lang ER, Otto WR et al. Molecular and cellular analysis of embryonic avian tongue development. Anat Embryol 2001; 204:179-187.
120. Reichert CB. Das Entwickelungsleben im Wirbelthierreich. 1838.
121. Kölliker A. Entwicklungsgeschichte des Menschen und der höheren Thiere. 1879; Engelmann, Leipzig.
122 Goette A. Entwickelungsgeschichte der Unke (Bombinator igneus) als Grundlage einer vergleichenden Morphologie der Wirbelthiere. 1875; Leopold Voss, Leipzig.
123. Minot CS. Human Embryology. 1892; Wood and Co, New York.
124. Murray PDF. The origin of the dermis. Nature 1928; 122:609.
125. Le Douarin NM. Particularites du noyau interphasique chez la caille japonaise (Coturnix c. Japonica). Utilisation de ses particularités comme "marquage biologique" dans les recherches sur les interactions tissulaires et les migrations cellulairs au cours de l'ontogenèse. Bull Biol Fr Bel 1969; 103:435-452.
126. Le Douarin NM. A biological cell labeling technique and its use in experimental embryology. Dev Biol 1973; 30:217-222.
127. Mauger A. The role of somitic mesoderm in the development of dorsal plumage in chick embryos. I. Origin, regulative capacity and determination of the plumage-forming mesoderm. J Embryol Exp Morphol 1972; 28:313-341.
128. Christ B, Jacob M, Jacob HJ. On the origin and development of the ventrolateral abdominal muscles in the avian embryo. An experimental and ultrastructural study. Anat Embryol 1983; 166:87-101.
129. Le Lievre CS, Le Douarin NM. Mesenchymal derivatives of the neural crest: analysis of chimaeric quail and chick embryos. J Embryol exp Morph 1975; 34:125-154.
130. Noden DM. The role of the neural crest in patterning of avian cranial skeletal, connective and muscle tissues. Dev Biol 1983; 96:144-165.
131. Couly GF, Coltey PM, Le Douarin NM. The developmental fate of the cephalic mesoderm in quail-chick chimeras. Development 1992; 114:1-15.
132. Olivera-Martinez I, Coltey M, Dhouailly D et al. Mediolateral somitic origin of ribs and dermis determined by quail-chick chimeras. Development 2000; 127:4611-4617.
133. Olivera-Martinez I, Thelu J, Teillet MA et al. Dorsal dermis development depends on a signal from the dorsal neural tube, which can be substituted by Wnt-1. Mech Dev 2001; 100:233-244.
134. Atit R, Sgaier SK, Mohamed OA et al. Beta-catenin activation is necessary and sufficient to specify the dorsal dermal fate in the mouse. Dev Biol 2006; 296:164-176.
135. Sengel P. Morphogenesis of skin. Cambridge University Press, 1976; Cambridge.
136. Brill G, Kahane N, Carmeli C et al. Epithelial-mesenchymal conversion of dermatome progenitors requires neural tube-derived signals: characterization of the role of neutrophin-3. Development 1995; 121:2583-2594.

137. Marcelle C, Lesbros C, Linker C. Somite patterning: a few more pieces of the puzzle. Results Probl Cell Differ 2002; 38:81-108.

138. Houzelstein D, Cheraud Y, Auda-Boucher G et al. The expression of the homeobox gene Msx1 reveals two populations of dermal progenitor cells originating from the somites. Development 2000; 127:2155-2164.

139. Buchner G, Broccoli V, Bulfone A et al. MAEG, an EGF-repeat containing gene, is a new marker associated with dermatome specification and morphogenesis of its derivatives. Mech Dev 2000; 98:179-182.

140. Olivera-Martinez I, Missier S, Fraboulet S et al. Differential regulation of the chick dorsal thoracic dermal progenitors from the medial dermomyotome. Development 2002; 129:4763-4772.

141. Duong TD, Erickson CA. MMP-2 plays an essential role in producing epithelial- mesenchymal transformations in the avian embryo. Dev Dyn 2004; 229:42-53.

142. Christ B, Huang R, Wilting J. The development of the avian vertebral column. Anat Embryol 2000; 202:179-194.

143. Sorrell JM, Caplan AI. Fibroblast heterogeneity: more than skin deep. J Cell Sci 2004; 117:667-675.

144. Wessells NK. Morphology and proliferation during early feather development. Dev Biol 1965; 12:131-153.

145. Li L, Cserjesi P, Olson EN. Dermo-1; a novel twist-related bHLH protein expressed in the developing dermis. Dev Biol 1995; 172:280-292.

146. Scaal M, Füchtbauer EM, Brand-Saberi B. Cdermo-1 expression indicates a role in avian skin development. Anat Embryol 2001; 203:1-7.

147. Scaal M, Pröls F, Füchtbauer EM et al. BMPs induce dermal markers and ectopic feather tracts. Mech Dev 2002; 110:51-60.

148. Sengel P. Pattern formation in skin development. Int J Dev Biol 1990; 34:33-50.

149. Dhouailly D. Dermo-epidermal interactions during morphogenesis of cutaneous appendages in amniotes. Front Matrix Biol 1977; 4:86-121.

150. Chuong CM. Molecular Basis of Epithelial Appendage Morphogenesis. R.G. Landes Company, 1998; Austin, TX.

151. Pispa J, Thesleff I. Mechanisms of ectodermal organogenesis. Dev Biol 2003; 262:195-205.

152. Wilting J, Christ B, Yuan L et al. Cellular and molecular mechanisms of embryonic haemangiogenesis and lymphangiogenesis. Naturwissenschaften 2003; 90:433-448.

153. Wilting J, Brand-Saberi B, Kurz H et al. Development of the embryonic vascular system. Cell Mol Biol Res 1995; 41:219-232.

154. Pardanaud L, Luton D, Prigent M et al. Two distinct endothelial lineages in ontogeny, one of them related to hemopoiesis. Development 1996; 122:1363-1371.

155. Kardon G, Campbell JK, Tabin CJ. Local extrinsic signals determine muscle and endothelial cell fate and patterning in the vertebrate limb. Dev Cell 2002; 3:533-545.

156. He L, Papoutsi M, Huang R et al. Three different fates of cells migrating from somites into the limb bud. Anat Embryol 2003; 207:29-34.

157. Huang R, Zhi Q, Patel K et al. Dual origin and segmental organisation of the avian scapula. Development 2000; 127:3789-3794.

158. Wang B, He L, Ehehalt F et al. The formation of the avian scapula blade takes place in the hypaxial domain of the somites and requires somatopleure-derived BMP signals. Dev Biol 2005; 287:11-18.

159. Tajbakhsh S, Rocancourt D, Buckingham M. Muscle progenitor cells failing to respond to positional cues adopt nonmyogenic fates in myf-5 null mice. Nature 1996; 384:266-270.

160. Solursh M, Fischer M, Meier S et al. The role of extracellular matrix in the formation of the sclerotome. J Embryol exp Morph 1979; 54:75-98.

161. Trelstad RL. Mesenchymal cell polarity and morphogenesis of chick cartilage. Dev Biol 1977; 59:153-163.

162. Hirano S, Hirako R, Kajita N et al. Morphological analysis of the role of the neural tube and notochord in the development of somites. Anat Embryol 1995; 192:445-457.

163. Burgess R, Rawls A, Brown D et al. Requirement of the paraxis gene for somite formation and musculoskeletal patterning. Nature 1996; 384:570-573.

164. Barnes GL, Alexander PG, Hsu CW et al. Cloning and characterization of chicken Paraxis: a regulator of paraxial mesoderm development and somite formation. Dev Biol 1997; 189:95-111.

165. Ostrovsky D, Sanger JW, Lash JW. Somitogenesis in the mouse embryo. Cell Differ 1988; 23:17-26.

166. Hatta K, Takagi S, Hajime F et al. Spatial and temporal expression pattern of N- cadherin cell adhesion molecules correlated with morphogenetic processes of chicken embryos. Dev Biol 1987; 120:215-227.

167. Duband J.-L, Dufour S, Hatta K et al. Adhesion molecules during somitogenesis in the avian embryo. J Cell Biol 1987; 104:1361.

168. Takeichi M. The cadherins: cell-cell adhesion molecules controlling animal morphogenesis. Development 1988; 102:630-655.

169. Dietrich S, Schubert FR, Gruss P et al. Altered Pax gene expression in mouse notochord mutants: the notochord is required to initiate and maintain ventral identity in the somite. Mech Dev 1993; 44:189-207.

170. Koseki H, Wallin J, Wilting J et al. A role of Pax-1 as a mediator of notochordal signals during the dorso-ventral specification of vertebrae. Development 1993; 119:649-660.

171. Peters H, Doll U, Niessing J. Differential expression of the chicken Pax-1 and Pax-9 gene. In situ hybridization and immunohistochemical analysis. Dev Dyn 1995; 203:1-16.

172. Walther C, Guenet J.-L, Simon D et al. Pax: A murine multigene family of paired box containing genes. Genomics 1991; 11:424-434.

173. Noll M. Evolution and role of Pax genes. Curr Opin Genet Dev 1993; 3:595-605.

174. Deutsch U, Dressler GR, Gruss P. Pax 1, a member of paired box homologous murine gene family, is expressed in segmented structures during development. Cell 1988; 53:617-626.

175. Ebensperger C, Wilting J, Brand-Saberi B et al. Pax-1, a regulator of sclerotome development is induced by notochord and floor plate signals in avian embryos. Anat Embryol 1995; 191:297-310.

176. Borycki AG, Strunk K, Savary R et al. Distinct signal/response mechanisms regulate pax 1 and QmyoD activation in sclerotomal and myotomal lineages of quail somites. Dev Biol 1997; 185:185-200.

177. Balling R, Neubüser A, Christ B. Pax genes and sclerotome development. Semin Cell Dev Biol 1996; 7:129-136.

178. Müller TS, Ebensperger C, Neubüser A et al. Expression of avian Pax-1 and Pax-9 in the sclerotome is controlled by axial and lateral tissues, but intrinsically regulated in pharyngeal endoderm. Dev Biol 1996; 178:403-417.

179. Borycki AG, Mendham L, Emerson CP Jr. Control of somite patterning by Sonic hedgehog and its downstream signal response genes. Development 1998; 125:777-790.

180. Buttitta L, Mo R, Hui CC et al. Interplays of Gli2 and Gli3 and their requirement in mediating Shh-dependent sclerotome induction. Development 2003; 130:6233-6243.

181. Johnson RL, Läufer E, Riddle RD et al. Ectopic expression of Sonic hedgehog alters dorsal-ventral patterning of somites. Cell 1994; 79:1165-1173.

182. Fan C.-M, Porter JA, Chiang C et al. Long-range sclerotome induction by sonic hedgehog: direct role of the amino-terminal claevage product and modulation by the cyclic AMP signaling pathway. Cell 1995; 81:457-465.

183. Marti E, Takada R, Bumcrot DA et al. Distribution of Sonic hedgehog peptides in the developing chick and mouse embryo. Development 1995; 121:2537-2547.

184. McMahon JA, Takada S, Zimmermann LB et al. Noggin-mediated antagonism of BMP signaling is required for growth and patterning of the neural tube and somite. Genes Dev 1998; 12:1438-1452.

185. Dockter JL. Sclerotome induction and differentiation. In: Somitogenesis, Part 2,(Ordahl CP. ed.) Academic Press, London, 2000; New York, 77.

186. Hoang BH, Thomas JT, Abdul-Karim FW et al. Expression pattern of two Frizzled-related genes, Frzb-1 and Sfrp-1 during mouse embryogenesis suggests a role for modulating action of Wnt family members. Dev Dyn 1998; 212:364-374.

187. Cauthen CA, Berdougo E, Sandler J et al. Comparative analysis of the expression patterns of Wnts and Frizzleds during early myogenesis in chick embryos. Mech Dev 2001; 104: 133-138.

188. Schubert FR, Mootoosamy RC, Walters EH et al. Wnt6 marks sites of epithelial transformations in the chick embryo. Mech Dev 2002; 114:143-148.

189. Rodriguez-Niedenführ M, Dathe V, Jacob HJ et al. Spatial and temporal pattern of Wnt-6 expression during chick development. Anat Embryol 2003; 206:447-451.

190. Brand-Saberi B, Ebensperger C, Wilting J et al. The ventralizing effect of the notochord on somite differentiation in chick embryos. Anat Embryol 1993; 188:239-245.

191. Wheelock M, Knudsen K. N-cadherin-associated proteins in chicken muscle. Differentiation 1991; 46:35-42.

192. Schmidt C, Stoeckelhuber M, McKinnell I et al. Wnt 6 regulates the epithelialisation process of the segmental plate mesoderm leading to somite formation. Dev Biol 2004; 271:198-209.

193. Hirsinger E, Duprez D, Jouve C et al. Noggin acts downstream of Wnt and Sonic Hedgehog to antagonize BMP4 in avian somite patterning. Development 1997; 124:4605-4614.

194. Amthor H, Conolly D, Patel K et al. The expression and regulation of follistatin and a follistatin-like gene (flik) during avian somite compartmentalization and myogenesis. Dev Biol 1996; 178:343-362.

195. Wilting J, Kurz H, Brand-Saberi B et al. Kinetics and differentiation of somite cells forming the vertebral column: studies on human and chick embryos. Anat Embryol 1994; 190:573-581.

196. Wilting J, Ebensperger C, Müller TS et al. Pax-1 in the development of the cervico-occipital transitional zone. Anat Embryol 1995; 192:221-227.

197. Huang R, Stolte D, Kurz H et al. Ventral axial organs regulate expression of myotomal Fgf-8 that influences rib development. Dev Biol 2003; 255:30-47.

198. Stolte D, Huang R, Christ B. Spatial and temporal pattern of Fgf-8 expression during chicken development. Anat Embryol 2002; 205:1-6.

199. Teillet M.-A, Watanabe Y, Jeffs P et al. Sonic hedgehog is required for survival of both myogenic and chondrogenic somitic lineages. Development 1998; 125:2019-2030.

200. Hebrok M, Wertz K, Füchtbauer EM. M-twist is an inhibitor of muscle differentiation. Dev Biol 1994; 165:537-544.

201. Füchtbauer E.-M. Expression of m-twist during post-implantation development of the mouse. Dev Dyn 1995; 204:316-322.

202. Spicer DB, Rhee J, Cheung WL et al. Inhibition of myogenic bHLH and MEF2 transcription factors by the bHLH protein twist. Science 1996; 272:476-1480.

203. Bialek P, Kern B, Yang X et al. A twist code determines the onset of osteoblast differentiation. Dev Cell 2004; 6:423-435.

204. Hornik C, Brand-Saberi B, Rudloff S et al. TWIST is an integrator of Shh, FGF and BMP signaling. Anat Embryol 2004; 209:31-39.

205. Furumoto TA, Miura N, Akasaka T et al. Notochord-dependent expression of MFH1 and Pax1 cooperates to maintain the proliferation of sclerotome cells during the vertebral column development. Dev Biol 1999; 210:15-29.

206. Sanders EJ. Cell death in the avian sclerotome. Dev Biol 1997; 192:551-563.

207. Huang R, Zhi Q, Wilting J. The fate of somitocoele cells in avian embryos. Anat Embryol 1994; 190:243-250.

208. Huang R, Zhi Q, Neubüser A et al. Function of somite and somitocoele cells in the formation of the vertebral motion segment in avian embryo. Acta Anat 1996; 155:231-241.

209. Sanders EJ, Parker E. Ablation of axial structures activates apoptotic pathways in somite cells of the chick embryo. Anat Embryol 2001; 204:389-398.

210. Schmidt C, Christ B, Patel K et al. Experimental induction of BMP-4 expression leads to apoptosis in the paraxial and lateral plate mesoderm. Dev Biol 1998; 202:253-263.

211. Monsoro-Burq AH, Duprez D, Watanabe Y et al. The role of bone morphogenetic proteins in vertebral development. Development 1996; 122:3607-3616.

212. Christ B. Die Entwicklung der Körperwandmetamerie, experimentelle Untersuchungen an Hühnerembryonen. Habilitationsschrift 1975; Ruhr-Universität Bochum.

213. Ebner V. von. Urwirbel und Neugliederung der Wirbelsäule. Sitzungsber Akad Wiss Wien 1888; III/97:194-206.

214. Rickmann M, Fawcett LW, Keynes RJ. The migration of neural crest cells and the growth of motor axons through the rostral half of the chick somite. J Embryol exp Morph 1985; 90:437.

215. Keynes RJ, Stern CD. Segmentation in the vertebrate nervous system. Nature 1984; 310:786-789.

216. Bronner-Fraser M. Analysis of the early stages of trunk neural crest migration in avian embryos using monoclonal antibody HNK-1. Dev Biol 1986; 115:44-55.

217. Teillet M.-A, Kalcheim C, Le Douarin NM. Formation of the dorsal root ganglion in the avian embryo: segmental origin and migratory behavior of neural crest progenitor cells. Dev Biol 1987; 120:329.

218. Hrabe de Angelis M, McIntyre J, Gossler A. Maintenance of somite borders in mice requires the Delta homologue DII1. Nature 1997; 386:717-721.

219. Del Amo FF, Smith DE, Swiatek PJ et al. Expression pattern of Motch, a mouse homolog of Drosophila Notch, suggests an important role in early post-implantation mouse development. Development 1992; 115:737-744.

220. Reaume AG, Conlon RA, Zirngibl R et al. Expression analysis of a Notch homologue in the mouse embryo. Dev Biol 1992; 154:377-387.

221. Bettenhausen B, Hrabe de Angelis M, Simon D et al. Transient and restricted expression during mouse embryogenesis of DiI1, a murine gene closely related to Drosophila Delta. Development 1995; 121:2407-2418.

222. Jen WC, Wettstein D, Turner D et al. The Notch ligand, X-Delta-1, mediates segmentation of the paraxial mesoderm in Xenopus embryos. Development 1997; 124:1169-1178.

223. del Barco Barrantes I, Elia AJ, Wunsch K et al. Interaction between Notch signalling and Lunatic fringe during somite boundary formation in the mouse. Curr Biol 1999; 9:470-480.

224. Rida PCG, Le Minh N, Jiang Y.-J. A Notch feeling of somite segmentation and beyond. Dev Biol 2004; 265:2-22.

225. Saga Y, Takeda H. The making of the somite. Molecular events in vertebrate segmentation. Nat Rev Genet 2001; 2:835-845.

226. Stockdale FE, Nikovits W Jr, Christ B. Molecular and cellular biology of avian somite development. Dev Dyn 2000; 219:304-321.

227. Bussen M, Petry M, Schuster-Gossler K et al. The T-box transcription factor Tbx18 maintains the separation of anterior and posterior somite compartments. Genes Dev 2004; 18:1209-1221.
228. Johnson J, Rhee J, Parsons SM et al. The anterior/posterior polarity of somites is disrupted in paraxis-deficient mice. Dev Biol 2001; 229:176.
229. Corning HK. Über die sogenannte Neugliederung der Wirbelsäule und über das Schicksal der Urwirbelhöhle bei Reptilien. Morph Jb 1881; 17:611-622.
230. Bagnall KM, Higgins S, Sanders EJ. The contribution made by a single somite to the vertebral column: experimental evidence in support for resegmentation using the chick-quail chimera model. Development 1988; 103:69-85.
231. Bagnall KM, Higgins S, Sanders EJ. The contribution made by a single somite to tissue within a body segment and assessment of their integration with similar cells from adjacent segments. Development 1989; 107:931-943.
232. Huang R, Zhi Q, Brand-Saberi B. New experimental evidence for somite resegmentation. Anat Embryol 2000a; 202:195-200.
233. Ewan KBR, Everett AW. Evidence for resegmentation in the formation of the vertebral column using the novel approach of retroviral-mediated gene transfer. Exp Cell Res 1992; 198:315-320.
234. Aoyama H, Asamoto K. The developmental fate of the rostral/caudal half of a somite for vertebra and rib formation: experimental confirmation of the resegmentation theory using chick-quail chimeras. Mech Dev 2000; 99:71-82.
235. Evans DJR. Contribution of somitic cells to the avian ribs. Dev Biol 2003; 256:114-126.
236. Kieny M, Mauger A, Sengel P. Early regionalization of somitic mesoderm as studied by the development of axial skeleton of the chick embryo. Dev Biol 1972; 28:142-161.
237. Jacob M, Christ B, Jacob HJ. Über die regionale Determination des paraxialen Mesoderms junger Hühnerembryonen. Verh Anat Ges 1975; 69:263-269.
238. Burke AC, Nelson CE, Morgan BA et al. Hox genes and the evolution of vertebrate axial morphology. Development 1995; 121:333-346.
239. Duboule D, Dollé P. The structural and functional organization of the murine HOX gene family resembles that of Drosophila homeotic genes. EMBO J 1989; 8:14971505.
240. Graham A, Papalopulu N, Krumlauf R. The murine and Drosophila homeobox clusters have common features of organization and expression. Cell 1989; 5:367-378.
241. Krumlauf R. Hox genes in vertebrate development. Cell 1994; 78:191.
242. Gruss P, Kessel M. Axial specification in higher vertebrates. Curr Opin Genet Dev 1991; 1:204-210.
243. Kessel M. Molecular coding of axial positions by Hox genes. Semin Dev Biol 1991; 2:367.
244. McGinnis W, Krumlauf R. Homeobox genes and axial pattering. Cell 1992; 68:283.
245. Kessel M, Gruss P. Homeotic transformations of murine vertebrae and concomitant alteration of Hox codes induced by retinoic acid. Cell 1991; 67:89-104.
246. Condie BG, Capecchi MR. Mice homozygous for a targeted disruption of Hoxs-3 (Hox-4.1) exhibit anterior transformations of the first and second cervical vertebrae, the atlas and axis. Development 1993; 119:579.
247. Kostic D, Capecchi MR. Targeted disruption of the murine Hoxa-4 and Hoxa-6 genes result in homeotic transformations of the vertebral column. Mech Dev 1994; 46:231-247.
248. Medina-Martinez O, Bradley A, Ramirez-Solis R. A large targeted deletion of Hoxb1-Hoxb9 produces a series of single segment anterior homeotic transformations. Dev Biol 2000; 222:71.
249. Lufkin T, Mark M, Hart CP et al. Homeotic transformation of the occipital bones of the skull by ectopic expression of a homeobox gene. Nature 1992; 359:835-841.
250. Kessel M, Balling R, Gruss P. Variations of cervical vertebrae after expression of a Hox-1.1 transgene mice. Cell 1990; 61:301.
251. Duboule D, Morata G. Colinearity and functional hierarchy among genes of the homeotic complexes. Trend Genet 1994; 10:358-364.
252. Boncinelli E, Simeone A, Acompora D et al. HOX gene activation by retinoic acid. Trends Genet 1991; 7:329.
253. Kessel M. Respecification of vertebral identities by retinoic acid. Development 1992; 115, 487.
254. Dubrulle J, McGrew MJ, Pourquié O. FGF signaling controls somite boundary position and regulates segmentation clock control of spatiotemporal Hox gene activation. Cell; 2001; 106:219-232. 106, 219.
255. Zakány J, Kmita M, Alarcon P et al. Localized and transient transcription of Hox genes suggest a link between patterning and the segmentation clock. Cell 2001; 106:207-217.
256. Burke AC. Hox genes and the global patterning of the somitic mesoderm. Curr Top Dev Biol 2000; 47:155.
257. Monsoro-Burq AH and Le Douarin NM. Duality of molecular signaling involved in vertebral chondrogenesis. In: Ordahl CP, ed. Somitogenesis, Part 2. London, New York: Academic Press, 2000:43-75.

258. Sudo H, Takahashi Y, Tonegawa A et al. Inductive signals from the somatopleure mediated by bone morphogenetic proteins are essential for the formation of the sternal component of avian ribs. Dev Biol 2001; 232:284-300.

259. Schweitzer R, Chyung JH, Murtaugh LC et al. Analysis of the tendon cell fate using Scleraxis, a specific marker for tendons and ligaments. Development 2001; 128:3855-3866.

260. Brent AE, Schweitzer R, Tabin CJ. A somitic compartment of tendon progenitors. Cell 2003; 113:235-248.

261. Hall BK. Chondrogenesis of the somitic mesoderm. Adv Anat Embryol Cell Biol 1977; 53:3.

262. Christ B, Wilting J. From somites to vertebral column. Ann Anat 1992; 174:23-32.

263. Christ B, Huang R, Wilting J. The development of the avian vertebral column. Anat Embryol 2000; 202:179-194.

264. Halata Z, Grim M, Christ B. Origin of spinal cord meninges sheaths of peripheral nerves and cutaneous receptors including Merkel cells. An experimental and ultrastructural study with avian chimeras. Anat Embryol 1990; 182:529-537.

265. Mittapalli VR, Huang R, Patel K et al. Arthrotome: a specific joint forming compartment in the avian somite. Dev Dyn 2005; 234:48-53.265.

266. Blechschmidt E. Die vorgeburtliche Entwicklungsstadien des Menschen. Karger, Basel-1961; New York.

267. Verbout AJ. The development of the vertebral column. Adv Anat Embryol Cell Biol 1985; 90:1-122.

268. Mansouri A, Yokota Y, Wehr R et al. Paired-related murine homeobox gene expressed in the developing sclerotome, kidney and nervous system. Dev Dyn 1997; 210:53-65.

269. Neidhardt L, Lispert A, Hermann BG. A mouse gene of the paired-related homeobox xlass expressed in the caudal somite compartment and in the developing vertebral column, kidney and nervous system. Dev Genes Evol 1997; 207:330-339.

270. Mansouri A, Voss AK, Thomas T et al. Uncx4.1 is required for the formation of the pedicles and proximal ribs and acts upstream of Pax-9. Development 2000; 127:2251-2258.

271. Leitges M, Neidhardt L, Haenig B et al. The paired homeobox gene Uncx4.1 specifies pedicles transverse processes and proximal ribs of the vertebral column. Development 2000; 127:2259-2267.

272. Schrägle J, Huang R, Christ B et al. Control of the temporal and spatial Uncx4.1 expression in the paraxial mesoderm of avian embryos. Anat Embryol 2004;

273. Christ B, Jacob HJ, Jacob M. Experimentelle Untersuchungen zur Entwicklung der brustwand beim Hühnerembryo. Experientia 1974; 30:1449-1451.

274. Henderson DJ, Conway SJ, Copp AJ. Rib truncations and fusions in the Sp2H mouse reveal a role for Pax3 in specification of the ventro-lateral and posterior parts of the somite. Dev Biol 1999; 209:143-158.

275. Tallquist MD, Weismann KE, Hellström M et al. Early myotome specification regulates PDGFA expression and axial skeleton development. Development 2000; 127:5059-5070.

276. Grass S, Arnold HH, Braun T. Alterations in somite patterning of Myf-5-deficient mice. A possible role for FGF-4 and FGF-6. Development 1996; 122:141-150.

277. Jacob M, Jacob HJ, Christ B. Die frühe Differentierung des chordanahen Bindegewebes. Raster- und transmissionselektronenmikroskopische Untersuchungen an Hühnerembryonen. Experientia 1975b; 31:1083-1086.

278. Smits O, Lefebre V. Sox 5 and Sox 6 are required for notochord extracellular matrix sheath formation, notochord cell survival and development of the nucleus pulposus of intervertebral discs. Development 2003; 130:1135-1148.

279. Töndury G. Entwicklungsgeschichte und Fehlbildungen der Wirbelsäule. Hippokrates, 1958; Stuttgart.

280. Chen Z.-F, Behringer RR. Twist is required in head mesenchyme for cranial neural tube morphogenesis. Genes Dev 1995; 9:686-699.

281. Wallin J, Wilting J, Koseki H et al. The role of Pax-1 in skeleton development. Development 1994; 120:1109-1121.

282. Miura N, Wanaka A, Tohyama M et al. MFH1, a new member of the forkhead domain family is expressed in developing mesenchyme. FEBS Lett 1993; 326:171-176.

283. Neubüser A, Koseki H, Balling R. Characterisation and developmental expression of Pax9, a paired-box-containing gene related to Pax1. Dev Biol 1995; 170:701-716.

284. Kaestner KH, Bleckmann SC, Monaghan AP et al. Clustered arrangement of winged helix 1 genes fkh-6 and MFH 1. Possible implications for mesoderm development. Development 1996; 122:1751-1758.

285. Winnier GE, Hargett L, Hogan BLM. The winged helix transcription factor MFH1 is required for proliferation and patterning of paraxial mesoderm in the mouse embryos. Genes Dev 1997; 11:926-940.

286. Behrens A, Haigh J, Mechta-Grigoriou F et al. Impaired intervertebral disc formation in the absence of Jun. Development 2003; 130:103-109.

287. Peters H, Wilm B, Sakai N et al. Pax1 and Pax9 synergistically regulate vertebral column development. Development 1999; 126:5399-5407.
288. Barnes GL, Hsu CW, Mariani BD et al. Chicken Pax-1 gene: Structure and expression during embryonic somite development. Differentiation 1996; 61:13-23.
289. Buchberger A, Schwarzer M, Brand T et al. Chicken winged-helix transcription factor cFKH-1 prefigures axial and appendicular skeletal structures during chicken embryogenesis. Dev Dyn 1998; 212:94-101.
290. Mankoo BS, Skuntz S, Hassegan I et al. The concerted action of Meox homeobox genes is required upstream of genetic pathways essential for the formation, patterning and differentiation of somites. Development 2003; 130:4655-4664.
291. Mankoo BS, Collins NS, Ashby P et al. Mox2 is a component of the genetic hierarchy controlling limb muscle development. Nature 1999; 400:69-73.
292. Kraus F, Haenig B, Kispert A. Cloning and expression analysis of the mouse T-box gene Tbx 18. Mech Dev 2001; 100:83-86.
293. Holtzer H. Morphogenetic influence of the spinal cord on the axial skeleton and musculature. Anat Rec 1951; 109:373-374.
294. Holtzer H. An experimental analysis of the development of the spinal column: the dispensability of the notochord. J Exp Zool 1952a; 121:573-591.
295. Holtzer H. An experimental analysis of the development of the spinal column. I. Response pf precartilage cells to size variations of the spinal cord. J Exp Zool 1952b; 121:121-148.
296. Holtzer H, Detwiler SR. An experimental analysis of the development of the spinal column. III, Induction of skeletogenous cells. J Exp Zool 1953; 123:335-366.
297. Holtzer H, Detwiler SR. The dependence of somitic differentiation on the neural axis. Anat Rec 1954; 118:390.
298. Lash J, Holtzer S, Holtzer H. An experimental analysis of the development of the spinal column. Exp Cell Res 1957; 13:292-303.
299. Lash JW, Hommes FA, Zilliken F. Induction of cell differentiation. I. The in vivo induction of vertebral cartilage with a low-molecular-weight tissue component. Biochim Biophys Acta 1962; 56:313-319.
300. Avery G, Chow M, Holtzer H. An experimental analysis of the development of the spinal column. V. Reactivity of chick somites. J Exp Zool 1956; 132:409-426.
301. Strudel G. L'action morphogène du tube nerveux et de la corde sur la differenciation de vertebrès et des muscles vertebaux chez l'embryon de poulet. Arch Anat Microsc Morphol Exp 1955; 44:209-235.
302. Strudel G. Induction de cartilage in vitro par l'extrait de tube nerveux et de chorde de l'embryon de poulet. Dev Biol 1962; 4:67-86.
303. Christ B. Experimente zur Lageentwicklung der Somiten. Verh Anat Ges 1970; 64:555-564.
304. Murtaugh LC, Chyung JH, Lassar AB. Sonic hedgehog promotes somitic chondrogenesis by altering the cellular response to BMP signaling. Genes Dev 1999; 13:225-237.
305. Murtaugh LC, Zeng I, Chyung JH et al. The chick transcriptional repressor Nkx3.2 acts downstream of Shh to promote BMP-dependent axial chondrogenesis. Dev Cell 2001; 1:411-422.
306. Lettice LA, Purdie LA, Carlson GJ et al. The mouse bagpipe gene controls development of axial skeleton, skull and spleen. Proc Natl Acad Sci USA 1999; 96:9655-9700.
307. Tribioli C, Lufkin T. The murine Bapx 1 homeobox gene plays a critical role in embryonic development of the axial skeleton and spleen. Development 1999; 126:5699-5711.
308. Akazawa H, Komuro I, Sugitani Y. Targeted disruption of the homeobox transcription factor Bapx1 results in lethal skeletal dysplasia with asplenia and gastroduodenal malformation. Genes Cell 2000; 5:499-513.
309. Rodrigo I, Bovolenta P, Mankoo BS et al. Meox homeodomain proteins are required for Bapx1 expression in the sclerotome and activate its transcription by direct binding to its promoter. Mol Cell Biol 2004; 24:2757-2766.
310. Rodrigo I, Hill RE, Balling R et al. Pax1 and Pax9 activate Bapx1 to induce chondrogenic differentiation in the sclerotome. Development 2003; 130:473-482.
311. Zeng L, Kempf H, Murtaugh LC et al. Shh establishes an Nkx3.2/Sox9 autoregulatory loop that is maintained by BMP signals to induce somitic chondrogenesis. Genes Dev 2002: 16,1990-2005.
312. Bell DM, Leung KK, Wheatley SC et al. SOX9 directly regulates the type-II collagen gene. Nat Genet 16:174-178.
313. Healy C, Uwanohgo D, Sharpe PT. Expression of the chicken Sox9 gene marks the onset of cartilage differentiation. Ann N Y Acad Sci 1996; 785:261-262.
314. Bi W, Huang W, Whitworth DJ et al. Happloinsufficiency of Sox9 results in defective cartilage primordia and premature skeletal mineralization. Proc Natl Acad Sci USA 2001; 98:6698-6703.
315. Watanabe Y, Le Douarin NM. A role for BMP-4 in the development of subcutaneous cartilage. Mech Dev 1996; 57: 69-78.

316. Iida K, Koseki H, Kakinuma H et al. Essential role of the winged helix transcription factor MFH-1 in aortic arch patterning and skeletogenesis. Development 1997; 124:4627-4638.

317. Freitas C, Rodrigues S, Charrier JB et al. Evidence for medial/lateral specification and positional information within the presomitic mesoderm. Development 2001; 128:5139-5147.

318. Pourquié O, Coltey M, Bréant C et al. Control of somite patterning by signals from the lateral plate. Proc Natl Acad Sci USA 1995; 92:3219-3223.

319. Huang R, Zhi Q, Wilting J et al. Sclerotomal origin of the ribs. Development 2000b; 127:527-532.

320. Wilting J, Brand-Saberi B, Huang R et al. The angiogenic potential of the avian somite. Dev Dyn 1995b, 202:165-171.

321. Eichmann A, Marcelle C, Bréant C et al. Two molecules related to the VEGF receptor are expressed in early endothelial cells during avian embryonic development. Mech Dev 1993; 42:33-48.

322. Nimmagadda S, Geetha Loganathan P, Wilting J et al. Expression pattern of VEGFR-2 (Quek1) during quail development. Anat Embryol 2004; 20:219-224.

323. Nimmagadda S, Geetha Loganathan P, Christ B et al. BMP-4 and noggin control embryonic blood vessel formation by antagonistic regulation of VEGFR-2 (Quek1) expression. Dev Biol 2005; 280:100-110.

324. Norris WE, Stern CD, Keynes RJ. Molecular differences between the rostral and caudal halves of the sclerotome in the chick embryo. Development 1989; 105:541.

325. Tosney KW. Cell and cell-interactions that guide motor axons in the developing chick embryo. Bio Essays 1991; 13:17-24.

326. Davies JA, Cook GM, Stern CD et al. Isolation from chick somites of a glycoprotein fraction that causes collapses of dorsal root ganglion growth cones. Neuron 1990; 4:11-19.

327. Wang HU, Anderson DJ. Eph family transmembrane ligands can mediate repulsive guidance of trunk neural crest cell migration and motor axon outgrowth. Neuron 1997; 18:383.

328. Henkemeyer M, Marengere LEM, McGlade J et al. Immunolocalization of the Nuk receptor tyrosine kinase suggests roles in segmental patterning of the brain and axonogenesis. Oncogene 1994; 9:1001-1008.

329. Ohta K, Nakamura M, Hirokawa K et al. The receptor kinase, Cek8, is transiently expressed on subtypes of motoneurons in the spinal cord during development. Mech Dev 1996; 54:59-69.

333. Robinson V, Smith A, Felnniken AM et al. Role of Eph receptors and ephrins in neural crest pathfinding. Cell Tissue Res 1997; 290:265-274

331. Tan S.-S, Crossin KL, Hoffman S et al. Asymmetric expression in somite of cytoactin and its proteoglycan ligand is correlated with neural crest cell distribution. Proc Natl Acad Sci USA 1987; 84:7981-7988.

332. Mackie EJ, Tucker RP, Halfter W et al. The distribution of tenscin coincides with pathway of neural crest cell migration. Development 1988; 102:237-256.

333. Layer PG, Alber A, Rathjen FG. Sequential activation of butyrylcholinesterase in motoneurons and myotoms preceding growth of motor axons. Development 1988; 102:387-396.

334. Goldstein RS, Teillet, MA, Kalcheim C. The microenvironment created by grafting multiple rostral half-somites is mitogenic for neural crest cells. Proc Natl Acad Sci USA 1990; 87:4476-4480.

335. Goldstein RS, Avivi C, Geffen R. Initial axial level-dependent differences in size of avian dorsal root ganglia are imposed by the sclerotome. Dev Biol 1995; 168:214-222.

336. Stern CD, Keynes RJ. Interactions between somite cells: the formation and maintenance of segment boundaries in the chick embryo. Development 1987; 99:261- 272.

337. Goldstein RS, Kalcheim C. Determination of epithelial half-somites in skeletal morphogenesis. Development 1992; 116:441-445.

338. Oakley RA, Tosney KW. Peanut agglutinin and chondroitin-6-sulfate are molecular markers for tissues that act as barriers to axon advance in the avian embryo. Dev Biol 1991; 47- 156.

339. Kerr RSE, Newgreen DF. Isolation and characterization of chondroitin sulfate proteoglycans from embryonic quail that influence neural crest cell behavior. Dev Biol 1997; 192:108-115.

340. Landolt RM, Vaugham L, Winterhalter KH et al. Versican is selectively expressed in embryonic tissues that act as barriers to neural crest cell migration an axon outgrowth. Development 1995; 121:2303.

341. Eickholdt BJ, Mackenzie SL, Graham A et al. Evidence for collapsin-1 functioning in the control of neural crest migration in both trunk and hindbrain regions. Development 1999; 126:2181-2189.

342. Debby-Brafmann A, Burstyn-Cohen T, Klar A et al. F-spondin is expressed in somite regions avoided by neural crest cells and mediates the inhibition of distinct somitic domains to neural crest migration. Neuron 1999; 22:475-488.

343. Böhme C. Lichtmikroskopische Untersuchungen über die Struktur des Leptomeninx encephalis bei Gallus domesticus. Z Anat Entwickl-Gesch 1973; 140:215-236.

344. Hochstetter F. Über die Entwicklung und Differenzierung der Hüllen des Rückenmarkers beim Menschen. Morph Jb 1934; 74:1-104.

345. O'Rahilly R, Müller F. The meninges in human development. J Neuropath exp Neurol 1986; 45:588-608.

346. Hanincec P, Grim M. Localization of dipeptidylpeptidase IV and alkaline phosphatase in developing spinal cord meninges and peripheral nerve coverings of the rat. Int J Devl Neuroscience 1990; 8:175-185.
347. Marin F, Nieto MA. Expression of chicken slug and snail in mesenchymal components of the developing central nervous system. Dev Dyn 2004; 230:144-148.
348. Kurz H, Gärtner T, Eggli PS et al. First blood vessels in the avian neural tube are formed by a combination of dorsal angioblast immigration and ventral sprouting of endothelial cells. Dev Biol 1996; 173:133-147.
349. Hogan KA, Ambler CA, Chapman DL et al. The neural tube patterns vessels developmentally using the VEGF signaling pathway. Development 2003; 131:1503-1513.
350. Brent AE, Braun T, Tabin CJ. Genetic analysis of interactions between the somitic muscle, cartilage and tendon cell lineages during mouse development. Development 2005; 132:515-528.
351. Smith TG, Sweetman D, Patterson M et al. Feedback interactions between MKP3 and ERK MAP kinase control scleraxis expression and the specification of rib progenitors in the developing chick somite. Development 2005; 132:1305-1314.
352. Folkman J, D'Amore P. Blood vessel formation: what is its molecular basis? Cell 1996; 87:1153-1155.
353. Parker LH, Schmidt M, Jin S.-W et al. The endothelial-cell-derived secreted factor Egfl7 regulates vascular tube formation. Nature 2004; 428:754.
354. Pardanaud L, Dieterlen-Lievré F. Emergence of endothelial and hemopoietic cells in the avian embryo. Anat Embryol 1993; 187:107-114.
355. Klessinger S, Christ B. Axial structures control laterality in the distribution pattern of endothelial cells. Anat Embryol 1996; 193:319-330.
356. Carmeliet P. Mechanisms of angiogenesis and arteriogenesis. Nat Med 2000; 6:389-395.
357. Hagedorn M, Balke M, Schmidt A et al. VEGF coordinates interaction of pericytes and endothelial cells during vasculogenesis and experimental angiogenesis. Dev Dyn 2004; 230:23-33.
358. Wilting J, Papoutsi M, Schneider M et al. The lymphatic endothelium of the avian wing is of somitic origin. Dev Dyn 2000; 217:271-278.
359. Wilting J, Papoutsi M, Othmanm-Hassan K et al. Development of the avian lymphatic system. Microsc Res Tech 2001; 55:81-91.
360. Yamaguchi TP, Dumont DJ, Conlon RA et al. fkl-1 and flt-related receptor tyrosine kinase is an early marker for endothelial cell precursor. Development 1993; 118:489-498.
361. Tosney KW. Cell and cell-interactions that guide motor axonas in the developing chick embryo. Bio Essays 1991; 13:17-24.
362. Ring C, Hassell J, Halfter W. Expression pattern of collagen IX and potential role in the segmentation of the peripheral nervous system. Dev Biol 1996; 180:41.

CHAPTER 2

Avian Somitogenesis:
Translating Time and Space into Pattern

Beate Brand-Saberi,* Stefan Rudloff and Anton J. Gamel

Abstract

Vertebrates have a metameric bodyplan that is based on the presence of paired somites. Somites develop from the segmental plate in a cranio-caudal sequence. At the same time, new material is added from Hensen's node, the primitive streak and the tail bud. In this way, the material residing in the segmental plate remains constant and comprises 12 prospective somites on each side. Prospective segment borders are not yet determined in the caudal segmental plate. Prior to segmentation, the cranial segmental plate undergoes epithelialization, which is controlled by signals from the neural tube and ectoderm. The bHLH transcription factor Paraxis is critically involved in this process. Formation of a new somite from the cranial end of the segmental plate is a highly controlled process involving complex cell movements in relation to each other. *Hox* genes specify regional identity of the somites and their derivatives. In the chicken a transposition of thoracic into cervical vertebrae has occurred as compared to the mouse. Transcription factors of the bHLH and homeodomain type also specify the cranio-caudal polarity and that of particular cell groups within the somites. According to segmentation models, somitogenesis is under the control of a "segmentation clock" in combination with a morphogen gradient. This hypothesis has recently found support from molecular data, especially the cycling expression of genes such as *cHairy1* and *Lunatic Fringe*, which depend on the Notch/Delta pathway of signal transduction. FGF8 has been described to be distributed along a cranio-caudal gradient. The first oscillating gene described shown to be independent of Notch is *Axin2,* encoding a negative regulator of the canonical Wnt pathway and a target of Wnt3a. Wnt3a and Axin2 show a similar distribution as FGF8 with high levels in the tailbud. The chick embryo has recently become accessible to molecular approaches such as overexpression by electroporation and RNA interference which can be expected to help elucidating some of the still open questions concerning somitogenesis.

Introduction

Metamerism as a Construction Plan

The basic organization underlying all vertebrate and many invertebrate embryos is metamerism. This pattern consists of a repeated sequence of segments and is the prerequisite for bodies capable of locomotion, because joints will form in a metameric array of elements at the intersections between the segments, although these may not be identical in position to the primary pattern of segments, due to secondary rearrangement ("resegmentation" described below). Hence, it appears that this construction principle has been conserved during evolution. In spite of the historical view of poly-phyletic development of segmentation, recent molecular evidence suggests that certain principles of metameric development may indeed be shared between protostomes and deuterostomes.[1]

*Corresponding Author: Beate Brand-Saberi—Department of Molecular Embryology, Institute for Anatomy and Cell Biology, Albertstrasse 23 79104 Freiburg, Germany.
Email: beate.brand-saberi@anat.uni-freiburg.de

Somitogenesis, edited by Miguel Maroto and Neil V. Whittock. ©2008 Landes Bioscience and Springer Science+Business Media.

However, in many multicellular organisms two different ways of segmentation must be distinguished: The first involves the regular partitioning of a pre-existing tissue, whereas the second arises dynamically as new tissue is being generated. Somitogenesis is an example of the second type of segmentation. For the vertebrate body, this segmentation determines the basic pattern of the musculoskeletal system of the trunk as well as the secondary segmental arrangement of the peripheral nervous system[2] and the adaxial blood vessels.

A number of classifications for somites have been developed, one of the first being that of Christ and Ordahl.[3] According to their nomenclature, the youngest somite pair is designated as I, the second last as II and so on. Extension of this terminology into the unsegmented portion of the paraxial mesoderm resulted in the introduction of negative Roman numerals, such as I for the forming somite and II for the next prospective one.[4] In a new nomenclature, the nascent somite is classified as S0; consequently the last somite is called SI and the next prospective somite S-I.[5]

For more convenience and to distinguish the two nomenclatures more easily, many authors use Arabic numerals, e.g. S1, S0, S-1, instead of Roman numerals. In this review, the latter nomenclature will be used, which is compared to the older one in Figure 1 and 3.

Gastrulation

In vertebrate embryos, gastrulation is a crucial event, which leads to the formation of the three germ layers, ectoderm, mesoderm and endoderm. The mesoderm yields the largest pool of cells in the vertebrate body, the material for the locomotory system including cartilage, bone, skeletal muscle and tendons, the heart and the greatest part of the blood and lymphatic vessels, as well as the blood cells and the urogenital system.

During gastrulation in the chick, mesendodermal progenitor cells ingress through the primitive streak and Hensen's node by delaminating from the epiblast, which is hence the source of all three germ layers.[6] Delamination occurs as a consequence of destabilization of cell-cell contacts such as cadherin-mediated adherens junctions. This can be achieved by catenin phosphorylation after activation of the receptor tyrosine kinase Met by its ligand Scatter factor/Hepatocyte Growth Factor (SF/HGF) that has been shown to be expressed in the primitive streak.[7]

Ingression from the epiblast is a continuous process that starts cranially and goes on as the primitive streak retreats caudally. At this time, first oscillations of gene expression are described.[8] which also play a role later during somitogenesis (see below). Upon its formation, the mesoderm soon partitions into four distinct compartments: the axial, paraxial, intermediate and lateral plate mesoderm. To understand the process of somitogenesis, it should be borne in mind that paraxial mesoderm is continuously formed as the primitive streak regresses and further caudal elongation of the embryonic axis occurs, when mesenchymal cells are added from the tailbud and form the segmental plate (a term traditionally used in chick embryos synonymous to the presomitic mesoderm in the mouse). As the paraxial mesoderm matures, it will give rise to paired metameric spheres: the somites.

Epithelialization of the Segmental Plate

During segmentation, the length of the segmental plate remains fairly constant comprising the material for 12 prospective somites, which means that the rate of somite formation must be somehow linked to proliferation. The formation of somites from the segmental plate involves a mesenchymal-to-epithelial transition (MET).

The mesenchymal state of the PSM is likely to be maintained by the repressive action of Snail1 (mouse) or Snail2 (chicken) of transcription factors on the expression of genes associated with an epithelial phenotype.[9,10] By binding to E-boxes in the promoter region of E-cadherin, for example, they not merely cause the downregulation of this gene, but furthermore, contribute to the accumulation of free β-catenin in the cell, which is important for transducing Wnt signals, as will be described later. Just prior to segmentation, this negative regulation subsides which can be observed by morphological changes that only occur in the cranial portion of the segmental plate. Here, the cells increase in number, become denser and are arranged in a simple epithelium at the periphery of the paraxial mesoderm.[11,12] During this process, the bHLH-transcription factor Paraxis starts to be expressed in the cranial half of the segmental plate. The cell adhesion molecule N-cadherin[12] is upregulated in the two to three prospective somites. Subsequently, the cadherin-associated intracellular protein β-catenin becomes strongly expressed from S-1 on.[13] β-catenin serves a dual role: A structural one by linking the intracellular portion of cadherin to associated molecules and the actin-cytoskeleton, as well as a role as a signalling molecule mediating in the canonical Wnt-pathway. In the latter, β-catenin becomes stabilized through the inhibitory effect of activated Dishevelled upon GSK-3β, translocates into the nucleus and coactivates the LEF-1/TCF transcription factor by displacing the corepressor protein Groucho.[14] Paraxis has been identified as a regulator of the epithelial organization in the paraxial mesoderm.[15] Murine mutants in the *paraxis* gene fail to form epithelialized somites, but the paraxial mesenchyme still segregates into sequential portions assumed to correspond to somites in normal mice. This knockout mutant shows that epithelialization and

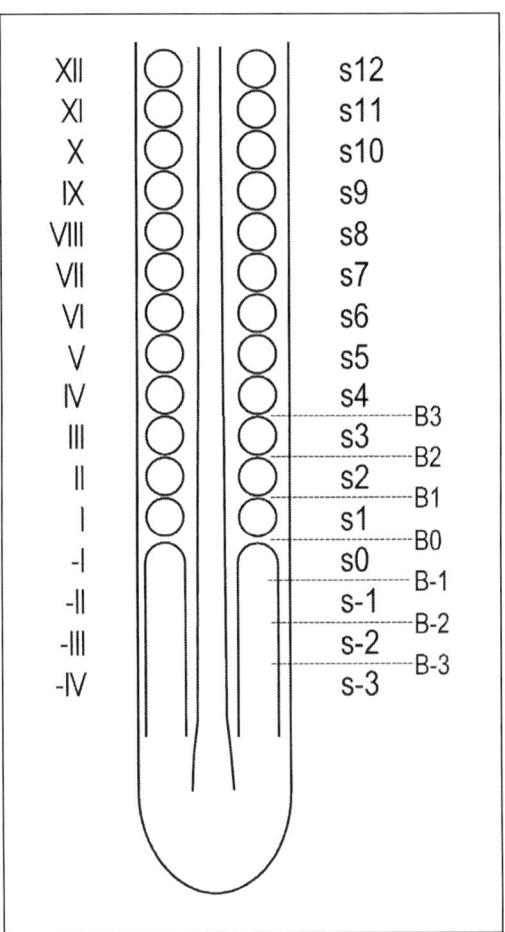

Figure 1. Highly regionalized or stage-dependent expression patterns and processes require precise terms to communicate. Here two widely used nomenclatures for somite staging in avian somitogenesis are compared. (Left side) Roman numerals are used for somites, to indicate segmental plate levels preceded by a minus sign. (Right side) Arabic numerals combined with S, as abbreviation for somite, using S0 for the nascent somite and negative numbers to indicate segmental plate levels. Intersomitic levels are indicated by B followed by the number of the following somite stage.

segmentation are independent events. Apart from segmentation, specification of somitic cells was normal as assessed by the sclerotome markers *Scleraxis, Pax1* and *Pax9* and the dermomyotomal markers *Pax3, Pax7,* and myogenic markers such as *Myogenin.*[16]

In this context, it is important to mention that some authors have described pre-existing groups of mesenchymal cells in the segmental plate, the so-called somitomeres.[12,17,18] The somitomeres can be observed by stereo scanning electron microscopy as concentric swirls of cells in the paraxial mesoderm.[19,20] The number of somitomeres corresponds to the number of prospective somites that are contained within the segmental plate. The borders between individual somitomeres are not well delineated and they do not represent compartments before somites are formed. As described below, a surprising degree of highly controlled cell rearrangement occurs in the progression from somitomere to somite. Furthermore, the somitomeres do not reflect a prepattern for the somitic segments, because the paraxial mesoderm has been shown by grafting experiments to be capable of regulating their allocation to particular somites.[21] This capacity has also been found when signalling molecules like FGF8 and Wnt3a were applied to the segmental plate, which can alter the size and number of the segments.[22] Determination of the cranio-caudal identity occurs at S-1/S0 and hence precedes the determination of dorso-ventral identity in the paraxial mesoderm.[16,23] The latter is stably determined only after S3.[24]

In spite of the fact that *paraxis*-mutants still show segmentation, this appears to be true mainly of sclerotome derivatives, while the dermomyotome—and as a result the myotome—show compromised development; especially *Pax3* expression was decreased. Thus, epithelialization may be instrumental to the normal patterning process.[25] The importance of cell adhesiveness and cell contacts has been recognized for some time.[26,27] It has been shown that cell-to-cell contacts are especially abundant in the cranial portion of the segmental plate in comparison to the caudal portion.[28,29] Targeted mutations in the N-cadherin gene revealed defects in somitogenesis with smaller and irregularly formed segments.[30]

Segmentation occurs in a highly controlled fashion in the transition zone between the first somitomere (S0) and the last somite (S1). This transition is accompanied by rapid changes in the cytoskeleton. Due to the dynamic and three-dimensional nature of this process, it can best be investigated by in vivo imaging. Kulesa and Fraser[31] have taken an admirable approach by labelling cells in the segmental plate of embryos in culture and following up the process of somite formation by time-lapse microscopy. From their studies, it became clear that the segment boundary forms in a specific way, which cannot easily be anticipated from histological sections. Epithelialization seems to first occur at the anterior and medial border of somite S0. It then continues in posterior direction and sculpts the newly formed somite by finishing at the posterior border. Here, a complex choreography of cells movements was disclosed as the new somite separates from the segmental plate. Instead of a straight clean cut across the segmental plate, the separation manoeuvre takes a form resembling a ball in a socket. Interestingly, not all cells contribute to S0, as the cells composing the wall of the socket retract posteriorly across the future segment border and curve inward, eventually building the anterior border of the next to form somite (S-1). Contemporaneously, the group of cells left behind moves anteriorly and assembles the posterior border of S0. Thus the sliding movement of these two groups of cells in opposite directions leads to the formation of the somitic border (Fig. 2). The intertwined nature of the separating somite/segmental plate boundary is unanticipated in the process of somitogenesis showing the importance derived from the integration of 4D data into the analysis of developmental processes.

These findings also have an important bearing on the interpretation of expression patterns of genes implied in the control of somitogenesis. Surprisingly from these in vivo data, most of the genes that are thought to be involved in somitogenesis show a straight expression border across the cranial segmental plate.

Some light has been shed on the generation of the medial and lateral restriction of the somites: BMPs are the prevailing signals in the lateral plate mesoderm. In experiments where the BMP antagonist Noggin was applied to the lateral plate, ectopic somites formed laterally to the normal paraxial mesoderm. The results show that the mesodermal compartments can be respecified by

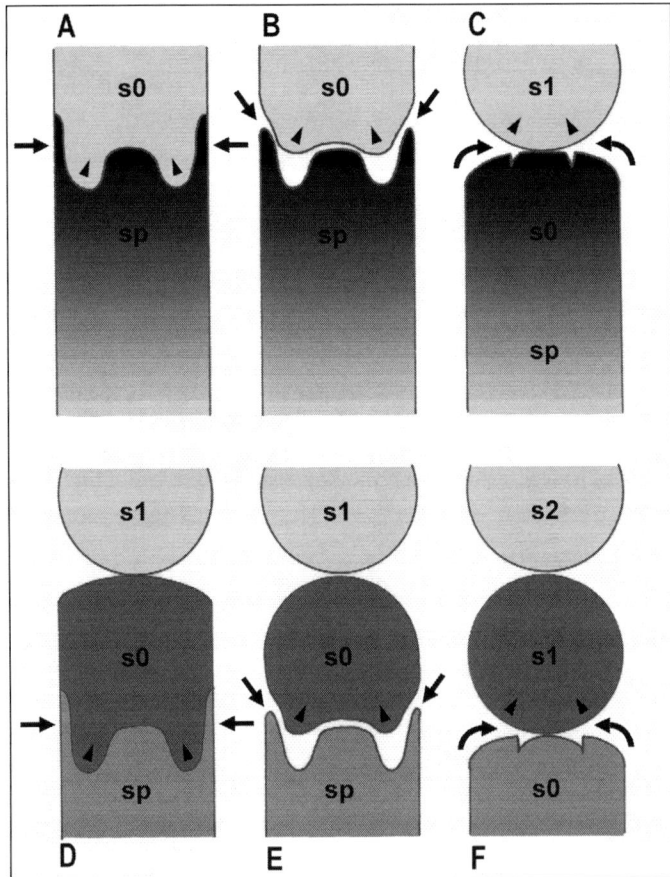

Figure 2. Schematic drawing to illustrate six phases of the complex process that forms a complete somite sphere from the segmental plate (shaded). A-C) The segmental plate (sp) first forms a kind of socket (arrows) for the nacent somite (S0). The separation from S0 is completed by withdrawel of the caudal cell masses of S0 (arrowheads) from the socket to form the new somite (S1). At the same time the overhanging lips of the socket retract and fold inwards to fill the gaps and to form the cranial epithelialized border of the next nascent somite (bent arrows) and (D-E) the next cycle ensues to complete the formation of the next somite (new S1). Drawing according to Kulesa and Fraser.[31]

ectopic signals.[32] During normal development, *Noggin* is expressed in the early segmental plate and also at the level of S0. Its function here could serve the formation of somites in the presence of high levels of BMPs emanating from the intermediate mesoderm and lateral plate (Fig. 3B).

The Anterior Posterior Polarization of the Paraxial Mesoderm

The findings reported in the previous paragraph indicate that the somitic boundary cells are likely to have special properties making them competent to undergo complex morphogenetic process.[31] In fact, grafting experiments between quail and chick segmental plates showed that the formation of a somitic boundary is established by a particular set of cells capable of organizing ectopic boundaries when grafted to a new position within the segmental plate. Those

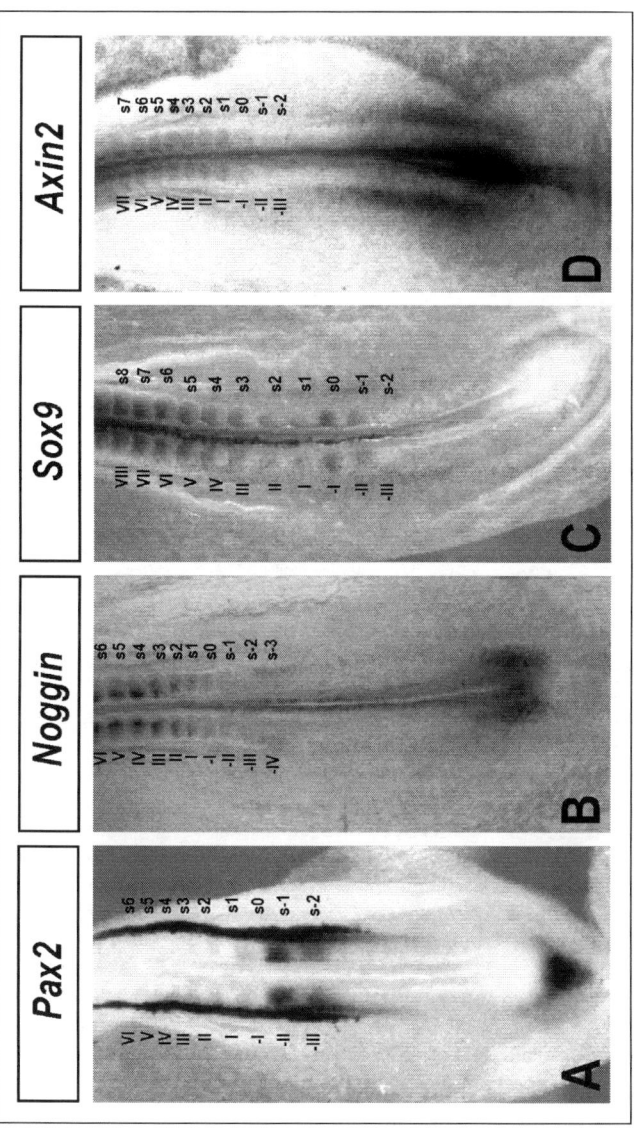

Figure 3. Examples of gene expression during avian somitogenesis indicating genes that are each involved in different processes and phases of somite development. Staging of somites and segmental plate mesoderm is given in both nomenclatures according to Figure.1. A) Pax2 expression in the segmental plate at somite levels S-1 and S-2. The cranial expression domain condenses before ceasing in the nascent somite (S0). Pax2 is also strongly expressed caudally at Hensen's node and laterally to the paraxial mesoderm in the Wolffian duct as well as in the underlying intermediate mesoderm. B) Noggin expression shows up first in the caudal paraxial mesoderm where it fades out towards the presomitic levels. Later the expression comes up again in the somites from stage S0 on and becomes manifest in the dorsomedial part of the more mature somites. C) Sox9, a marker for neural crest cells and cartilage formation, here expressed in the differentiating sclerotome of the cranial somites from somite S5 on, is also expressed at presomitic levels (S0) and (S-1). D) Snapshot of Axin2 expression pattern. The expression can be found in the caudal segmental plate mesoderm whereas the cranial part of the segmental plate shows no expression. In the more mature somites from somite stage (S4) on, Axin2 is also expressed in the medial halves of the somites.

boundary-forming cells have an instructive effect on the cells anterior to them and cause these to epithelialize and separate.[33] In this connection, two important molecules have been identified which mediate the property of boundary induction: Notch and Lunatic fringe. Notch represents together with Delta (in the chick Delta-like1 and Delta-like3) a transmembrane receptor-ligand system, which has a key role in somitogenesis. Signalling occurs when Notch on the one cell binds Delta on another cell resulting in a conformational change in the intracellular portion of Notch (NIC). This enables the protease Presenilin-1 to cleave NIC leading to the entry of that cleaved portion into the nucleus where it binds to the CSL (CBF1, SuCH & Lag1 group = suppressor of hairless in *Drosophila*)-transcription factor.[34] By NIC binding to CSL in the control region of target genes, a repressor is removed and activation of transcription is enabled, mainly by histone acetyltransferases.[35] Lunatic fringe, on the other hand, is a glycosyltransferase that has been shown to modulate Notch activity,[36] and is itself a target of Notch. This relationship causes a negative feedback loop leading to a repeated boundary formation.[34,37-39] But not only the existence of boundary cells suggests a regionalization of the segmental plate. Upon rotation by 180° along the anterior-posterior axis, segmentation occurs caudally instead of cranially, implying that only the cranial portion of the segmental plate has acquired the potential to generate somites.[11]

Two aspects are to be considered in anterior-posterior polarization of the paraxial mesoderm: The first is the cranio-caudal gradient of maturity that results from the inherent nature of mesoderm formation by caudal addition of material. The second aspect is the large scale patterning of the vertebrate body by the Hox-code. The first aspect will be addressed in a separate paragraph of this chapter dealing with cycling genes or oscillation. The second aspect is of major importance for the regionalization of the trunk and for the initiation areas of limb bud development. Hence there are high similarities between different vertebrate taxa, but also some characteristics that distinguish the chicken from other vertebrates.

Genes affecting the pattern of body segments of different regions were first described in *Drosophila*,[40] but soon turned out to be evolutionary highly conserved and could be identified throughout the animal kingdom.[41-43] The Hox genes are coding for homeobox-containing transcription factors which result in so-called homeotic transformations when affected by mutations. Vertebrate *Hox* genes are organized in four clusters (A, B, C and D), which map to different chromosomes and go back to one original cluster in lower organisms as a result of gene-duplication. Each cluster contains 13 groups of genes that are arranged along the chromosome in a sequential fashion (A1 towards the 3' end and A13 towards 5'end). Corresponding genes in each cluster are called paralogues. Gene expression of the members in all paralogous groups shows a patterned sequence that corresponds to their arrangement on the chromosome. This remarkable feature is termed colinearity.

Although expression of *Hox* genes is first initiated during formation of the primitive streak, they fulfil their patterning function after gastrulation when a sharp cranial border of expression is established for each gene.[44,45] The "code" is based on the combination of *Hox* genes expressed in each particular body region and translated into regional differences between vertebrae and intrinsic back muscles and other regional characteristics such as the development of the limb girdles.[46] In the chick, there exist four somites that become part of the head (4.5 to be precise due to resegmentation), 14 somites form the cervical region, seven form the thoracic region, four somites form the lumbar region and finally nine somites from the sacral region. In comparison to the mouse, this means a massive transposition between the cervical (seven somites) and the thoracic (13 somites) region. The transposition is paralleled by posteriorization of the expression of the paralogues 6-9 (reviewed in ref. 47).

In spite of an overwhelming body of evidence for the action of the Hox code, there is still very little information concerning their mode of action. A crucial problem is that up to now no target genes of Hox transcription factors have been identified in vertebrates, whereas a number of candidate genes have already been described in Drosophila.[48] For the *Hoxb* genes, it has recently been demonstrated that their colinear expression in the epiblast regulates the movements of cells during gastrulation.[49] In this way, *Hoxb* genes exert a direct control on the establishment of spatial colinearity. These important findings imply that the *Hoxb* genes control targets that have a function in the modulation of migratory properties of cells.

Resegmentation of the Somitic Derivatives

The function of the musculo-skeletal system is only possible, if muscle and skeletal elements are arranged out of register resulting in a bridging of joints by muscle. As muscle and cartilage arise from the same entity, the somite, this can only be achieved by a positional rearrangement of these two elements. This process has been postulated as early as 1850 and was described as "Neugliederung" or "resegmentation".[50] Remak proposed that the caudal part of one somite fuses with the cranial part of the next somite. Later reports support this view.[51-53] The morphological bases for this are the presence of an intrasegmental boundary and the capability of cells to spread around and rearrange. Both requirements are met in the sclerotome. The intrasegmental boundary was first observed by von Ebner (1888; von Ebner's fissure[54]). It divides the sclerotome into a cranial and a caudal half. Sclerotome cells are highly motile mesenchymal cells, which fuse across the primary segment boundary to form the anlage of one vertebra. As a consequence, von Ebner's fissure demarcates the intervertebral fissure. A number of experimental studies in avian embryos confirm this process. Quail-chick grafting of single somites show that each somite contributes to one half of two adjacent vertebral bodies and the intervening vertebral disc.[55-58] Resegmentation does not only apply to the vertebral body, but also to the neural arches, spinous and transverse processes and the ribs. In contrast, the insertions and origins of segmental muscles are formed from one somite so that the myotomes of one somite link the cranial and caudal sclerotome halves. Hence, one somite represents the precursor of a motion segment (i.e., facing halves of two adjacent vertebrae, vertebral disc and segmental muscle).

Hence, segments of the vertebral column resemble segments of *Drosophila*, which are formed by the caudal half of one parasegment and the cranial half of the adjacent parasegment. This means that the status of the somite equals that of an insect parasegment. The similarity between the two processes in different phyla suggests a common segmented ancestor.[1,59]

Regionalization of Somites and Segmental Plate

Homeobox-containing genes may also specify cell behaviour in particular regions. For example the expression of *Lbx1*, a homologue of the *Drosophila* gene *ladybird* in vertebrates,[60] (reviewed in ref. 61), is restricted to the regions that give rise to individually migrating myogenic precursor cells detaching from the lateral dermomyotome. This occurs at limb bud levels (somites 16-21 for the wing bud; somites 26-32 for the leg bud) and at the cranio-cervical level (somites 2-6).

Especially the transition zone from S-1 to S0 is a region of dynamic changes in expression of many genes, some of which were not anticipated to be expressed in that area. A candidate gene for regulating cytoskeletal changes that are involved in somitogenesis is *EphA4*.[62] It encodes an ephrin receptor expressed at the level of S0 and S-1 through S-3. Ephrin-mediated signalling links to activation of members of the Rho small GTPase family[59] and has been implicated in neuronal pathfinding.[63,64] *EphA4* expression depends on signals from the ectoderm known to be also important for epithelialization of the segmental plate.[62] However, targeting the *EphA4* gene has not lead to a somite phenotype, possibly by redundancy of more than one Ephrin receptor involved in this process.[65,66]

Uncx4.1, a homeodomain-containing transcription factor, is expressed in each caudal half of the prospective and definite somites from S-1 on.[67] Therefore, Uncx4.1 is commonly used as a marker for proper formation of the caudal half of the somite. Up to now only *Pax9* is identified as a downstream target of Uncx4.1.[68] Since *Pax9* is expressed considerably later in development and the phenotype of the Uncx4.1$^{-/-}$ mouse is much more severe than the Pax9$^{-/-}$ phenotype, Uncx4.1 is likely to control additional targets in the presomitic mesoderm as well as later in the sclerotome forming the vertebral column.

The *Meso* genes are also involved in polarization of the paraxial mesoderm, but in contrast to Uncx4.1, the transactivating *cMeso1* is predominantly expressed in the cranial portion of the prospective somites. Its closely related neighboring gene *cMeso2* is expressed in the caudal portion of the prospective somites. *cMeso2* lacks the transactivation domain and may rather act as a repressor, possibly for *cMeso1*.[69,70]

One of the chief prerequisites for the realization of the vertebrate bodyplan is to keep premature differentiation at bay. The tendency to differentiate is present in mesodermal cells from gastrulation on and particularly applies to myogenesis. This has lead to the idea that signals involved in myogenesis are permissive rather than instructive. The myogenic bias increases with maturation and becomes especially high at somitogenesis, possibly due to an increase in cell contacts.[71,72] The expression of a number of genes such as *Pax2* and *Sox9* in the segmental plate and early somites of avian embryos may appear less enigmatic by interpreting their presence in this context. The HMG gene *Sox9* is a main switch gene for cartilage development. It is expressed in the sclerotome and initiates cartilage formation in the vertebral column.[73,74] Nevertheless in chicken, *Sox9* is expressed in the presomitic mesoderm (S0 and S-1) in a highly dynamic manner (Fig. 3C). It is at present not clear whether this is only a sign of early initiation of cartilage differentiation comprising the plasticity of the paraxial mesoderm or whether Sox9 has distinct functions in somite formation.

Beside *Pax,3* a marker of myogenic specification which is expressed throughout the segmental plate and later in the dorsal half of the somites,[75] *Pax2* is the second known paired-box gene that is expressed in the chick presomitic mesoderm.[76,77] The sevenpass transmembrane receptors of the frizzled family mediate Wnt signals. Receptor binding can be competitively inhibited by secreted frizzled related proteins (sFRP). Recently sFRP2 was identified as a target of Pax2.[78] In this way *Pax2* may, by inhibition of Wnt signalling, contribute to somite formation and prevent differentiation of mesoderm cells ahead of time. Further support for the importance of regulating Wnt signals comes from investigating sFRP1/2 double mutants.[79] Herein, the observed mutant embryos disclose defects in AP axis formation in the thoracic region including an anterior shift in the expression of the *Hox* code, as well as, randomized, incomplete segmentation in the interlimb region. Inactivation of sFRPs in the mutants leads to an activation of *Wnt3a* in the tail bud, which causes perturbations in Notch-related oscillating genes like *Lfng* or *Hes7* that eventually account for the abnormal somitogenesis and malformations.

Wnts from the neural tube (Wnt1, Wnt3a and Wnt4) are known to trigger robust muscle differentiation in the somitic paraxial mesoderm.[80] Already in the cranial segmental plate the muscle regulatory factor *MyoD* is under the control of Wnt5b. At the same level the *frizzled receptors 1, 2* and *7* are highly expressed. Premature muscle differentiation is furthermore prevented by the influence of BMPs.[81]

Finally, *Axin2* is dynamically expressed in the segmental plate and later in the medial portion of the somites. Axin2 has been shown to block the canonical Wnt-pathway and is at the same time involved in the segmentation clock.[22]

Oscillations in Gene Expression Underlying Somitogenesis

The formation of somites from the segmental plate is a highly controlled, repetitive process that occurs at a periodicity of 90 minutes in the chicken embryo. One obvious advantage of this is the synchronization of the two sides of the body. However, the generation of this periodicity has been a phenomenon that has fascinated many scientists for a long time and is still the object of intense investigation.

Segmentation has been considered to be an intrinsic property of the paraxial mesoderm, because somites can form from isolated segmental plate in vitro in the absence of neural tube and notochord.[82,83] The ectoderm has later been shown to account for epithelialization and dermomyotome development.[84]

To explain the periodicity of somite formation, several theoretical models have been put forward,[85-88] among which the "clock and wavefront" model by Cooke and Zeeman (1976) has received new attention as a result of the discovery of genes that are periodically expressed, the so-called cycling genes (reviewed in ref. 5). The "clock and wavefront" model postulates the existence of a molecular clock in the segmental plate and a wavefront of cell maturation enabling somites to develop in the cranial portion of the segmental plate. The wavefront can also be interpreted as an anterior-posterior gradient, the slope of which determines the speed of segmentation.[89] Thus, that model has also been referred to as the "clock and gradient" model.[90,91]

The first gene to be described to undergo cyclic expression was the basic helix-loop-helix (bHLH) transcriptional repressor c-hairy1 in the chick.[92] The periodicity of expression coincides with the cycles of somite formation, i.e., 90 minutes in the chick. The expression pattern is highly dynamic and can be divided into three phases (Fig. 4). Initially, *c-hairy1* is expressed in a broad domain in the posterior segmental plate including the tail bud (Fig. 4A). In the next phase, *c-hairy1* expression sweeps anteriorly and finally contracts to a single domain at the level of S-1 (third phase, Fig. 4B,C) before the expression cycle restarts from the tail bud. In the newly formed somites, *c-hairy1* expression is restricted to the most caudal portions (Fig. 4C,A).

The described expression pattern is unaffected by the presence or absence of the tail bud or any of the surrounding tissues, like neural tube, notochord and ectoderm etc. *c-hairy1* also oscillates in explant cultures of the segmental plate. It was an explant system that first provided evidence for the cyclic nature of *c-hairy1* expression. Embryos of particular stages were cut apart in the mediosagittal plane. One half was fixed immediately and subjected to in situ hybridization for *c-hairy,1* while the other half was cultivated for different time periods (15 to 90 minutes). The shift in expression between the left and the right side of the same embryo indicated that gene expression was periodically changing with each cycle of somite formation.[92]

Oscillating genes have hence been described in many other vertebrates during somitogenesis. In avian embryos, *Lunatic fringe* (Lfng)[93,94] was another one of high interest, because it encodes a modulator of Notch activity and is a target of Notch,[36,93-95] which is also true for *c-hairy1* and *Delta*.[8,92,96] Thus all oscillating genes mentioned so far are somehow related to the Notch pathway and consequently cycle in phase with Notch. The first oscillating gene described that is independent of Notch is *Axin2*, a negative regulator of the canonical Wnt pathway.[22]

A requirement for a superimposed oscillator that could act independently of Notch can be deduced from the fact that cycling gene expression is still observed in mutants of the Notch pathway of mouse and zebrafish.[25,36,97-111] Surprisingly, although all mutants show defects in segmentation, cyclic gene expression was not totally absent, but rather uncoordinated. This suggests that the Notch pathway could exert a synchronizing function on cell-intrinsic oscillations and does not necessarily account for their generation. Two signalling molecules have been described to be distributed in a graded fashion along the anterior-posterior axis of the embryo: Wnt3a and FGF8.[22,112,113] Both molecules are present at high levels at the posterior end of the embryo. The domain of high FGF8 expression shows a high correlation with the region of the paraxial mesoderm that is still labile in determination, whereas the domain of downregulated *fgf8* expression equals the "determination zone".[112] Wnt3a has been shown to be involved in elongation of the anterior-posterior axis.[114,115] Recent evidence suggests a role for Wnt-signalling in segmentation.[22] A link between the gradient and the clock is provided by the fact that Wnt3a activates the transcription of its own inhibitor, *Axin2*. Axin2 in turn, leads to a destabilization of β-catenin by activation of GSK3 and interrupts the canonical Wnt-pathway. Thus it regulates its own transcription via a negative feedback loop. When Axin2 is degraded, Wnt signalling is reassumed. In this way, periodic cycles are generated that can be translated into a metameric pattern of somites. Notably, *Axin2* cycles out of phase with *Lfng*. Furthermore, the hypomorph *Wnt3a* mutant, *vestigial tail*, shows absence of *Lfng* oscillation and strong downregulation of *fgf8* expressed in a similar fashion in the chick embryo, where its involvement in segmentation can be examined directly. Dact1, another Wnt/beta-catenin antagonist[116] that binds to and regulates Dishevelled, an integral component of the canonical Wnt-pathway, also cycles in the PSM along with Axin2.[117] The presence of several LEF-1/TCF binding sites in its promoter and first intron make Dact1 a plausible candidate for cooperating with Axin2 in the regulation of periodic Wnt-signalling. Finally, a link between Wnt-signalling and Notch-signalling is conceivable at the level of Dishevelled, which may inhibit Notch signalling and is itself being negatively influenced by Axin2 and Dact1.[22,117] Thus, whenever Wnt signalling is acting, Notch signalling will be switched off and vice versa.

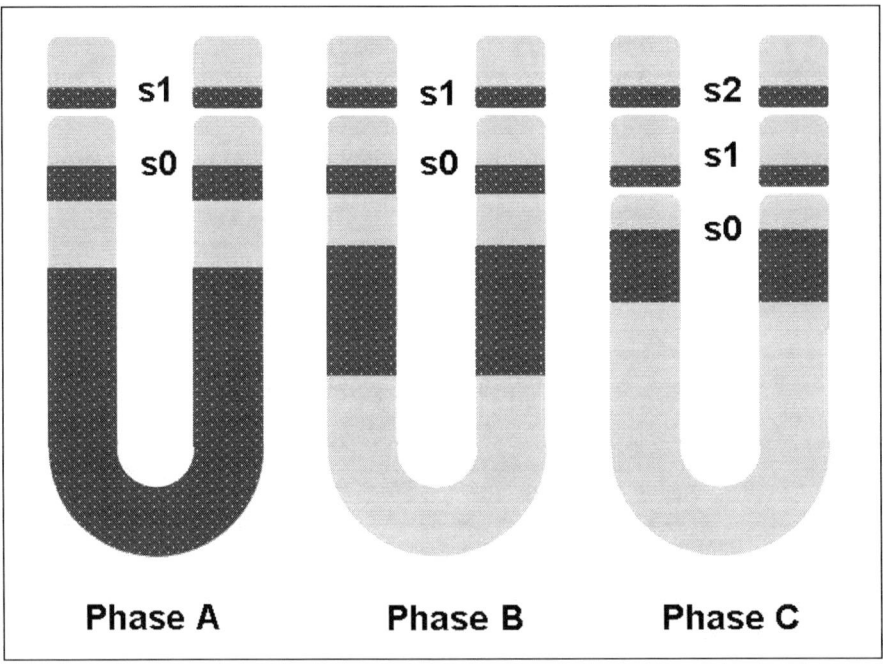

Figure 4. Schematic overview summarizing the cycling expression pattern of chairy1 in chick paraxial mesoderm over a period of 90 minutes. A) Initial expression of chairy1 in the posterior segmental plate. B) Expression shifts towards the cranial part of the segmental plate and C) condenses posteriorly to the next nascent somite and in its caudal half. Expression pattern then starts over to phase (A), where expression in the posterior segmental plate comes up again and expression in the nascent and then newly formed somite is restricted to the most caudal portion.

Recently, a third mode of oscillation was described for the transcriptional repressors Snail1 or Snail2 in mouse or chicken embryos, respectively.[118] Although *Snail1* cycles in phase with Notch target genes, it does not depend on Notch signalling, as its dynamic expression pattern is still active in Notch pathway mutants. Instead, *Snail1* depends on the integrity of the Wnt pathway and FGF8 signalling. This unique regulation of the *Snail* genes brings together all three main signalling pathways in the PSM (Notch, Wnt and FGF), integrating the clock mechanism with the wavefront progression. Moreover, the repressive function of Snail on epithelial markers and the liberation of beta-catenin from the cadherins in the first place could make the PSM competent to respond to the signals of the segmentation clock.[118] Overexpression of *Snail2* in the chicken embryo with the help of in ovo electroporation blocks segmentation by suppressing *Lfng* and cMeso1 expression and prevents MET, which is marked by the absence of *Paraxis* in transfected areas. With dropping below the threshold for FGF8 at the determination front, the negative influence of the *Snail* genes on segmentation and MET subsides, which stabilizes the Notch response at the most anterior PSM resulting in border formation and epithelialization.

Conclusion and Future Considerations

As we have seen, somitogenesis involves various interesting aspects of how changes in the behaviour of individual cells are used to create a coordinated pattern of striking symmetry. The avian embryo lends itself especially to manipulations that help us to elucidate the process of somitogenesis, because it is easily accessible. This will be possible to an even greater extent in the future, due to the release of the chicken genome and the establishment of new in ovo technique such as electroporation and RNA interference. An important aspect in somitogenesis is the synchronization of morphogenetic events, which is achieved through Notch-signalling. The morphological changes leading to somite formation are exerted by the acquisition of cell-contacts and epithelialization of the segmental plate. Positional information is mediated by the Hox code, which in turn depends on the segmentation clock and controls the ingression of cells during gastrulation. Periodically oscillating gene expression in the segmental plate and the evidence for gradients of signalling molecules along the anterior-posterior axis were first discovered in the chick and have recently provided a molecular basis for the postulated "clock and gradient" that accounts for the fascinating process of translating time and space into pattern: somitogenesis.

Acknowledgements

We are grateful to Randy Johnson, Paul Kulesa and Olivier Pourquié for comments on the manuscript, to Eric-Jens Burlefinger for preparation and assembly of the figures and to Ulrike Uhl for technical help with the manuscript. Original work shown in some of the illustrations was supported by the DFG.

References

1. De Robertis EM. Evolutionary biology. The ancestry of segmentation. Nature 1997; 387(6628):25-6.
2. Bronner-Fraser M, Stern C. Effects of mesodermal tissues on avian neural crest cell migration. Dev Biol 1991; 143(2):213-7.
3. Christ B, Ordahl CP. Early stages of chick somite development. Anat Embryol (Berl) 1995; 191(5):381-96.
4. Gamel AJ, Brand-Saberi B, Christ B. Halves of epithelial somites and segmental plate show distinct muscle differentiation behavior in vitro compared to entire somites and segmental plate. Dev Biol 1995; 172(2):625-39.
5. Pourquie O, Tam PP. A nomenclature for prospective somites and phases of cyclic gene expression in the presomitic mesoderm. Dev Cell 2001; 1(5):619-20.
6. Kochav S, Ginsburg M, Eyal-Giladi H. From cleavage to primitive streak formation: a complementary normal table and a new look at the first stages of the development of the chick. II. Microscopic anatomy and cell population dynamics. Dev Biol 1980; 79(2):296-308.
7. Selleck MA, Stern CD. Fate mapping and cell lineage analysis of Hensen's node in the chick embryo. Development 1991; 112(2):615-26.
8. Jouve C, Palmeirim I, Henrique D et al. Notch signalling is required for cyclic expression of the hairy-like gene HES1 in the presomitic mesoderm. Development 2000; 127(7):1421-9.
9. Barrallo-Gimeno A, Nieto MA. The Snail genes as inducers of cell movement and survival: implications in development and cancer. Development 2005; 132(14):3151-61.
10. Cano A, Perez-Moreno MA, Rodrigo I, Locascio A, Blanco MJ, del Barrio MG, Portillo F, Nieto MA. The transcription factor snail controls epithelial-mesenchymal transitions by repressing E-cadherin expression. Nat Cell Biol 2000; 2(2):76-83.
11. Keynes RJ, Stern CD. Mechanisms of vertebrate segmentation. Development 1988; 103(3):413-29.
12. Tam PP, Trainor PA. Specification and segmentation of the paraxial mesoderm. Anat Embryol(Berl) 1994; 189(4):275-305.
13. Marcelle C, Lesbros C, Linker C. Results and problems in cell differentiation. In: Brand-Saberi B, ed. Vertebrate Myogenesis. Heidelberg, Berlin: Springer-Verlag, 2002:81-108.
14. Tolwinski NS, Wieschaus E. A nuclear function for armadillo/beta-catenin. PLoS Biol 2004; 2(4): E95.
15. Burgess R, Cserjesi P, Ligon KL, et al. Paraxis: a basic helix-loop-helix protein expressed in paraxial mesoderm and developing somites. Dev Biol 1995; 168(2):296-306.
16. Burgess R, Rawls A, Brown D, et al. Requirement of the paraxis gene for somite formation and musculoskeletal patterning. Nature 1996; 384(6609):570-3.
17. Jacobson AG, Meier S. Somites in developing embryos. In: Bellairs R, Ede DA, Walsh DA, eds. New York: Plenum, 1980.
18. Jacobson AG. Somitomeres: mesodermal segments of vertebrate embryos. Development 1988; 104 Suppl:209-20.
19. Meier S. Development of the chick embryo mesoblast. Formation of the embryonic axis and establishment of the metameric pattern. Dev Biol 1979; 73(1):24-45.

20. Tam PP, Meier S, Jacobson AG. Differentiation of the metameric pattern in the embryonic axis of the mouse. II. Somitomeric organization of the presomitic mesoderm. Differentiation 1982; 21(2):109-22.

21. Packard DS, Jr. The influence of axial structures on chick somite formation. Dev Biol 1976; 53(1):36-48.

22. Aulehla A, Wehrle C, Brand-Saberi B, et al. Wnt3a plays a major role in the segmentation clock controlling somitogenesis. Dev Cell 2003; 4(3):395-406.

23. Saga Y, Hata N, Koseki H, et al. Mesp2: a novel mouse gene expressed in the presegmented mesoderm and essential for segmentation initiation. Genes Dev 1997; 11(14):1827-39.

24. Christ B, Brand-Saberi B, Grim M, et al. Local signalling in dermomyotomal cell type specification. Anat Embryol(Berl) 1992; 186(5):505-10.

25. Conlon RA, Reaume AG, Rossant J. Notch1 is required for the coordinate segmentation of somites. Development 1995; 121(5):1533-45.

26. Bellairs R, Curtis AS, Sanders EJ. Cell adhesiveness and embryonic differentiation. J Embryol Exp Morphol 1978; 46:207-13.

27. Bellairs R, Sanders EJ, Portch PA. Behavioural properties of chick somitic mesoderm and lateral plate when explanted in vitro. J Embryol Exp Morphol 1980; 56:41-58.

28. Beloussov LV, Naumidi, II. Cell contacts and rearrangements preceding somitogenesis in chick embryo. Cell Differ 1983; 12(4):191-204.

29. Cheney CM, Lash JW. An increase in cell-cell adhesion in the chick segmental plate results in a meristic pattern. J Embryol Exp Morphol 1984; 79:1-10.

30. Radice GL, Rayburn H, Matsunami H, et al. Developmental defects in mouse embryos lacking N-cadherin. Dev Biol 1997; 181(1):64-78.

31. Kulesa PM, Fraser SE. Cell dynamics during somite boundary formation revealed by time-lapse analysis. Science 2002; 298(5595):991-5.

32. Tonegawa A, Takahashi Y. Somitogenesis controlled by Noggin. Dev Biol 1998; 202(2):172-82.

33. Sato Y, Yasuda K, Takahashi Y. Morphological boundary forms by a novel inductive event mediated by Lunatic fringe and Notch during somitic segmentation. Development 2002; 129(15):3633-44.

34. Lecourtois M, Schweisguth F. Indirect evidence for Delta-dependent intracellular processing of notch in Drosophila embryos. Curr Biol 1998; 8(13):771-4.

35. Wallberg AE, Pedersen K, Lendahl U, et al. p300 and PCAF act cooperatively to mediate transcriptional activation from chromatin templates by notch intracellular domains in vitro. Mol Cell Biol 2002; 22(22):7812-9.

36. Evrard YA, Lun Y, Aulehla A, et al. lunatic fringe is an essential mediator of somite segmentation and patterning. Nature 1998; 394(6691):377-81.

37. Dale JK, Malapert P, Chal J, et al. Oscillations of the snail genes in the presomitic mesoderm coordinate segmental patterning and morphogenesis in vertebrate somitogenesis. Dev Cell 2006; 10(3):355-66.

38. Schroeter EH, Kisslinger JA, Kopan R. Notch-1 signalling requires ligand-induced proteolytic release of intracellular domain. Nature 1998; 393(6683):382-6.

39. Struhl G, Adachi A. Nuclear access and action of notch in vivo. Cell 1998; 93(4):649-60.

40. Lewis EB. A gene complex controlling segmentation in Drosophila. Nature 1978; 276(5688):565-70.

41. Akam M. Hox and HOM: homologous gene clusters in insects and vertebrates. Cell 1989; 57(3):347-9.

42. Kappen C, Ruddle FH. Evolution of a regulatory gene family: HOM/HOX genes. Curr Opin Genet Dev 1993; 3(6):931-8.

43. Duboule D. The function of Hox genes in the morphogenesis of the vertebrate limb. Ann Genet 1993; 36(1):24-9.

44. Gaunt SJ. Expression patterns of mouse Hox genes: clues to an understanding of developmental and evolutionary strategies. Bioessays 1991; 13(10):505-13.

45. Gaunt SJ, Strachan L. Forward spreading in the establishment of a vertebrate Hox expression boundary: the expression domain separates into anterior and posterior zones, and the spread occurs across implanted glass barriers. Dev Dyn 1994; 199(3):229-40.

46. Kessel M, Gruss P. Homeotic transformations of murine vertebrae and concomitant alteration of Hox codes induced by retinoic acid. Cell 1991; 67(1):89-104.

47. Burke AC. Hox Genes and the global patterning of the somitic mesoderm. In: Ordahl C, ed. Somitogenesis. San Diego: Academic Press, 2000:155ff.

48. Graba Y, Aragnol D, Pradel J. Drosophila Hox complex downstream targets and the function of homeotic genes. Bioessays 1997; 19(5):379-88.

49. Iimura T, Pourquie O. Collinear activation of Hoxb genes during gastrulation is linked to mesoderm cell ingression. Nature 2006; 442(7102):568-71.

50. Remak R. Untersuchungen über die Entwicklung der Wirbelthiere. Erste Lieferung über die Entwicklung des Hühnchens im Ei. Berlin: Reimer, 1850.
51. Schultze O. Ueber embryonale und bleibende Segmentierung. Verh. Anat. Ges. 1896.
52. Williams LW. The somites of the chick. Am J Anat 1910; 11:55-100.
53. Sensenig EC. The early development of the human vertebral column. Contr Embryol Carnegie Inst Publ 1949; 33(583):21-41.
54. von Ebner V. Urwirbel und Neugliederung der Wirbelsäule. Sitzungsber. Akad. Wiss. Wien Math. Naturwiss. Kl. Abt. 1888; 97(3):194-206.
55. Beresford B. Brachial muscles in the chick embryo: the fate of individual somites. J Embryol Exp Morphol 1983; 77:99-116.
56. Bagnall KM, Higgins SJ, Sanders EJ. The contribution made by a single somite to the vertebral column: experimental evidence in support of resegmentation using the chick-quail chimaera model. Development 1988; 103(1):69-85.
57. Huang R, Zhi Q, Neubuser A, et al. Function of somite and somitocoele cells in the formation of the vertebral motion segment in avian embryos. Acta Anat(Basel) 1996; 155(4):231-41.
58. Huang R, Zhi Q, Schmidt C, et al. Sclerotomal origin of the ribs. Development 2000; 127(3):527-32.
59. Holland SJ, Gale NW, Gish GD, et al. Juxtamembrane tyrosine residues couple the Eph family receptor EphB2/Nuk to specific SH2 domain proteins in neuronal cells. EMBO J 1997; 16(13):3877-88.
60. Parkyn G, Mootoosamy RC, Cheng L, et al. Hypaxial muscle development. In: Brand-Saberi B, ed. Vertebrate Myogenesis. Heidelberg, Berlin: Springer-Verlag, 2002:127-41.
61. Brohmann H, Jagla K, Birchmeier C. The role of Lbx1 in migration of muscle precursor cells. Development 2000; 127(2):437-45.
62. Schmidt C, Christ B, Maden M, et al. Regulation of Epha4 expression in paraxial and lateral plate mesoderm by ectoderm-derived signals. Dev Dyn 2001; 220(4):377-86.
63. Holland SJ, Peles E, Pawson T, et al. Cell-contact-dependent signalling in axon growth and guidance: Eph receptor tyrosine kinases and receptor protein tyrosine phosphatase beta. Curr Opin Neurobiol 1998; 8(1):117-27.
64. O'Leary DD, Wilkinson DG. Eph receptors and ephrins in neural development. Curr Opin Neurobiol 1999; 9(1):65-73.
65. Dottori M, Hartley L, Galea M, et al. EphA4(Sek1) receptor tyrosine kinase is required for the development of the corticospinal tract. Proc Natl Acad Sci U S A 1998; 95(22):13248-53.
66. Helmbacher F, Schneider-Maunoury S, Topilko P, et al. Targeting of the EphA4 tyrosine kinase receptor affects dorsal/ventral pathfinding of limb motor axons. Development 2000; 127(15):3313-24.
67. Schrägle J, Huang R, Christ B, et al. Control of the temporal and spatial Uncx4.1 expression in the paraxial mesoderm of avian embryos. Anat Embryol(Berl) 2004; 208(4):323-32.
68. Mansouri A, Voss AK, Thomas T, et al. Uncx4.1 is required for the formation of the pedicles and proximal ribs and acts upstream of Pax9. Development 2000; 127(11):2251-8.
69. Buchberger A, Bonneick S, Klein C, et al. Dynamic expression of chicken cMeso2 in segmental plate and somites. Dev Dyn 2002; 223(1):108-18.
70. Buchberger A, Seidl K, Klein C, et al. cMeso-1, a novel bHLH transcription factor, is involved in somite formation in chicken embryos. Dev Biol 1998; 199(2):201-15.
71. Holt CE, Lemaire P, Gurdon JB. Cadherin-mediated cell interactions are necessary for the activation of MyoD in Xenopus mesoderm. Proc Natl Acad Sci U S A 1994; 91(23):10844-8.
72. George-Weinstein M, Gerhart J, Blitz J, et al. N-cadherin promotes the commitment and differentiation of skeletal muscle precursor cells. Dev Biol 1997; 185(1):14-24.
73. Healy C, Uwanogho D, Sharpe PT. Regulation and role of Sox9 in cartilage formation. Dev Dyn 1999; 215(1):69-78.
74. Zeng L, Kempf H, Murtaugh LC, et al. Shh establishes an Nkx3.2/Sox9 autoregulatory loop that is maintained by BMP signals to induce somitic chondrogenesis. Genes Dev 2002; 16(15):1990-2005.
75. Williams BA, Ordahl CP. Pax-3 expression in segmental mesoderm marks early stages in myogenic cell specification. Development 1994; 120(4):785-96.
76. Suetsugu R, Sato Y, Takahashi Y. Pax 2 expression in mesodermal segmentation and its relationship with EphA4 and Lunatic-fringe during chicken somitogenesis. Gene Expr Patterns 2002; 2(1-2):157-61.
77. Suetsugu R, Sato Y, Takahashi Y. Pax 2 expression in mesodermal segmentation and its relationship with EphA4 and Lunatic-fringe during chicken somitogenesis. Mech Dev 2002; 119 Suppl 1:S155-9.
78. Brophy PD, Lang KM, Dressler GR. The secreted frizzled related protein 2(SFRP2) gene is a target of the Pax2 transcription factor. J Biol Chem 2003; 278(52):52401-5.
79. Satoh W, Gotoh T, Tsunematsu Y, et al. Sfrp1 and Sfrp2 regulate anteroposterior axis elongation and somite segmentation during mouse embryogenesis. Development 2006; 133(6):989-99.

80. Wagner J, Schmidt C, Nikowits W, Jr., Christ B. Compartmentalization of the somite and myogenesis in chick embryos are influenced by wnt expression. Dev Biol 2000; 228(1):86-94.
81. Linker C, Lesbros C, Stark MR, Marcelle C. Intrinsic signals regulate the initial steps of myogenesis in vertebrates. Development 2003; 130(20):4797-807.
82. Bellairs R. The Development Of Somites In The Chick Embryo. J Embryol Exp Morphol 1963; 11:697-714.
83. Packard DS, Jr., Zheng RZ, Turner DC. Somite pattern regulation in the avian segmental plate mesoderm. Development 1993; 117(2):779-91.
84. Sosic D, Brand-Saberi B, Schmidt C, Christ B, Olson EN. Regulation of paraxis expression and somite formation by ectoderm- and neural tube-derived signals. Dev Biol 1997; 185(2):229-43.
85. Cooke J, Zeeman EC. A clock and wavefront model for control of the number of repeated structures during animal morphogenesis. J Theor Biol 1976; 58(2):455-76.
86. Kerszberg M, Wolpert L. A clock and trail model for somite formation, specialization and polarization. J Theor Biol 2000; 205(3):505-10.
87. Meinhardt H. Hierarchical inductions of cell states: a model for segmentation in Drosophila. J Cell Sci Suppl 1986; 4:357-81.
88. Stern CD, Fraser SE, Keynes RJ, Primmett DR. A cell lineage analysis of segmentation in the chick embryo. Development 1988; 104 Suppl:231-44.
89. Cooke J. Control of somite number during morphogenesis of a vertebrate, Xenopus laevis. Nature 1975; 254(5497):196-9.
90. Slack JMW. The problems of early development and the means of their solution. In: From Egg to Embryo: Determinative Events in Early Development. Cambridge: Cambridge Univ Press, 1983:214ff.
91. Slack JMW. Regional Specification in Early Development. Cambridge: Cambridge Univ Press, 1991.
92. Palmeirim I, Henrique D, Ish-Horowicz D, Pourquie O. Avian hairy gene expression identifies a molecular clock linked to vertebrate segmentation and somitogenesis. Cell 1997; 91(5):639-48.
93. Aulehla A, Johnson RL. Dynamic expression of lunatic fringe suggests a link between notch signaling and an autonomous cellular oscillator driving somite segmentation. Dev Biol 1999; 207(1):49-61.
94. McGrew MJ, Dale JK, Fraboulet S, Pourquie O. The lunatic fringe gene is a target of the molecular clock linked to somite segmentation in avian embryos. Curr Biol 1998; 8(17):979-82.
95. Forsberg H, Crozet F, Brown NA. Waves of mouse Lunatic fringe expression, in four-hour cycles at two-hour intervals, precede somite boundary formation. Curr Biol 1998; 8(18):1027-30.
96. Jiang YJ, Aerne BL, Smithers L, Haddon C, Ish-Horowicz D, Lewis J. Notch signalling and the synchronization of the somite segmentation clock. Nature 2000; 408(6811):475-9.
97. Holley SA, Geisler R, Nusslein-Volhard C. Control of her1 expression during zebrafish somitogenesis by a delta-dependent oscillator and an independent wave-front activity. Genes Dev 2000; 14(13):1678-90.
98. Holley SA, Julich D, Rauch GJ, Geisler R, Nusslein-Volhard C. her1 and the notch pathway function within the oscillator mechanism that regulates zebrafish somitogenesis. Development 2002; 129(5):1175-83.
99. Itoh M, Kim CH, Palardy G, Oda T, Jiang YJ, Maust D, Yeo SY, Lorick K, Wright GJ, Ariza-McNaughton L, Weissman AM, Lewis J, Chandrasekharappa SC, Chitnis AB. Mind bomb is a ubiquitin ligase that is essential for efficient activation of Notch signaling by Delta. Dev Cell 2003; 4(1):67-82.
100. van Eeden FJ, Granato M, Schach U, Brand M, Furutani-Seiki M, Haffter P, Hammerschmidt M, Heisenberg CP, Jiang YJ, Kane DA, Kelsh RN, Mullins MC, Odenthal J, Warga RM, Nusslein-Volhard C. Genetic analysis of fin formation in the zebrafish, Danio rerio. Development 1996; 123:255-62.
101. Henry CA, Urban MK, Dill KK, Merlie JP, Page MF, Kimmel CB, Amacher SL. Two linked hairy/Enhancer of split-related zebrafish genes, her1 and her7, function together to refine alternating somite boundaries. Development 2002; 129(15):3693-704.
102. Oates AC, Ho RK. Hairy/E(spl)-related(Her) genes are central components of the segmentation oscillator and display redundancy with the Delta/Notch signaling pathway in the formation of anterior segmental boundaries in the zebrafish. Development 2002; 129(12):2929-46.
103. Sieger D, Tautz D, Gajewski M. The role of Suppressor of Hairless in Notch mediated signalling during zebrafish somitogenesis. Mech Dev 2003; 120(9):1083-94.
104. Gajewski M, Sieger D, Alt B, Leve C, Hans S, Wolff C, Rohr KB, Tautz D. Anterior and posterior waves of cyclic her1 gene expression are differentially regulated in the presomitic mesoderm of zebrafish. Development 2003; 130(18):4269-78.
105. Geling A, Steiner H, Willem M, Bally-Cuif L, Haass C. A gamma-secretase inhibitor blocks Notch signaling in vivo and causes a severe neurogenic phenotype in zebrafish. EMBO Rep 2002; 3(7):688-94.
106. Oka C, Nakano T, Wakeham A, de la Pompa JL, Mori C, Sakai T, Okazaki S, Kawaichi M, Shiota K, Mak TW, Honjo T. Disruption of the mouse RBP-J kappa gene results in early embryonic death. Development 1995; 121(10):3291-301.

107. Bessho Y, Miyoshi G, Sakata R, Kageyama R. Hes7: a bHLH-type repressor gene regulated by Notch and expressed in the presomitic mesoderm. Genes Cells 2001; 6(2):175-85.
108. Hrabe de Angelis M, McIntyre J, 2nd, Gossler A. Maintenance of somite borders in mice requires the Delta homologue DII1. Nature 1997; 386(6626):717-21.
109. Kusumi K, Sun ES, Kerrebrock AW, Bronson RT, Chi DC, Bulotsky MS, Spencer JB, Birren BW, Frankel WN, Lander ES. The mouse pudgy mutation disrupts Delta homologue Dll3 and initiation of early somite boundaries. Nat Genet 1998; 19(3):274-8.
110. Wong PC, Zheng H, Chen H, Becher MW, Sirinathsinghji DJ, Trumbauer ME, Chen HY, Price DL, Van der Ploeg LH, Sisodia SS. Presenilin 1 is required for Notch1 and DII1 expression in the paraxial mesoderm. Nature 1997; 387(6630):288-92.
111. Zhang N, Gridley T. Defects in somite formation in lunatic fringe-deficient mice. Nature 1998; 394(6691):374-7.
112. Dubrulle J, McGrew MJ, Pourquie O. FGF signaling controls somite boundary position and regulates segmentation clock control of spatiotemporal Hox gene activation. Cell 2001; 106(2):219-32.
113. Dubrulle J, Pourquie O. fgf8 mRNA decay establishes a gradient that couples axial elongation to patterning in the vertebrate embryo. Nature 2004; 427(6973):419-22.
114. Takada S, Stark KL, Shea MJ, Vassileva G, McMahon JA, McMahon AP. Wnt-3a regulates somite and tailbud formation in the mouse embryo. Genes Dev 1994; 8(2):174-89.
115. Greco TL, Takada S, Newhouse MM, McMahon JA, McMahon AP, Camper SA. Analysis of the vestigial tail mutation demonstrates that Wnt-3a gene dosage regulates mouse axial development. Genes Dev 1996; 10(3):313-24.
116. Fisher DA, Kivimae S, Hoshino J, Suriben R, Martin PM, Baxter N, Cheyette BN. Three Dact gene family members are expressed during embryonic development and in the adult brains of mice. Dev Dyn 2006; 235(9):2620-30.
117. Suriben R, Fisher DA, Cheyette BN. Dact1 presomitic mesoderm expression oscillates in phase with Axin2 in the somitogenesis clock of mice. Dev Dyn 2006; 235(11):3177-83.
118. Dale JK, Maroto M, Dequeant ML, Malapert P, McGrew M, Pourquie O. Periodic notch inhibition by lunatic fringe underlies the chick segmentation clock. Nature 2003; 421(6920):275-8.

CHAPTER 3

Genetic Analysis of Somite Formation in Laboratory Fish Models

Christoph Winkler* and Harun Elmasri

Abstract

The repeated appearance of somites is one of the most fascinating aspects of vertebrate embryogenesis. Recent studies identified complex regulatory circuits that provide the molecular basis for the "clock and wave front" model, postulated almost 30 years ago by Cooke and Zeeman. The highly coordinated process of somite formation involves several networks of molecular cascades including the Delta/Notch, Wnt, FGF and retinoid signalling pathways. Studies in mouse, Xenopus and especially chicken over the last decade have helped to understand the role and interactions of these pathways in somitogenesis. More recently, this has been supplemented by experiments in zebrafish. This animal model offers the possibility of performing large scale mutagenesis screens to identify novel factors and pathways involved in somitogenesis. Molecular cloning of zebrafish somite mutants mainly resulted in genes that belong to the Delta/Notch pathway and therefore underlined the importance of this pathway during somitogenesis. The fact that other pathways have not yet been identified by genetic screening in this species was assumed to be caused by functional redundancy of duplicated genes in zebrafish. In 2000, a large-scale mutagenesis screen has been initiated in Kyoto, Japan using the related teleost medaka (*Oryzias latipes*). In this screen, mutants with unique phenotypes have been identified, which have not been described in zebrafish or mouse. In this chapter, we will review the progress that has been made in understanding the molecular control of somite formation in zebrafish and will discuss recent efforts to screen for novel phenotypes using medaka somitogenesis mutants.

Introduction

Somites are transient embryonic structures of the paraxial mesoderm that give rise to all striated muscle of the adult body, the axial skeleton and dermis during later embryogenesis. They form repeatedly from the unsegmented presomitic mesoderm (PSM) in a highly coordinated temporal and spatial manner. The periodicity underlying this process is controlled by a molecular oscillator, the segmentation clock, which is active in the tailbud of all vertebrate embryos. As outlined in detail below, this clock is coupled to a so called wave front activity, which determines the position, where the next somite will emerge. It has been one of the most fascinating fields in modern developmental biology to identify the molecular nature of both the oscillator and the wave front.

Why Fish?

Over the last decade, the zebrafish has attracted considerable attention as an animal model to study gene functions during vertebrate development and disease. It provides several experimental advantages that offer a combination of molecular and embryological strategies to approach

*Corresponding Author: Christoph Winkler—Department of Biological Sciences, National University of Singapore, Singapore. Email: dbswcw@nus.edu.sg

Somitogenesis, edited by Miguel Maroto and Neil V. Whittock. ©2008 Landes Bioscience and Springer Science+Business Media.

developmental questions. Zebrafish produce a large number of eggs that are externally fertilized and develop outside the mother. Development is extremely fast and takes only 2.5 days from fertilization to hatching of the free-swimming larvae. The embryos are highly transparent and are big enough to allow micromanipulations like injection of DNA, RNA or antisense oligos into early embryos or transplantation of cells and beads soaked with chemicals. These features together with a relatively short generation time of approx. 3 months provided the basis for efficient mutagenesis

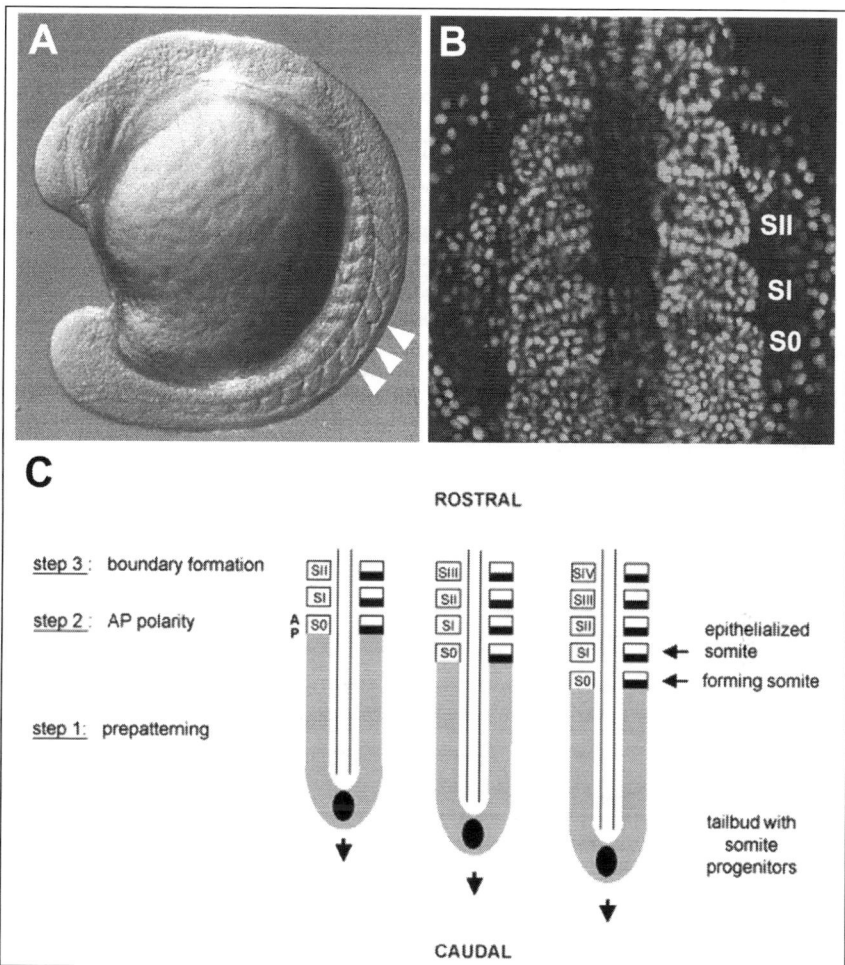

Figure 1. Somitogenesis in zebrafish. A) Lateral view of a living embryo at 16 hours post fertilization (hpf). Somites appear as prominent, chevron-shaped structures in the trunk (marked by arrowheads). B) Confocal microscopy of a living zebrafish embryo expressing a histon2a:RFP fusion protein in nuclei of transgenic cells (picture courtesy of Matthias Schäfer). Dorsal view of trunk region with anterior to the top showing presomitic and somitic mesoderm. S0 marks the next somitomere to be epithelialized. Note the regular positioning of epithelial and loose distribution of mesenchymal cells in formed somites. C) Schematic diagram of a zebrafish tailbud, dorsal view. The presomitic mesoderm is shown in grey. As in higher vertebrates, three steps of somite formation (i.e., PSM prepatterning, establishment of anterior-posterior (AP) polarity and somite boundary formation) can be distinguished. Nomenclature of somitomeres is as described in reference 4.

screens in zebrafish to identify recessive embryonic lethal phenotypes. Besides several small scale screens in many laboratories worldwide, also large scale screens have been conducted in Boston and Tübingen.[1-3] These screens identified thousands of mutants showing deficiencies in a variety of developmental processes. Over the last years, a large number of the mutated loci were identified and many more are expected to follow after the recent completion of the zebrafish genome sequencing project (see www.ensembl.org/Danio_rerio/; The Wellcome Trust Sanger Institute). By now, also several mutations that affect somitogenesis have been identified by positional cloning.

Zebrafish to Screen Mutants

According to their phenotype, zebrafish somitogenesis mutants can be classified into two general groups. One group consists of the so called *you*-type mutants, including *you-too* (*yot*; *gli2*),[5] *sonic-you* (*syu*; *shh*),[6] *chameleon* (*con*; *dispatched homolog 1*),[7] and *u-boot* (*ubo*; *prdm1*).[8] Molecular cloning revealed that all *you*-type mutants are deficient in Hedgehog signalling. Instead of showing a typical chevron shape (Fig. 1A), somites of homozygous *you*-type mutants exhibit a U-shape. While segmentation appears normal in these mutants, somite patterning is impaired. Among other phenotypes, this leads to defective muscle differentiation (reviewed in ref. 9). The second group of zebrafish somite mutants was named *fss*-type mutants (Table 1) and includes *fused somites* (*fss*),[10] *after eight* (*aei*),[11] *deadly seven* (*des*),[12] *beamter* (*bea*),[13] *mind bomb* (*mib*),[14] *spadetail* (*spt*),[15] and *before eight* (*bfe*).[16,17] Except for *fss*, *spt* and *bfe*, all other mutants in this group are exclusively affected in the Delta/Notch signalling pathway.

Genetically Dissecting the Clock in Zebrafish: The Delta/Notch Somitogenesis Mutants

Already several decades ago, theoretical models have been developed that postulated the existence of a clock in the presomitic mesoderm of vertebrates.[18] Investigations of the zebrafish *delta/notch* somitogenesis mutants contributed significantly to the molecular dissection of the transcriptional oscillator that reflects the segmentation clock. The first gene identified in a zebrafish somite mutant was *deltaD* that carries a premature stop codon in *aei* mutants.[11] Homozygous *aei* embryos generate 7 to 9 pairs of regular somites in the anterior trunk, but completely lack all

---

Table 1. The zebrafish fss-type somitogenesis mutants

Mutant	Gene	Somite Phenotype	Other Phenotypes	Reference
after eight (aei)	*deltaD*	forms only first 7-9 somite pairs	none	11
deadly seven (des)	*notch1a*	forms only first 7-9 somite pairs	none	12
beamter (bea)	*deltaC*	forms only first 3-4 somite pairs	none	13
mind bomb (mib)	*RING ubiquitin ligase*	Irregular formed somite boundaries in posterior segments	Increased neural progenitor numbers	14
fused somites (fss)	*tbx24*	No somite boundaries	none	10
spadetail (spt)	*tbx16*	No trunk somites	Normal notochord, otherwise defective trunk mesoderm	15
before eight (polypterus)	*integrin a5*	Somite boundaries absent in anterior two to ten somites	Defective head cartilage	16,17

posterior somite boundaries (Fig. 2A,B). Except neuronal hyperplasia,[19] no defective structures other than somites were reported. *aei* mutants are viable and develop into fertile adults. In this mutant, transcription of several downstream components of the Delta/Notch cascade, including the hairy-related genes *her1*,[11] *her7*,[20,21] and *hey1* (Fig. 2C,D)[22] is impaired. In the PSM, the oscillating expression of *her1* and *her7*, as well as of *deltaC* is strongly perturbed.[11,20,21] These genes instead often show a characteristic "salt and pepper" mode of expression, rather than a coordinated wave that sweeps the PSM in caudal to rostral directions. Similar phenotypes were also observed in the *deadly-seven* (*des*) mutant.[12] The finding that *notch1a* is deficient in this mutant further added to the view that Delta/Notch signalling is required for the formation of the posterior zebrafish somites. It furthermore demonstrated the importance of this pathway for the induction and maintenance of oscillatory gene expression in the PSM (reviewed in ref. 23). This was also supported by the

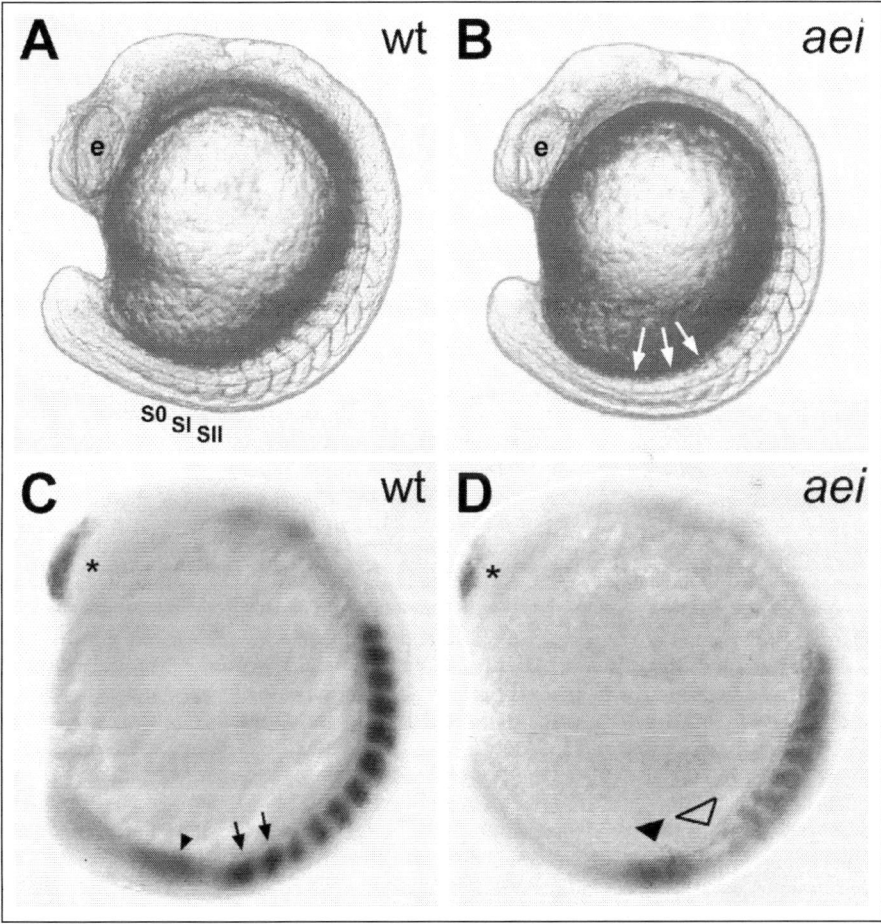

Figure 2. The zebrafish *after eight* (*aei*) mutant carrying a deficiency in the *deltaD* gene. While the wild-type (wt) embryo in (A) exhibits 15 fully epithelialized somites, a homozygous *aei* mutant at the same stage shows only 8-9 pairs of fully developed somites. The remaining paraxial mesoderm remains unsegmented (white arrows). Expression of the Delta/Notch target genes *her1* (arrowheads) and *hey1* (arrows) in zebrafish wt (C) and *aei* mutant at the 12-somite stage is shown (for details see ref. 22). e, eye; asterisk marks *hey1* expression in the forebrain.

characterization of a γ-ray induced zebrafish mutant with a large deletion that includes both, the *her1* and *her7* loci.[24] Interestingly, in all these mutants only posterior somites were affected. It was suggested that genetic redundancy might account for the normal formation of the more robust anterior somites in these mutants,[20] rather than Delta/Notch not being involved in anterior somitogenesis. Interestingly, however, even the knock-down of zebrafish *Suppressor-of-Hairless* (*Su(H)*), a central component downstream of general Notch signalling, did not result in a block of anterior somite formation.[25] Although the observed defects were stronger than for example in *aei* and *des* mutants, the first 5 to 7 pairs of somites showed only mildly affected phenotypes. Thus, it was suggested that additional genetic pathways may be active in the specification of the most anterior somites.[25]

All *fss*-type somite mutants show perturbations in the expression of the cycling genes *her1*, *her7* and *deltaC*.[11,12,20] Also other Delta/Notch targets in the PSM are affected, like for example *hey1* and *her11*.[22,26] This is consistent with Delta/Notch signalling being either the output or a central component of the oscillator. In fact, studies in zebrafish demonstrated that Delta/Notch activity itself establishes the oscillator. Overexpression of the Notch1a receptor results in an up-regulation of *her1* transcription.[27] On the other hand, overexpression of *her1* down-regulates transcription of *deltaC* and *deltaD* suggesting a negative feedback loop in this pathway. This was supported by Morpholino antisense experiments, where the knock-down of *her1* or *her7* resulted in an up-regulation of the corresponding *her1* and *her7* transcription.[12,20] However, it should be noted that the view of a direct auto-regulatory repression of *her1* and *her7* as central part of this model has been challenged by work from M. Gajewski et al.[21] Using intron-specific riboprobes, the authors demonstrated an inhibition of endogenous *her1* expression in *her1* Morpholino injected embryos. Consequently, they postulated an activating rather than repressing activity of Her1 (and possibly Her7). It, however, can not be excluded that Her1 activates a thus far unknown repressor, which integrates into the proposed inhibitory loop.

Taken together, experiments in zebrafish led to the hypothesis that oscillating Delta/Notch activity is at the core of the segmentation clock. Delta/Notch signalling is initiated in the posterior PSM by thus far unknown mechanisms. Through the consecutive induction of transcriptional activators and repressors auto-regulatory feedback loops are established. The resulting alternating gene activities are then coordinated between neighbouring cells, which finally results in oscillating transcription patterns that sweep the PSM in caudal to rostral direction in each cycle of somite formation.[28,29]

According to a model described in ref. 18, Delta ligands activate Notch receptors, which subsequently results in the activation of the target genes *her1* and *her7*. These encode transcriptional repressors. After protein accumulation, they subsequently down-regulate their own transcription, as well as that of Delta ligands. After degradation of Her1 and Her7 repressor proteins, *delta* genes are switched on again which results in the recurrent activation of the Delta/Notch pathway. The critical aspect in this circuit is the translational delay in repressor production and turnover rate during receptor decay.[30] This consequently results in a periodic switch-on and switch-off of target gene transcription in PSM cells. As Delta ligands non-autonomously activate receptors on adjacent cells, this oscillating activity is also transmitted to the neighbouring cells. This is believed to result in a highly coordinated wave of transcription that sweeps the PSM. Elegant transplantation and temperature shift experiments have shown that Delta/Notch interactions are also crucial for the synchronization of the oscillating transcription between neighbouring cells and minimize developmental noise caused by stochastic gene expression.[28,29]

Dissecting the Wave Front in Zebrafish: FGF Signalling and Tbx24

The oscillating waves of gene expression generated by the Delta/Notch feedback loop are terminated at a position in the anterior PSM, where the next pair of somites will form. In the early models, this has been attributed to a so called "wave front" activity. Insight into the molecular control of this wave front came from the analysis of the *fused somites* (*fss*) zebrafish mutant.[10] Molecular cloning of this mutant identified a novel T-box containing transcription factor gene, named *tbx24*.[10] Other than in Delta/Notch mutants, no somite boundaries are formed along the entire anterior-posterior body axis and therefore all somites are affected in *fss* mutants. Based on several observations, it was postulated that the *fss/tbx24* gene is instrumental in establishing the wave front activity. Other than in *delta/notch* mutants, in *fss* the anterior *her1* stripe is completely missing, while the two posterior stripes appear normally in the PSM. In addition, expression of *mespa* and *papc*, two important regulators of AP somite polarity and boundary formation,[31] is absent in the anteriormost PSM of *fss* mutants. The *tbx24* gene is broadly expressed in the intermediate and anterior PSM. It was suggested that *tbx24* regulates the maturation of PSM cells, at the same time when oscillating Delta/Notch activity ceases in the anteriormost PSM.[23] *Tbx24* is required for stabilization of oscillating *her* expression in the anterior PSM. It furthermore controls the induction of segmentation genes, which regulate boundary formation and epithelialization. Interestingly, no *tbx24* ortholog has been identified so far in higher vertebrates. Thus, it is possible that this mechanism is unique to teleost fish.

Dubrulle and coworkers showed that a gradient of FGF activity is required to control the proper size of somites in chicken.[32] Transplantation of beads soaked with FGF8 into the PSM of chicken embryos resulted in somites with small size. In contrast, beads soaked with the FGF signalling inhibitor SU5402 increased the size of somites. FGF8 forms a mRNA and protein gradient with highest levels in the caudal part of the tailbud.[32] In 2001, Sawada et al reported that FGF8 regulates somite size also in zebrafish.[33] They showed that also in zebrafish, treatment with SU5402 increases somite size. In treated embryos, *her1* oscillations were prematurely stopped at more caudal positions. Furthermore, expression of markers in the epithelializing somites, like *mespa* and *papc*, was shifted caudally. On the other hand, ectopic FGF activity shifted these expression domains to the anterior.[33] Based on this, it was concluded that FGF signalling determines the position of the presomitic maturation front and consequently the place, where somite boundaries are formed. FGF signalling mediated through phosphorylated MAPK is highest in the caudal PSM. It shows intermediate to low levels at a position, where new somite boundaries are formed. According to this model, a distinct threshold concentration of FGF8 activity determines the position, where the wave front is active. There it stabilizes Notch oscillations and induces the maturation of presomitic cells. As the embryo continues to grow, the tailbud as source of FGF8 signal regresses caudally. In parallel to this, the area of the FGF8 threshold concentration that is critical for maturation is also gradually shifted caudally. The wave front activity, on the other hand, is mediated through *tbx24*, which at the RNA level is broadly expressed in the anterior and intermediate zebrafish PSM. *Tbx24* positive cells in the intermediate PSM are also exposed to high FGF8 protein levels and thus remain in an immature and proliferative state.

The parallel findings in zebrafish and chicken strongly suggest that FGF function during somitogenesis might be conserved among vertebrates. Unfortunately, however, no zebrafish FGF mutant with somitogenesis defects is known. The FGF8 deficient mutant *acerebellar* (*ace*) shows a severe defect in mid-hindbrain boundary formation, yet has only a very mild, if any, somite phenotype.[34] Interestingly, a simultaneous gene knock down of *fgf8* and *fgf24* by injecting Morpholino antisense oligonucleotides severely affects somite formation.[35] This supports the idea that closely related gene duplicates function redundantly and may prevent the appearance of somitic phenotypes in mutagenesis screens in a given mutant.

An important question with regard to proper positioning of the maturation front is, whether and how oscillatory Delta-Notch signalling is interconnected with the FGF pathway. Two recent studies have shed light on this.[36,37] It was shown that *her13.2* is regulated by FGF signalling and on the other hand interacts with Delta-Notch controlled *her13.1*. Consequently, a simultaneous knock-down of both *her13.1* and *her13.2* leads to a complete absence of somite boundaries along the entire body axis.

More recently, Kawamura and coworkers described *ripply1* as an additional factor that regulates the transition from unsegmented presomitic mesoderm into segmented somites in zebrafish.[38] *Ripply1* was identified in a screen for genes expressed in the tailbud. Its transcripts are found in the anterior PSM and the most recently formed somites. It encodes a nuclear protein associated with the corepressor Groucho. Gene knock-down of *ripply1* results in the repression of somite boundary formation, the ectopic up-regulation of PSM specific genes in anterior trunk regions and the loss of rostro-caudal polarity in the somitomeres. Thus, it was concluded that *ripply1* functions in the termination of the segmentation program in the anterior PSM and maintains rostro-caudal polarity of the forming somites.[38]

Other Pathways Implicated in Somitogenesis

Recently, Wnt signalling has been implicated as a crucial pathway regulating somite formation in higher vertebrates.[39] Wnt3a forms a gradient across the PSM with highest levels in the posterior tailbud. Similar to the situation with FGF signalling, it was suggested that a certain threshold level of Wnt3a activity drives final maturation of presomitic cells and their integration into epithelializing somites. It was shown that the Wnt antagonist *Axin2*, a downstream target in the Wnt pathway, shows oscillating expression in the PSM of mice.[39] In *Wnt3a* deficient *vestigial tail* (*vt*) mutant embryos, *Axin2* and *Fgf8* expression are strongly down-regulated, while in parallel expression of the Delta/Notch component *lunatic fringe* (*lfng*) is up-regulated. This results in larger somites and indicates that *Wnt3a* acts upstream of both Delta/Notch and FGF signalling. Subsequently, it was suggested that the Wnt pathway regulates the oscillatory activation of Delta/Notch signalling.[40,41] Thus, Wnt signalling appears to integrate the Delta/Notch driven clock with the FGF gradient and therefore plays a central role in setting up both, the segmentation clock and wave front activity.[42]

Unfortunately, no Wnt signalling mutant with somite defects has been identified in zebrafish. *Wnt5a* is mutated in the zebrafish *pipetail* mutant, which shows defects in tail formation but does not reveal an obvious somite phenotype.[43] Thus, the role of Wnt signals during zebrafish somitogenesis remains elusive. To this end, *axin2* and its co-ortholog *axin1/masterblind* have been described in zebrafish and medaka, but it remains unclear, whether they show oscillating expression in the PSM (Elmasri and Winkler, unpublished; www.zfin.org; see ref. 44). Nevertheless, several *wnt* genes are expressed in the tailbud, including *wnt3a*.[45,46] It will be interesting in the future to determine their function during somite formation. Unfortunately, however, there so far is no report that would indicate the involvement of Wnt signalling during zebrafish somitogenesis.

While the role of Wnts during zebrafish somitogenesis remains unclear, there is strong evidence that the role of retinoid signalling is conserved during somite formation in teleost fish (for review see ref. 47). A recent report has shown that somites form asymmetrically in the absence of retinoic acid signalling.[48] This demonstrates that retinoids in zebrafish control left-right asymmetry and are indispensable for the bilaterally symmetric formation of somites.

Medaka: A Model Complementary to Zebrafish

The Japanese teleost medaka (*Oryzias latipes*) has recently emerged as experimental model complementary to the zebrafish (see Medaka Special Issue of Mechanisms of Development 2004, Vol. 121 and refs. therein). Like zebrafish, medaka is oviparous and produces large numbers of embryos on a daily basis throughout the year. The embryos are covered by a rigid chorion that is transparent enough to allow observation of the rapid embryonic development. Depending on the incubation temperature, it takes between 7 to 12 days until the embryos hatch. The size of the embryos permits manipulations like microinjections and cell transplantations. Functional analysis by RNA injections, gene knock-down by Morpholino antisense approaches and several transgenic approaches including e.g., endonuclease- and transposon-mediated transgenesis have been established over the last years.[49,50] In addition, embryonic stem (ES) cells are available that can be used to study cell differentiation processes in vitro.[51,52] With approx. 900 Mb (zebrafish: 1700 Mb), the medaka genome is roughly a third of the size of the human genome. Recently, whole genome sequencing has been completed and Ensembl genebuild information can be retrieved from publicly available databases (see www.ensembl.org/Oryzias_latipes/index.html; dolphin.lab.nig.ac.jp/medaka/; medaka.utgenome.org/). Several important developmental processes have been intensively studied in the medaka at the molecular level, including eye development, sex determination and somitogenesis.[53-58]

Somite Formation in the Teleost Medaka

Morphologically, somite formation is remarkably similar in medaka and zebrafish with the difference that medaka forms a total of 35 somite pairs, while zebrafish embryos have 30 somites. Also, it takes approximately 60 minutes for one medaka somite to form at 26°C, while a zebrafish somite cycle is significantly shorter (30 min at 28°C; 51).[59,60] The expression of known somitogenesis marker genes shows several similarities between medaka and zebrafish. For example, *Lunatic fringe* shows oscillating expression in mice and chicken, but importantly shows no dynamic mode of PSM expression in both zebrafish and medaka.[57,61-64] In addition, whereas the *hairy/Enhancer-o f-split*-related gene *hey2* shows oscillating transcription in mice and chicken, the co-ortholog *hey1* is dynamically expressed in the PSM of both zebrafish and medaka.[22,57,65] However, there are also differences in gene expression between both fish species. *her7*, a hairy-related gene with central function in the somitogenesis clock, shows a characteristic wave of oscillating transcription in the zebrafish PSM.[20,21] Its expression starts at the caudal end of the tailbud and subsequently sweeps the PSM with at least two discrete bands moving rostrally. At the same time, the caudal-most expression decreases until it is completely vanished.[20] This is in sharp contrast to the situation in medaka, where the caudal expression of the *her7* ortholog appears to remain stable and only one band of *her7* transcripts detaches from the caudal domain and moves rostrally (Fig. 3).[57] The situation of only one stripe detaching in the rostral PSM has also been observed for other *her* paralogs in medaka.[58]

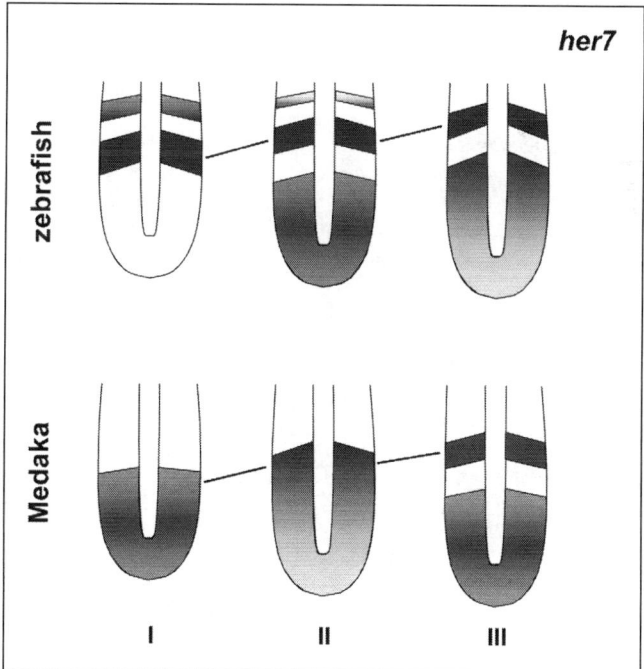

Figure 3. Schematic view of dynamic *her7* regulation during different phases of one cycle of somite formation in zebrafish and medaka. Periodic repression of *her7* in interstripe regions and the caudalmost domain of the PSM is evident in zebrafish, but not in medaka (for details see refs. 20,21,57,58).

Although the basic mechanisms of the molecular segmentation clock appear similar in many vertebrates, these observations suggest that there are subtle variations in the regulation of distinct clock elements in different vertebrates and even between related fish species.

Medaka Somitogenesis Mutants

Stimulated by the observation that different pairs of duplicated genes exist in distantly related fish species,[66-68] a large scale mutagenesis screen in medaka has been initiated in Kyoto, Japan in 2000.[69] Several mutants affecting a large variety of developmental processes were obtained. These showed phenotypes that so far have not been recorded in the zebrafish screens. This supported the hypothesis that variable degrees of functional redundancy exist in different fish species. A variety of mutants with defective somite formation has been identified in this screen.[56] Nine of these mutants were thus far characterized in more detail. Based on their morphological appearance and on the expression of diagnostic marker genes, these mutants could be classified into two distinct groups. Group 1 contains mutants that show severe defects in PSM formation and prepatterning. This is evident by the irregular and non-oscillating expression of *her7* and absent or reduced expression of the anterior PSM marker *mesp*. Group 2 mutants on the other hand show regular PSM prepatterning and dynamic PSM marker expression, but develop severe morphological somite boundary and AP polarity defects.

While several early morphological criteria are very similar in medaka and zebrafish somitogenesis mutants, there are also important differences. The vast majority of *fss*-type zebrafish mutants is homozygous viable. The mutants overcome their severe early embryonic defect and develop into mature and fertile adults. In contrast in medaka, all identified mutations are embryonic lethal.

Table 2. The medaka somitogenesis mutants

Mutant	Somitic Phenotypes	Head Phenotypes
Group 1: Mutations affecting tailbud formation and PSM prepatterning		
bremser (bms)	No somites formed, only partial boundaries	Arrested eye and forebrain morphogenesis
planlos (pll)	Partial formation of only 1-2 anterior somites	Arrested eye and forebrain morphogenesis
schnelles ende (sne)	Only the first two somite pairs form, tailbud reduced	Normal
samidare (sam)	No segmentation of posterior trunk, anterior six pairs of somites form normally	Arrested eye and forebrain morphogenesis
doppelkorn (dpk)	Individually fused somites, irregular somite size	Arrested eye and forebrain morphogenesis
Group 2: Mutations affecting AP-polarity and epithelialization		
kurzer (krz)	Variable somite shapes and sizes	Arrested eye and forebrain morphogenesis
orgelpfeifen (opf)	Mild morphological phenotype with slightly irregular somite shapes and variable mediolateral extension	Arrested eye and forebrain morphogenesis
fussel (fsl)	Individually fused somites, variable somite sizes	Necrotic
zahnluecke (zlk)	Irregular sizes and shapes of anterior somites, posterior somites partially missing	Arrested eye and forebrain morphogenesis

Homozygous mutant embryos usually die when their wild-type siblings have reached late somite stages. The only exception is *schnelles ende* (*sne*) that forms no posterior somites and survives up to 4 days after hatching. The early embryonic lethality most likely is not due to the somite defects, which in some mutants is manifested by the fusion of single somites or slightly altered somite size. Instead, embryonic death is likely to be caused by other deficiencies mainly affecting the developing head. In the severely affected *planlos* (*pll*) mutant, for instance, a strongly increased level of apoptosis was observed in the head region. In contrast, no significant alterations in cell death were obvious in the PSM and somites. In general, all medaka somite mutants show necrosis and retarded development of head structures or an arrest of eye and forebrain morphogenesis, except *schnelles ende* (*sne*) that forms a normal head. This is in contrast to the described zebrafish somite mutants, which exhibit no other apparent malformations.

Therefore, the situation in medaka somite mutants with pronounced early embryonic lethality and deficiencies in organs other than somites closely resembles the situation in mouse somite mutants. Knock-out mice deficient for Delta/Notch components, including *Delta1*, *Lfng* and *Hes7* die during embryogenesis or shortly after birth and show severe defects in somites and other embryonic tissues including the nervous system.[70-72]

Medaka Somite Mutants with PSM Prepatterning Defects

Several group 1 mutants show phenotypic defects that resemble those seen in zebrafish mutants. In mutants like *pll*, *sne* and *sam* the anterior somites are formed normally, but posterior somite formation is impaired and boundaries are not formed. While the overall morphology of these mutants appears very similar to zebrafish *aei*, *des* and *bea* mutants, there are important differences. The medaka *samidare* (*sam*) mutant resembles the zebrafish *after eight* (*aei*) carrying a mutation in *deltaD*.[11] In both mutants, the first 7-9 somites develop normally, but no posterior somite boundaries are formed. *her7* shows a typical "salt and pepper" expression pattern in *aei* and *sam* mutants, which is characteristic for defects in the Delta/Notch pathway.[11,12,20] In contrast to *aei*, *sam* mutants also show defects in the mid-hindbrain boundary (MHB) region. Such MHB phenotypes were described in the zebrafish FGF8 mutant *acerebellar* (*ace*).[34] Importantly, however, *ace* has hardly any somitic defect. It was postulated that this could be due to functional redundancy by additional FGF ligands expressed in the tailbud of zebrafish. Consistent with this idea, simultaneous knock-down of *fgf8* and *fgf24* in zebrafish results in severely malformed somites.[35] Therefore, it is tempting to speculate that the medaka *sam* gene is a component of the FGF signalling pathway, which has been shown to regulate somite and MHB formation also in medaka (see ref. 73; H.E. and C.W., unpublished data). As Delta/Notch target gene expression is affected in *sam* embryos, we postulate that the affected component lies upstream of the Delta/Notch circuit.

The medaka *doppelkorn* (*dpk*) mutant shows a unique somite phenotype. Morphologically, *dpk* mutants develop fused somites or somites with expanded size. On the expression level, the clock gene *her7* is not downregulated in the anterior PSM. Instead, it expands towards rostral areas. The expression passes the *mesp* domain in the anterior PSM and can be found in the segmented paraxial mesoderm (Fig. 4).[56] We therefore speculate that *dpk* is a molecular component of the wave front, which is required to stabilize clock genes in the maturation zone of the anterior PSM. Interestingly, expression of the central regulator of AP polarity in somites, *mesp* (see below), appears not to be affected in these mutants. Thus, it seems likely that *dpk* is different from the zebrafish *fss/tbx24* mutant,[10] which is the only wave front mutant known in zebrafish so far and which lacks *mesp* expression. It is also unlikely that *dpk* encodes the ortholog of zebrafish *ripply1*, as in *ripply1* deficient zebrafish embryos expression of both *her1* and *mespb* is anteriorly expanded.[38] In conclusion, it is tempting to speculate that the *dpk* mutation uncouples establishment of AP polarity from the wave front that terminates the oscillations of clock genes.

Positional cloning of these mutations is under way. Based on their unique phenotypes, the molecular identification of the mutated genes is likely to reveal novel components in Delta/Notch signalling or even pathways, that were thus far not implicated in teleost somitogenesis, like for example Wnt signalling.

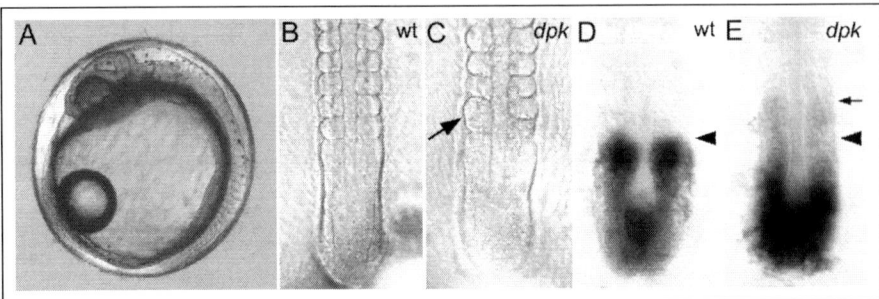

Figure 4. Somite formation in wildtype (wt) and *doppelkorn* (*dpk*) mutant medaka embryos. A) Lateral view of medaka embryo at 2.5 days post ferilization (dpf). (B, C) Dorsal view of the tailbud region at the 10 somite stage (38hpf). Note irregular somite size and fused somites (marked by arrow) in *dpk* mutants. D, E. Expression of *her7* (arrowheads) and *mesp* (arrowheads) in wt and *dpk* embryos at the 10 somite stage. The arrowhead marks expression of *mesp* (overlapped by strong *her7* staining) in the anterior PSM. In *dpk*, *her7* expands rostrally (arrow).

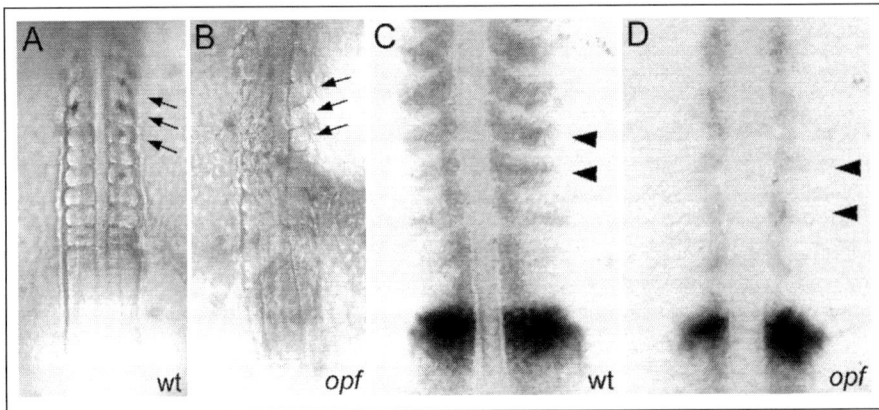

Figure 5. Somite defects in the medaka AP polarity mutant *orgelpfeifen* (*opf*). A, B) Dorsal views of trunk regions at the 10 somite stage, anterior is to the top. Note irregular somite shape and reversed orientation of somite boundaries (marked by arrows) in *opf* mutants. C) *myf5* expression in the posterior halves of individual wild-type somites (arrowheads). D) *myf5* expression in absent in *opf* somites (arrowheads) and is only evident in the adaxial mesoderm. Expression of *mesp* (dark domain at bottom), a central regulator of somite AP polarity in higher vertebrates, appears only slightly affected in *opf* mutants.

Medaka Mutants with Defective Somite Polarity

The anterior-posterior (AP) identity of somitic cells is already established in the PSM, as shown by elegant transplantation experiments in chicken.[74] In zebrafish *fss*-type mutants, AP polarity is strongly affected and leads to an irregular expression of markers that otherwise are restricted to either the rostral or caudal half of each somite. However, these AP defects are not directly caused by the respective gene mutations. They rather represent secondary defects, since early aspects of the somitogenesis clock are affected in all of these mutants. Therefore, these mutants will not identify the pathways regulating the specification of somitic AP identity. Genetic analyses in the mouse revealed a complex interplay of various feedback loops in the anterior PSM that lead to the restricted expression of genes either in the anterior or posterior half of the maturating somitomeres.[75] These loops most importantly involve the action of the Delta ligands, Dll1 and Dll3, which exhibit non-overlapping expression and nonredundant function in the anterior PSM and *Mesp* genes, which modulate Notch signalling.[75,76] Additional factors implicated in this complex network in the anterior PSM remain to be identified, however.

In medaka, several mutants have been identified that show regular expression of the cyclic gene *her7* in the PSM, but defective AP specification of somitic cells.[56] These mutants therefore are unique tools to study the establishment of AP identity in the presence of a normally working somite clock and to identify the molecular players involved in this process. Intriguingly, both types of possible AP deficiencies could be identified. The medaka mutants *kurzer* (*krz*) and *fussel* (*fsl*) show expanded *myf5* expression domains, indicative for a possible posteriorization of somitomeres. Concomitantly, expression of *lfng* in the rostral somite halves is reduced in these mutants. On the other hand, in the *orgelpfeifen* (*opf*) mutant, the *lfng* domain is strongly expanded, while *myf5* is almost completely lost in the developing somites (Fig. 5).[56] This strongly suggests that in this mutant, somites are anteriorized. Morphologically, all three mutants show similar phenotypes. The affected somites show irregular size and shape and in some cases also a fusion of somites is observed. Like all other medaka somite mutants, *krz*, *fsl* and *dpk* are also embryonic lethal. Thus, these patterning defects can not be followed until later stages, when for example effects on the axial skeleton could be observed. However, future mutant screens will hopefully identify hypomorphic alleles of these mutants. These should survive long enough to analyze the effects of anteriorization or posteriorization, respectively, on later steps of somite differentiation.

Taken together, the group 2 medaka mutants show unique phenotypes that so far have not been reported in zebrafish, mice or chicken. Thus, they promise to identify essential components of the molecular machinery that confers AP identity to somitomeres, a process that so far has not been accessible to mutant approaches in fish. In zebrafish and higher vertebrates, *mesp* genes have been shown to be instrumental regulators of somite AP polarity. For example, *Mesp2* knock-out mice show posteriorized somites, while somites of knock-out mice deficient for the Notch effector *Presenilin1* exhibit anteriorized characters.[77] It will be interesting to analyze in the future, whether any of the medaka group 2 gene products interacts with Mesp or Delta/Notch signalling, or possibly represents a component of a novel pathway involved in AP specification.

Conclusions and Outlook

During the last decades, theoretical models have been generated to decipher the complex and highly coordinated process of segmentation in the vertebrate body. It was only recently that the molecular mechanisms regulating somite formation are becoming unravelled. These studies identified oscillatory circuits and gradients of signalling molecules. Their expression and activity are remarkably consistent with the "clock and wave front" model initially postulated by Cooke and Zeemann nearly thirty years ago.[18] Because of its experimental advantages, the zebrafish model contributed significantly to our understanding of the regulatory feedback loops that establish oscillating expression of Delta/Notch pathway components in the PSM. The zebrafish offers the unique opportunity of combining mutant analysis with the application of Morpholino based knock-down approaches and transgenic strategies. Zebrafish mutants from large-scale mutagenesis screens in Boston and Tuebingen were extremely helpful to dissect the molecular basis of somite formation and to support findings obtained in higher vertebrate model organisms. With the emergence of medaka as a genetic model system complementary to the zebrafish, it appears likely that even more mutants in additional pathways will be discovered. The phenotypic differences observed between somite mutants of zebrafish and medaka predict that novel components might be identified. The ultimate goal will be to approach a saturating level of mutagenesis and target every signalling cascade implicated in somitogenesis. Toward this end, mutant screens in fish are presently designed that employ advanced technologies like e.g., insertional mutagenesis with modified retroviral vectors or transposable elements that will facilitate the subsequent cloning of the mutated loci.[50,78] With the advanced genome sequencing projects in both zebrafish and medaka, molecular identification of mutants will further be accelerated. Functional analyses in these mutants will help to dissect the different steps of somite formation and study the complex interactions of various signalling cascades. This will not only help to understand how somites form in fish, but will also significantly contribute to our understanding of somitogenesis in higher vertebrates.

Acknowledgements

We thank Matthias Schäfer for providing the picture in Figure 1B. We wish to thank Martin Gajewski, Daniel Liedtke and Manfred Gessler for critical comments and suggestions on the manuscript and Manfred Schartl for his constant support. We are extremely grateful to Hisato Kondoh and Makoto Furutani-Seiki for giving us the opportunity to participate in the ERATO screen and to cooperate on the medaka mutant project. We apologize to all colleagues, whose work was not cited due to space limitations.

References

1. Driever W, Solnica-Krezel L, Schier AF et al. A genetic screen for mutations affecting embryogenesis in zebrafish. Development 1996; 123:37-46.
2. Haffter P, Granato M, Brand M et al. The identification of genes with unique and essential functions in the development of the zebrafish, Danio rerio. Development 1996; 123:1-36.
3. Knaut H, Werz C, Geisler R et al. Tübingen 2000 Screen Consortium. A zebrafish homologue of the chemokine receptor Cxcr4 is a germ-cell guidance receptor. Nature 2003; 421:279-82.
4. Pourquie O, Tam PP. A nomenclature for prospective somites and phases of cyclic gene expression in the presomitic mesoderm. Dev Cell 2001; 1:619-20.
5. Karlstrom RO, Talbot WS, Schier AF. Comparative synteny cloning of zebrafish you-too: mutations in the Hedgehog target gli2 affect ventral forebrain patterning. Genes Dev 1999; 13:388-93.
6. Schauerte HE, van Eeden FJM, Fricke C et al. Sonic hedgehog is not required for the induction of medial floor plate cells in the zebrafish. Development 1998; 125:2983-93.
7. Nakano Y, Kim HR, Kawakami A et al. Inactivation of dispatched 1 by the chameleon mutation disrupts Hedgehog signalling in the zebrafish embryo. Dev Biol 2004; 269:381-92.
8. Baxendale S, Davison C, Muxworthy C et al. The B-cell maturation factor Blimp-1 specifies vertebrate slow-twitch muscle fiber identity in response to Hedgehog signaling. Nat Genet 2004; 36:88-93.
9. Stickney HL, Barresi MJ, Devoto SH. Somite development in zebrafish. Dev Dyn 2000; 219:287-303.
10. Nikaido M, Kawakami A, Sawada A et al. Tbx24, encoding a T-box protein, is mutated in the zebrafish somite-segmentation mutant fused somites. Nat Genet 2002; 31:195-9.
11. Holley SA, Geisler R, Nüsslein-Volhard C. Control of her1 expression during zebrafish somitogenesis by a Delta-dependent oscillator and an independent wave-front activity. Genes Dev 2000; 14:1678-90.
12. Holley SA, Julich D, Rauch GJ et al. Her1 and the notch pathway function within the oscillator mechanism that regulated zebrafish somitogenesis. Development 2002; 129:1175-83.
13. Julich D, Hwee Lim C, Round J et al. Tubingen 2000 Screen Consortium. beamter/deltaC and the role of Notch ligands in the zebrafish somite segmentation, hindbrain neurogenesis and hypochord differentiation. Dev Biol 2005; 286:391-404.
14. Itoh M, Kim CH, Palardy G et al. Mind bomb is a ubiquitin ligase that is essential for efficient activation of Notch signaling by Delta. Dev Cell 2003; 4:67-82.
15. Griffin KJ, Amacher SL, Kimmel CB et al. Molecular identification of spadetail: regulation of zebrafish trunk and tail mesoderm formation by T-box genes. Development 1998; 125:3379-88.
16. Julich D, Geisler R, Holley SA. Tubingen 2000 Screen Consortium. Integrinalpha5 and delta/notch signaling have complementary spatiotemporal requirements during zebrafish somitogenesis. Dev Cell 2005; 8:575-86.
17. Koshida S, Kishimoto Y, Ustumi H et al. Integrinalpha5-dependent fibronectin accumulation for maintenance of somite boundaries in zebrafish embryos. Dev Cell 2005; 8:587-98.
18. Cooke J, Zeeman EC. A clock and wavefront model for control of the number of repeated structures during animal morphogenesis. J Theor Biol 1976; 58:455-76.
19. Jiang YJ, Brand M, Heisenberg CP et al. Mutations affecting neurogenesis and brain morphology in the zebrafish, Danio rerio. Development 1996; 123:205-16.
20. Oates AC, Ho RK. Hairy/E(spl)-related (Her) genes are central components of the segmentation oscillator and display redundancy with the Delta/notch signaling pathway in the formation of anterior segmental boundaries in the zebrafish. Development 2002; 129:2929-46.
21. Gajewski M, Sieger D, Alt B et al. Anterior and posterior waves of cyclic her1 gene expression are differentially regulated in the presomitic mesoderm of zebrafish. Development 2003; 130:4269-78.
22. Winkler C, Elmasri H, Klamt B et al. Characterization of hey bHLH genes in teleost fish. Dev Genes Evol 2003; 213:541-53.
23. Holley SA, Takeda H. Catching a wave: the oscillator and wavefront that create the zebrafish somite. Semin Cell Dev Biol 2002; 13:481-88.

24. Henry CA, Urban MK, Dill KK et al. Two linked hairy/Enhancer of split-related zebrafish genes, her1 and her7, function together to refine alternating somite boundaries. Development 2002; 129:3693-704.
25. Sieger D, Tautz D, Gajewski M. The role of Suppressor of Hairless in Notch mediated signaling during zebrafish somitogenesis. Mech Dev 2003; 120:1083-94.
26. Sieger D, Tautz D, Gajewski M. her11 is involved in the somitogenesis clock in zebrafish. Dev Genes Evol 2004; 214:393-406.
27. Takke C, Campos-Ortega JA. her1, a zebrafish pair-rule like gene, acts downstream of notch signalling to control somite development. Development 1999; 126:3005-14.
28. Horikawa K, Ishimatsu K, Yoshimoto E et al. Noise-resistant and synchronized oscillation of the segmentation clock. Nature 2006; 441:719-23.
29. Jiang YJ, Aerne BL, Smithers L et al. Notch signaling and the synchronization of the somite segmentation clock. Nature 2000; 408:475-9.
30. Lewis J. Autoinhibition with transcriptional delay: a simple mechanism for the zebrafish somitogenesis oscillator. Curr Biol 2003; 13:1398-408.
31. Sawada A, Fritz A, Jiang YJ et al. Zebrafish Mesp family genes, mesp-a and mesp-b are segmentally expressed in the presomitic mesoderm and Mesp-b confers the anterior identity to the developing somites. Development 2000; 127:1691-702.
32. Dubrulle J, McGrew MJ, Pourquie O. FGF signaling controls somite boundary position and regulates segmentation clock control of spatiotemporal Hox gene activation. Cell 2001; 106:219-32.
33. Sawada A, Shinya M, Jiang YJ et al. Fgf/MAPK signaling is a crucial positional cue in somite boundary formation. Development 2001; 128:4873-80.
34. Reifers F, Bohli H, Walsh EC et al. Fgf8 is mutated in zebrafish acerebellar (ace) mutants and is required for maintenance of midbrain-hindbrain boundary development and somitogenesis. Development 1998; 125:2381-95.
35. Draper BW, Stock DW, Kimmel CB. Zebrafish fgf24 functions with fgf8 to promote posterior mesodermal development. Development 2003; 130:4639-54.
36. Kawamura A, Koshida S, Hijikata H et al. Zebrafish hairy/enhancer of split protein links FGF signaling to cyclic gene expression in the periodic segmentation of somites. Genes Dev 2005; 19:1156-61.
37. Sieger D, Ackermann B, Winkler C et al. her1 and her13.2 are jointly required for somitic border specification along the entire axis of the fish embryo. Dev Biol 2006; 293:242-51.
38. Kawamura A, Koshida S, Hijikata H et al. Groucho-associated transcriptional repressor ripply1 is required for proper transition from the presomitic mesoderm to somites. Dev Cell 2005; 9:735-44.
39. Aulehla A, Wehrle C, Brand-Saberi B et al. Wnt3a plays a major role in the segmentation clock controlling somitogenesis. Dev Cell 2003; 4:395-406.
40. Galceran J, Sustmann C, Hsu SC et al. LEF1-mediated regulation of Delta-like1 links Wnt and Notch signaling in somitogenesis. Genes Dev 2004; 18:2718-23.
41. Hofmann M, Schuster-Gossler K, Watabe-Rudolph M et al. WNT signaling, in synergy with T/TBX6, controls Notch signaling by regulating Dll1 expression in the presomitic mesoderm of mouse embryos. Genes Dev 2004; 18:2712-17.
42. Aulehla A, Herrmann BG. Segmentation in vertebrates: clock and gradient finally joined. Genes Dev 2004; 18:2060-67.
43. Rauch GJ, Hammerschmidt M, Blader P et al. Wnt5 is required for tail formation in the zebrafish embryo. Cold Spring Harb Symp Quant Biol 1997; 62:227-34.
44. Heisenberg CP, Houart C, Take-Uchi M et al. A mutation in the Gsk3-binding domain of zebrafish Masterblind/Axin1 leads to a fate transformation of telencephalon and eyes to diencephalon. Genes Dev 2001; 15:1427-34.
45. Krauss S, Korzh V, Fjose A et al. Expression of four zebrafish wnt-related genes during embryogenesis. Development 1992; 116:249-59.
46. Buckles GR, Thorpe CJ, Ramel MC et al. Combinatorial Wnt control of zebrafish midbrain-hindbrain boundary formation. Mech Dev 2004; 121:437-47.
47. Brent AE. Somite formation: Where Left meets right. Current Biology 2005; 15:468-70.
48. Kawakami Y, Raya A, Raya RM et al. Retinoic acid signalling links left-right asymmetric patterning and bilaterally symmetric somitogenesis in the zebrafish embryo. Nature 2005; 435:165-71.
49. Thermes V, Grabher C, Ristoratore F et al. I-SceI meganuclease mediates highly efficient transgenesis in fish. Mech Dev 2002; 118:91-8.
50. Grabher C, Henrich T, Sasado T et al. Transposon-mediated enhancer trapping in medaka. Gene 2003; 322:57-66.
51. Hong Y, Winkler C, Schartl M. Production of medakafish chimeras from a stable embryonic stem cell line. Proc Natl Acad Sci 1998; 95:3679-84.

52. Bejar J, Hong Y, Schartl M. Mitf expression is sufficient to direct differentiation of medaka blastula derived stem cells to melanocytes. Development 2003; 130:6545-53.
53. Del Bene F, Tessmar-Raible K, Wittbrodt J. Direct interaction of geminin and Six3 in eye development. Nature 2004; 427:745-49.
54. Nanda I, Kondo M, Hornung U et al. A duplicated copy of DMRT1 in the sex-determining region of the Y chromosome of the medaka, Oryzias latipes. Proc Natl Acad Sci 2002; 99:11778-83.
55. Schartl M. A comparative view on sex determination in medaka. Mech Dev 2004; 121:639-45.
56. Elmasri H, Winkler C, Liedtke D et al. Mutations affecting somite formation in the medaka (Oryzias latipes). Mech Dev 2004; 121:659-71.
57. Elmasri H, Liedtke D, Lücking G et al. her7 and hey1, but not lunatic fringe show dynamic expression during somitogenesis in Medaka (Oryzias latipes). Gene Expression Pattern 2004; 4:553-9.
58. Gajewski M, Elmasri H, Girschick M et al. Comparative analysis of her genes during fish somitogenesis reveals a mouse/chick-like mode of oscillation in medaka. Development, Genes and Evolution 2006; 216:315-32.
59. Iwamatsu T. Stages of normal development in the medaka Oryzias latipes. Mech Dev 2004; 121:605-18.
60. Kimmel CB, Ballard WW, Kimmel SR et al. Stages of embryonic development of the zebrafish. Dev Dyn 1995; 203:253–310.
61. Dale JK, Maroto M, Dequeant ML et al. Periodic notch inhibition by lunatic fringe underlies the chick segmentation clock. Nature 2003; 421:275–8.
62. Serth K, Schuster-Gossler K, Cordes R et al. Transcriptional oscillation of lunatic fringe is essential for somitogenesis. Genes Dev 2003; 17:912-25.
63. Leve C, Gajewski M, Rohr KB et al. Homologues of c-hairy1 (her9) and lunatic fringe in zebrafish are expressed in the developing central nervous system, but not in the presomitic mesoderm. Dev Genes Evol 2001; 211:493-500.
64. Prince VE, Holley SA, Bally-Cuif L et al. Zebrafish lunatic fringe demarcates segmental boundaries. Mech Dev 2001; 105:175-80.
65. Leimeister C, Dale K, Fischer A et al. Oscillating expression of c-hey2 in the presomitic mesoderm suggests that the segmentation clock may use combinatorial signaling through multiple interacting bHLH factors. Dev Biol 2000; 227:91-103.
66. Winkler C, Schäfer M, Duschl J et al. Functional divergence of two zebrafish midkine growth factors following fish-specific gene duplication. Genome Res 2003; 13:1067-81.
67. Amores A, Suzuki T, Yan YL et al. Developmental roles of pufferfish hox clusters and genome evolution in ray-fin fish. Genome Res 2004; 14:1-10.
68. Naruse K, Tanaka M, Mita K et al. A medaka gene map: the trace of ancestral vertebrate proto-chromosomes revealed by comparative gene mapping. Genome Res 2004; 14:820-8.
69. Furutani-Seiki M, Sasado T, Morinaga C et al. A systematic genome-wide screen for mutations affecting organogenesis in Medaka, Oryzias latipes. Mech Dev 2004; 121:647-58.
70. Hrabe de Angelis M, McIntyre J 2nd, Gossler A. Maintenance of somite borders in mice requires the Delta homologue DII1. Nature 1997; 386:717-21.
71. Evrard YA, Lun Y, Aulehla A et al. Lunatic fringe is an essential mediator of somite segmentation and patterning. Nature 1998; 394:377-81.
72. Bessho Y, Sakata R, Komatsu S et al. Dynamic expression and essential functions of Hes7 in somite segmentation. Genes Dev 2001; 15:2642-7.
73. Carl M, Wittbrodt J. Graded interference with FGF signalling reveals its dorsoventral asymmetry at the mid-hindbrain boundary. Development 1999; 126:5659-67.
74. Aoyama H, Asamoto K. Determination of somite cells: independence of cell differentiation and morphogenesis. Development 1988; 104:15-28.
75. Takahashi Y, Inoue T, Gossler A et al. Feedback loops comprising Dll1, Dll3 and Mesp2 and differential involvement of Psen1 essential for rostrocaudal patterning of somites. Development 2003; 130:4259-68.
76. Takahashi Y, Kitajima S, Inoue T et al. Differential contributions of Mesp1 and Mesp2 to the epithelialization and rostro-caudal patterning of somites. Development 2005; 132:787-96.
77. Takahashi Y, Koizumi K, Takagi A et al. Mesp2 initiates somite segmentation through the Notch signalling pathway. Nat Genet 2000; 25:390-6.
78. Amsterdam A, Hopkins, N. Retroviral-mediated insertional mutagenesis in zebrafish. Methods Cell Biol 2004; 77:3-20.

CHAPTER 4

Old Wares and New:
Five Decades of Investigation of Somitogenesis in *Xenopus laevis*

Duncan B. Sparrow*

Abstract

Somites are regular repeated structures formed in pairs on either side of the anterior-posterior axis of developing vertebrate embryos which give rise to all skeletal muscle of the body, the axial skeleton, the tendons and the dorsal dermis. Beginning in the middle of last century, somite formation has been extensively studied in the South African clawed frog (*Xenopus laevis*) using traditional embryological techniques. Recently, modern molecular methods have been applied to this system, producing substantial insights into the underlying molecular mechanisms driving these morphological events. In this review I discuss these new results in the context of the early embryological observations, looking at all levels of the process of somite formation, from the initial prepatterning of the presomitic mesoderm to the morphomechanical events required for the separation of each somite from the precursor tissue.

Introduction

Amphibian somite patterning and morphogenesis has been studied for many decades, thanks to the ease of embryological manipulation of embryos that can develop externally in a simple salt solution at room temperature. *Xenopus laevis* is the prevailing and best-understood amphibian model system. Many of the conclusions reached 20 years ago using the classical techniques of embryology are now being revisited with modern molecular techniques. In particular, techniques are now available to inject synthetic mRNAs to study the effects of targeted gene (and mutation) overexpression; morpholino oligonucleotides can be used to specifically "knockdown" specific transcripts; and transgenesis techniques can be used to either rapidly map transcriptional control elements in a promoter or as an additional way of targeting ectopic gene expression. As with many other developmental processes, somitogenesis appears to involve evolutionarily conserved regulatory mechanisms, although there may be differences at the fine morphological level. Therefore mouse, chick and zebrafish studies have also provided invaluable supplemental information. Thus, the *Xenopus* system is potentially a very useful tool for extending our understanding of how somitogenesis occurs.

Note: Standard embryo stages as defined by reference 1 will be used throughout this chapter. *Xenopus* embryos develop at a rate dependent on environmental temperature, thus to give the non-expert reader an idea of the timing of these stages, approximate times postfertilisation (hours post fertilisation; hpf) will be given assuming development at optimal temperature (23°C).

*Duncan B. Sparrow—Developmental Biology Program, Victor Chang Cardiac Research Institute, 384 Victoria Street, Darlinghurst, NSW 2010, Australia.
Email: d.sparrow@victorchang.edu.au

Somitogenesis, edited by Miguel Maroto and Neil V. Whittock. ©2008 Landes Bioscience and Springer Science+Business Media.

What Are Somites?

In vertebrates, all skeletal muscle of the body, the axial skeleton, the tendons and the dorsal dermis are derived from structures called somites. These are paired blocks of mesenchymal mesoderm located on either side of the neural tube that begin to form after gastrulation by a regular wave of segmentation in a rostral to caudal direction throughout the trunk and tail. The majority of the *Xenopus* somite is myotomal (Fig. 1), consisting of cells that will form muscle. In contrast to mammals, these myotomal cells initially differentiate into uninucleate myotubes and it is not until the onset of metamorphosis that these fuse to form multinucleate muscle fibres.[2,3] A small dorsolateral part of the somite is the dermatome (the precursor of the dermis), which forms a thin

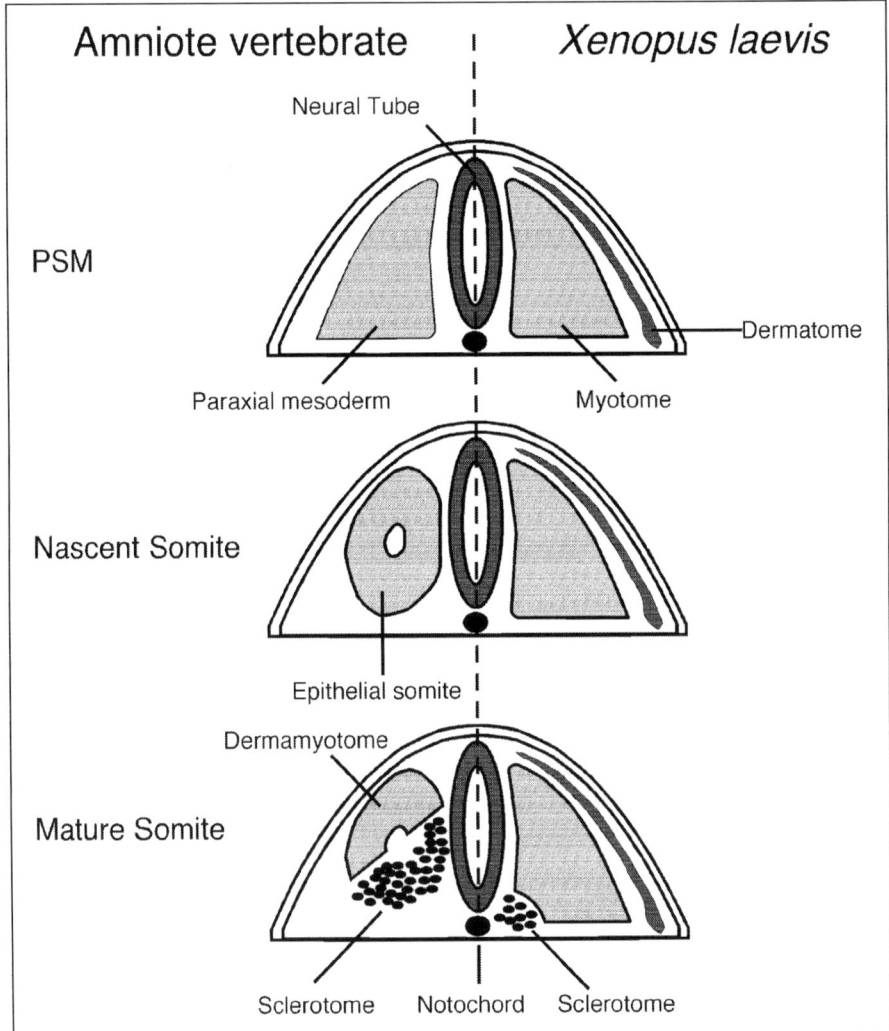

Figure 1. A schematic representation of transverse sections of an embryo, comparing the structure of the PSM and somites in amniote vertebrates (left of neural tube) and *Xenopus laevis* (right). The first section is taken through the unsegmented PSM, the second from the most recently formed somite and the last from a mature somite.

sheet covering the myotome. Whether this sheet is segmented in the same manner as the rest of the somite remains in dispute.[4,5] Finally the sclerotome (precursor of the ribs and vertebrae) is the smallest compartment of the somite, made up of cells located adjacent to the notochord that only become distinct from the myotome after stage 30 (~35 hpf).[5] These latter two compartments have largely been ignored by the research community, so relatively little is known about how they change as development proceeds (reviewed in ref. 6).

The developmental process of myogenesis in *Xenopus* has important differences to that of amniote vertebrates (such as mouse or chick, reviewed in ref. 7). Zygotic transcription of the myogenic factors Myf-5 and MyoD are first detectable during gastrulation (stages 9.5 and 10.5 respectively, ~9-11 hpf),[8,9] and they continue to be strongly expressed throughout the presomitic mesoderm (PSM) and somites as development proceeds. Muscle differentiation markers such as cardiac actin are also activated very early in development in the PSM. In mammals, the expression of myogenic factors and subsequent differentiation only occurs much later in development, once somites have segmented and epithelialized.

Where Do Somites Come From?

The precursor tissue from which somites are derived is an unsegmented sheet of mesenchyme called presomitic mesoderm (PSM; also called the paraxial mesoderm or the segmental plate). The PSM arises from the involuting marginal zone (IMZ) from which all mesoderm ultimately derives. This corresponds to a band of cells that surround the blastula embryo, lying between the prospective ectoderm at the animal pole and the prospective endoderm at the vegetal pole. In *Xenopus*, the IMZ consists of two cell layers: an epithelium of prospective endoderm; and the underlying deep prospective mesoderm (including somitic and notochordal mesoderm). During gastrulation, these cells move over the blastoporal lip (the equivalent of the node in amniotes such as chick and mouse) into the interior of the gastrulating embryo (a process known as involution). The prospective head, heart, lateral plate and ventral body wall mesoderm all undergo involution early in gastrulation and then actively migrate towards the anterior of the embryo.[10] Subsequently, the precursors of the somitic and notochordal mesoderm involute late in gastrulation. Once inside the embryo, these cells are not migratory. Instead, as gastrulation is completed, this tissue narrows and elongates along the anterior-posterior axis of the embryo in a process known as convergence/extension (reviewed in ref. 6).

Studies of cultured explants taken at stage 12.5 (late gastrula, ~14 hpf) containing both dorsal mesoderm and neural plate have revealed that it is only the most posterior 20-25% of this tissue that actually extends at any one given time. This process is not driven by cell division, since there is very little cell division in the presomitic mesoderm at this stage; nor is it driven by chances in cell shape. Instead it occurs by two distinct forms of cell intercalation. Firstly, around stage 15 (mid-neurula, ~18 hpf) some cells move in small groups from the endodermal epithelium to join the deep mesenchymal layer, a process called radial intercalation behaviour.[6] Once in the PSM, they de-epithelialize. This is confined to a narrow band of cells at the posterior edge of the presomitic mesoderm. The mixing of additional cells brings about rapid extension along the anterior-posterior axis without much change in the overall width of the tissue. More rostrally along the PSM a different process, called medio-lateral intercalation occurs. In this process, cells within the same layer rearrange, resulting in an overall slight elongation together with a narrowing in the tissue in general. These cells are myotomal and unlike amniote vertebrates, they already express markers of muscle differentiation (see above). At the same time these cells also change shape (although this is in opposition to the overall anterior-posterior elongation of the tissue). At stage 12.5-13 (~14-15 hpf), the cells of the presomitic mesoderm are increasingly rounded or polygonal as one moves caudally. However, beginning at stage 14 (16 hpf), these cells lengthen in the transverse axis until they lie in ranks with their long axis perpendicular to the notochord (Fig. 2). This transformation is initiated at the anterior of the PSM and spreads towards the posterior. This wave of cellular change (also known as the "zone of extension") precedes the wave of segmentation by five or six prospective somites.

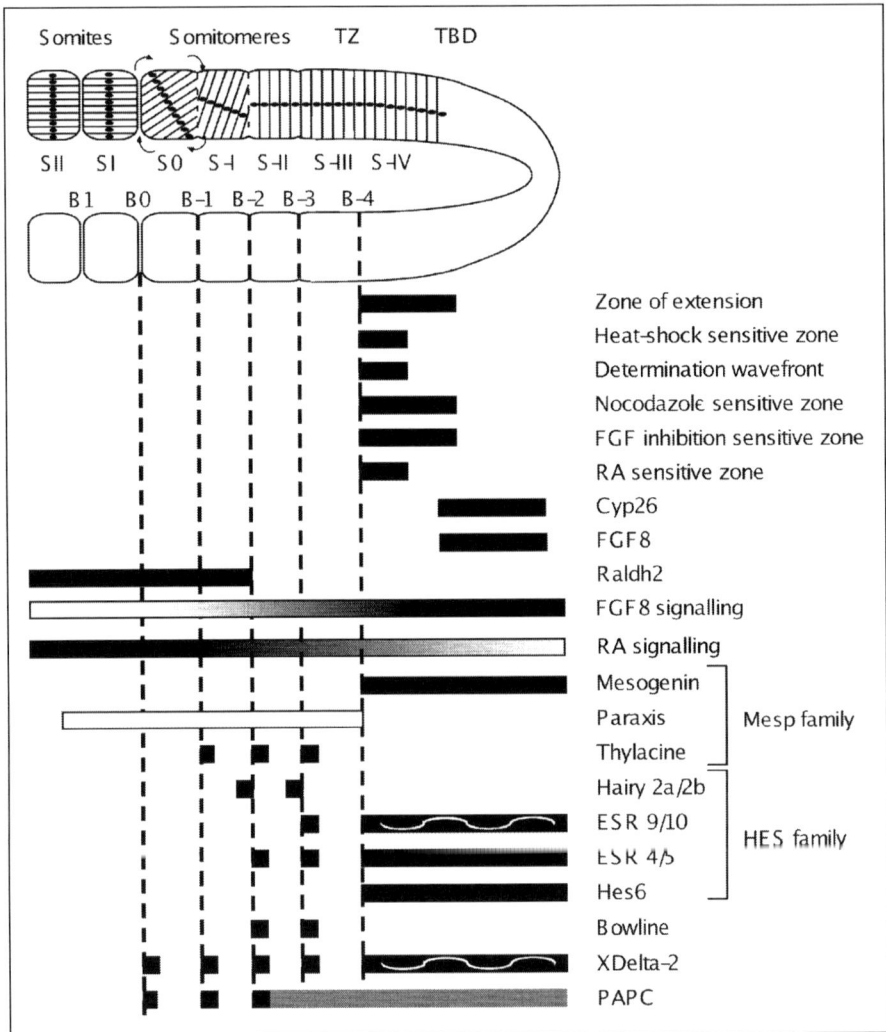

Figure 2. A schematic representation of a horizontal section taken through the paraxial mesoderm of a *Xenopus* embryo at stage 26. The top half of the embryo shows the arrangement of cells and nuclei undergoing the morphological rotation during somitogenesis. Standard nomenclature for mature and prospective somites and boundaries is indicated.[12] Below the schematic embryo are shown the various embryological regions and gene expression domains described in the text. Gradients of FGF8 and RA signaling are indicated by a gradient of shading. In the case of *ESR9, ESR10* and *X-Delta-2*, cyclical expression in the TBD is indicated by a wavy white line.

Structure of the PSM

Although the PSM morphologically appears as a loose featureless mesenchymal sheet, early stereo scanning electron microscope (SEM) analysis of the PSM from a variety of vertebrates suggested the presence of a loose arrangement of cells in squat bilaminar disks throughout the PSM, corresponding to a prepattern of prospective segments called "somitomeres" (reviewed in ref. 11). These were proposed to be fully established immediately after the allocation of the progenitor cells from the primitive streak or tail bud to the PSM. For the last decade or so, this observation has been controversial and largely discounted. Recent gene expression patterns showing discrete striped domains in the apparently unsegmented PSM suggests that prospective somites are patterned before segmentation occurs (see below). In the recent literature, "somitomere" is used interchangeably with "prospective somite". However the segmental expression of these molecular markers is limited to the rostral PSM and no gene has yet been found that is segmentally expressed in the caudal PSM. Therefore there is still no good evidence that "somitomeres" (in the original sense of the word) exist.

The nomenclature for segmental domains within the somites and PSM of all vertebrate species has been standardized as outlined in ref. 12 (see Fig. 2 for representation of the *Xenopus* PSM). The most recently formed complete somite is somite SI. The intrasomitic fissure between somite SI and the PSM is border 0 (B0). The PSM is then divided into prospective somites (or somitomeres), beginning rostrally with somite S0, S-I, S-II etc. Prospective boundaries are labelled beginning with –1 for the caudal boundary of somite S0. In *Xenopus*, somites S0 to S-II are referred to as the somitomere region; somite S-III and -IV are the transition zone (TZ); and somites S-V and higher are called the tailbud domain (TBD).[13] Embryological studies (reviewed in ref. 14) suggest that the PSM of *Xenopus* contains a total of nine prospective somites at any one time. This value is constant because the rate of recruitment of cells at the posterior end of the PSM is the same as the rate of exit of cells at the anterior end. This suggests that the TBD contains cells fated to form at least four somites in total. In other species, this number varies, for example mouse PSM contains 6 prospective somites[15] and chick 12.[16]

Morphological Descriptions of Segmentation in *Xenopus*

In *Xenopus*, segmentation of the presomitic mesoderm begins at stage 17 (mid-to-late neurula, ~19 hpf). The first five or six somites form in rapid succession and thereafter arise at a constant rate of approximately one somite every 45-120 minutes depending on environmental temperature.[4,17] A total of 46 somites form to the end of the tail, followed by some very small somites that are added continuously to the tail process that grows throughout larval life.[17] The physical process of segmentation has been extensively analysed by histology[4] and SEM[5,18] of whole embryos; and live video microscopy of cultured explants of dorsal mesoderm.[19] These allow the synthesis of an overall picture of the morphological processes involved (reviewed in ref. 6).

The intersomitic furrow forms first in the lateral region and then moves medially until it reaches the notochord.[19] Immediately prior to the formation of the fissure, multiple "near fissures" may be observed in the region where the boundary is expected to form, however only one of these straightens and intensifies to become the definitive furrow.[19] At the same time, cells that will come to lie on opposite sides of the boundary begin to move past one another, with cells anterior to the prospective boundary moving towards the notochord (when viewed from above) and posterior cells moving laterally. As the fissure extends medially, the cells in the rostral compartment continue morphological movements that now resemble a rotation of the entire somite block. Cells closest to the somitic furrow continue to move towards the notochord, while cells next to the previously formed somite boundary move laterally and those close to the notochord move rostrally. This results in the 90 degree rotation as described by Hamilton[4] (see also Fig. 2). Thus elongated mononucleated myotomal cells in the presomitic mesoderm lie with their long axis perpendicular to the anterior-posterior axis of the embryo, but after segmentation are parallel to it and span the entire width of the somite. When the dorso-ventral aspects of the somites are considered, initial histological analysis[4] suggested that the somitic cells rotate as a single block. However the SEM data[5] shows that the cells in the middle layer of the somite (nearest the notochord) rotate first, followed by the cells above and below. In general, SEM reveals that this rearrangement of cells is accompanied by the formation, extension and retraction of cell protrusions. This suggests that rotation occurs through individual cell movement.[5,19] However it is not clear whether the furrow arises as a consequence of cell movements, or whether it represents a controlled de-adhesion step required before cell are able to move. The potential molecular mechanisms controlling this process will be discussed at the end of this review.

A Comparison of *Xenopus* Segmentation with That of Amniote Vertebrates

In amniotes such as chick and mouse, there are some similarities but many differences in the mechanical processes of segmentation. An elegant study by Kulesa and Fraser[20] using time-lapse microscopy of somite boundary formation in live chick embryos illustrates this perfectly. Use of cell labelling shows that the initial stages of PSM formation in the chick are very similar to those in *Xenopus*. In particular, cells in the most posterior part of the PSM (in the region of Hensen's node, the equivalent to the blastoporal lip) actively exchange neighbours and disperse, spreading throughout large regions of the segmental plate. However, by the time cells have come within 4-5 prospective somites away from the forming somite (S0), their movements are reduced to less than a single somite length. However, there are marked differences in the actual process of segmentation. In chick, the initial tissue separation and fissure formation occurs at the same time on both the medial and lateral edges of the PSM. The fissure is not straight; instead the boundary between the segmenting somite and the PSM initially forms a "ball-and-socket" shape, with the somite forming the "ball" and the PSM the "socket". As the somitic cells pull away from the segmental plate, cells at the anterior lateral edge of the PSM appear to fold medially, filling the socket. These differences in morphological movements at the forming boundary are likely to be related to the differences in final somite structure: in *Xenopus* newly formed somites are cuboidal or chevron shaped and consist mainly of myotome overlaid by a separate thin sheet of dermatome, whereas in chick (like other amniotes) the somites form epithelial balls immediately after segmentation is complete (see Fig. 1). Only as the somite matures do these epithelial balls of tissue begin to differentiate and form myotomal, dermatomal and sclerotomal compartments. Given the differences in structure of the newly formed somites, it is likely that different cell movements may be required to achieve these final structures.

What Controls Where the Somitic Furrow Forms?

Over the past 30 years, many theoretical models have been proposed to explain how the regular pattern of somite boundary formation is achieved. The simplest model is that there is a cell-counting mechanism that triggers the formation of a somitic furrow after a regular number of cells. In *Xenopus* this is clearly not the case. The phenomemon of somite rotation and the lack of appreciable cell division in myotomal cells, means that the number of cells counted across a somite in horizontal section is a direct measure of the number of cells in the presomitic mesoderm that were recruited into that somite. For the initial 10-12 somites, this is an average of 9-10 cells. However as somitogenesis continues, the number of cells recruited declines, with the terminal somites of the tail tip containing as few as two cells.[17] Since somites are forming at a constant rate, this precludes the existence of a cell-counting mechanism (which would, by necessity, produce somites of identical cell number throughout the trunk and tail).

The most durable idea of how somitogenesis is achieved is the "Clock and Wavefront" model.[21,22] This proposes that all cells within the PSM oscillate between two internal states (the "clock"). These cells are locally coordinated such that cells positioned relatively closely in the embryo oscillate near-synchronously. The period of the oscillation corresponds to the time observed between the segmentation of successive somites. At the same time, a slowly moving wavefront progresses in a rostral to caudal direction through the PSM. As the wavefront passes a group of cycling cells, those in the permissive phase stop oscillating and begin to differentiate. The interaction between the oscillation and the wavefront produces a temporal separation between the adjacent groups of cells, allowing them to begin somite formation at different times in a regular metameric fashion. This model makes three predictions. Firstly, the anterior PSM, although morphologically unsegmented, nevertheless should contain a molecular prepattern sufficient for the regular formation of somites. Secondly, the posterior PSM should contain a "clock" that oscillates with a constant period. Lastly, a "wavefront" should pass slowly through the PSM in a rostral-caudal direction, acting to halt the oscillation of the "clock" and trigger the start of differentiation. There is now molecular evidence of each of these three required elements and these will be discussed in this order in the next three sections of this review.

Evidence for Segmental Prepatterning of the PSM

One of the first indications of the existence of a molecular prepattern within the rostral PSM was the observation that the gene *X-Delta-2* was expressed in discrete stripes in the morphologically unsegmented PSM of *Xenopus*.[23] This expression pattern consists of four discrete stripes in the rostral PSM and a broad expression domain in the TBD. As mentioned above, the *Xenopus* PSM has a unique arrangement, with mononucleate myotomal cells spanning the whole PSM and a regular number of these cells being recruited into each somite. Thus it is relatively easy to plot the exact location of gene expression patterns in the PSM by cutting horizontal sections through the embryo and counting the myotomal cell nuclei back from the most recently formed somitic fissure (B0). Using such sections, it was shown that *X-Delta-2* is expressed in the anterior halves of prospective somites S0, S-I, S-II and S-III (Fig. 2). Subsequently, more genes have been described with similar stripes of expression in the PSM of *Xenopus* (Table 1). These can be classified into two groups: those with stripes plus TBD expression (such as *X-Delta-2, ESR4* and *ESR5*) and those with stripes but no TBD expression (*Thylacine, Hairy2a/2b* and *Bowline*). All of these genes are expressed in the anterior halves of prospective somites (with the exception of *Hairy2a/2b*), but there is considerable variation in which prospective somites have expression (Fig. 2). In general, orthologues of these genes in other vertebrates, where identified, have similar patterns of PSM expression. A number of other genes are also segmentally expressed in this domain, but their expression patterns have not been precisely defined (Table 1).

Table 1. Segmentally expressed genes in *Xenopus:* A list of genes shown to be expressed in discrete stripes in the PSM. The first section is comprised of genes for which the precise localisation within the PSM has been determined, whilst the second section contains genes where the expression domain has not been well characterized. In both cases an indication is given of protein function.

Gene	Protein Function	Reference
Thylacine	Mesp family transcription factor	24
Mesogenin/Mespo	Mesp family transcription factor	65
Paraxis	Mesp family transcription factor	65
Hairy2a/b	HES family transcriptional repressor	23
ESR4	HES family transcriptional repressor	25
ESR5	HES family transcriptional repressor	25
ESR9	HES family transcriptional repressor	40
ESR10	HES family transcriptional repressor	40
Hes6	HES family transcriptional repressor	101
X-Delta-2	Notch ligand	23
PAPC	Protocadherin	31
Bowline	Groucho/TLE interactor	102
Genes with poorly described segmental expression		
Hey-1	HES family transcriptional repressor	103
Hairy1	HES family transcriptional repressor	32
Lfng	Glycosyltransferase	29
Nrarp	Notch signaling inhibitor	30
EphA4	Eph receptor	100
EphrinB2	Eph ligand	97
Dickkopf 1	BMP signaling inhibitor	104
Xblimp-1	Zn finger transcription factor	105

These expression patterns might be either a readout of a molecular prepattern, or a direct reflection of the processes involved in setting up the pattern. The roles of these genes in the segmentation process have been investigated by expressing them ectopically in the developing *Xenopus* embryo. In the case of somitogenesis, synthetic transcripts can be directed into the prospective mesoderm by injection into the equatorial region of a single blastomere of a two cell stage embryo, which is then allowed to develop until around stage 26 (early tadpole; 30 hpf). This results in an embryo in which one half ectopically expresses the injected transcript, whilst the other acts as a non-injected control. Using this system to individually overexpress the segmentally expressed genes *X-Delta-2, Thylacine 1, ESR5* and *Hes6* [23-26] results in gross abnormalities of somite segmentation limited to the injected side of the embryo. This confirms that these genes have a role in setting up the segmental prepattern and are not merely downstream effectors. A clue to the underlying mechanisms driving somitogenesis has been provided by the observation that most of the genes shown to have segmental PSM expression are either components of the Notch signaling pathway or downstream targets thereof.

The Notch Signaling Pathway

The Notch signaling pathway regulates cell fate decisions during embryonic development in all metazoans (reviewed in ref. 27). Notch proteins are a family of single-pass membrane receptors that are expressed in many cell types in temporally and spatially restricted patterns. Their ligands, members of the Delta and Serrate/Jagged families, are also membrane-bound. Because both ligand and receptor are on the cell surface, the activation of the Notch signaling pathway requires close cell-cell interaction. Ligand binding results in proteolytic cleavage of the receptor, allowing the Notch intracellular domain (NICD) to be translocated to the nucleus. There, the NICD interacts with members of the CSL family of DNA binding proteins, converting them from repressors to activators and switching on a wide array of downstream target genes. This signal is terminated through ubiquitylation of the NICD, followed by proteasome-dependent degradation. The basic pathway is modulated at a wide variety of levels including negative transcriptional feedback, posttranslation modification and transcript and protein stability. Of these, the most relevant to somitogenesis is the activity of the Fringe family of glycosyltransferases, which act in the Golgi apparatus to change the glycosylation patterns of the Notch extra-cellular domain and alter the ligand binding specificity of the receptor (see below). Using injection of synthetic transcripts, Notch signaling can be blocked in a number of different ways. These include expression of dominant negative forms of *XSu(H)* (a *Xenopus* member of the CSL family), a dominant negative version of the ligand *X-Delta-2*, or the *Enhancer of split*-related transcription repressors *ESR4* and *ESR5*. [23-25] Alternatively the pathway can be activated, for example by the expression of constitutively active forms of *XSu(H)*. In both cases disruption of Notch signaling in the PSM leads to poor and irregular boundary formation, suggesting that Notch signaling is a key component of the segmentation process. Further studies have revealed that Notch signaling is required for both the establishment of the segmental prepattern in the TZ, as well as maintaining and elaborating the pattern as differentiation commences in the somitomere region. Studies of null mutant mice and zebrafish, as well as ectopic expression experiments in chick, confirm these conclusions. The Notch1 receptor itself is expressed throughout the PSM and somites in *Xenopus*,[28] so activation of signaling must be controlled by ligands (*X-Delta-2*)[23,25] and modulated by modifiers of the pathway (*Lfng* and *nrarp*)[29,30] and downstream targets (*Thylacine* and *HES* family genes).[23-25]

Regulation of Segmental Gene Expression in Xenopus

In the *Xenopus* PSM, the Notch signaling pathway is first activated in a broad domain of cells in the TBD. However, beginning in the TZ, this resolves into an ON or OFF state, corresponding to anterior and posterior prospective somite halves respectively. Notch signaling then acts to maintain this segmental prepattern in the somitomere region. The initial striped pattern is achieved by a dynamic mechanism that represses the expression of Notch pathway genes in the posterior half-segment. This repression depends on a negative feedback loop in which *ESR5* (a direct target of Notch signaling) acts as a transcriptional repressor to shut off the Notch signal.[25] *ESR5* also acts in the anterior half-segment to maintain the expression of Notch target genes (such as *Thylacine* and *X-Delta-2*).[23-25] Of particular interest, *Thylacine* (a transcriptional activator of the Mesp class of bHLH DNA binding proteins) is involved in the subsequent specification of anterior identity, for example by activation of genes such as PAPC[31] required to maintain the segmental integrity needed for the formation of the somitic fissure (see below).

How this might work at the molecular level has been investigated with respect to the promoters of *Hairy2a*[32] and *Thylacine*.[24] *Hairy2a*, a member of the *Hairy/enhancer of split* (*HES*) family of transcriptional repressors, is expressed in two stripes in the PSM, in the posterior halves of prospective somites S-I and S-II (Fig. 2). In the case of *Thylacine*, although an enhancer element sufficient to drive PSM-specific expression has been mapped at a gross scale, investigation of the precise promoter elements required for segmental expression have not yet been reported.[13] The *Hairy2a* promoter has been investigated in more detail. In this case, the stripes of expression are controlled by two co-operating mechanisms. Firstly the proximal promoter contains two Su(H) binding sites in a head-to-head arrangement immediately upstream of the TATA box and transcriptional initiation site. This is very similar to the Su(H) paired sites (SPS) of E(spl) genes of *Drosophila* that are directly activated by Notch signaling.[33] The *Hairy2a* SPS site includes a variant of the conserved hexamer site that has been proposed to bind a general activator of transcription.[32] Upstream of the SPS site are two N boxes, which are putative binding sites for the HES family of transcriptional repressors. Secondly, the 3' UTR contains a 25bp element that destabilizes the mRNA, resulting in a transcript with a short half-life. Neither of these sequences alone is sufficient to recapitulate the striped pattern of *Hairy2a* expression, although together they can reconstitute the full stripe pattern. Remarkably, both of these elements appear highly conserved in the promoters and 3' UTR of other members of the *HES* family, even in other vertebrate species[32,34] (DBS unpublished data). Element swapping experiments show that these conserved sequences are functional when placed in the context of the *Hairy2a* gene and this points to a high degree of evolutionary conservation of control of gene expression in this family of genes. However, there are differences in the precise localisation of the stripes between different family members, with *c-hairy-1* class genes (of which *Hairy2a* is a member) being restricted to the posterior half segment and the *c-hairy-2* group (including *ESR4* and *ESR5*) being restricted to the anterior half segment. In addition, there is also variation in which prospective somites gene expression is observed. This suggests that elements in the proximal promoter apart from the SPS site are required to fine tune the expression of each family member.

Cycling Genes—Evidence of a "Clock"?

At the same time that *X-Delta-2* was described, evidence for the existence of an oscillator at work in the PSM was provided by studies of the *c-hairy-1* gene in chick.[35] Within individual cells of the PSM, this gene is expressed in a rhythmic ON/OFF manner that begins during gastrulation and is maintained throughout somitogenesis until ceasing when the cells approach the site of segmentation. Pulses of gene expression are followed by degradation of transcripts to undetectable levels, with the length of the complete ON/OFF cycle corresponding to the time it takes to form a complete somite. Thus, it is likely that the total number of cycles an individual cell experiences in the PSM prior to being incorporated in a somite is equal to the number of prospective somites within the PSM. Although the length of each ON/OFF cycle remains constant, the length of the ON period grows progressively shorter as the cell experiences more pulses. In addition, there is a slight phase shift between adjacent cells, such that as each new cell is added to the posterior end of the PSM, it is slightly delayed in the initiation of the cycling when compared to its anterior neighbour. Overall, this pattern of pulsatory gene expression generates the illusion of a continuous progressive wave of gene expression that moves from the posterior of the embryo to the anterior. As it progresses rostrally, the apparent band of gene expression becomes narrower, eventually coming to rest at the boundary of the forming somite (S0). In the case of *c-hairy-1*, expression is maintained in the caudal half of each somite after it has separated from the PSM. This illusory wave is a manifestation of the "clock" and should not be confused with the "wavefront", which

Table 2. Genes known to be expressed in an oscillating pattern in the PSM. Genes known to be expressed cyclically are shown arranged into three groups on the basis of the final location of the stripe of expression in a prospective half-segment. The species in which cycling has been demonstrated is indicated. The presence of a somitogenesis phenotype in null mutant embryos is also indicated, where known.

Gene	Species	Stripe Location	Null Mutant Somitogenesis Phenotype?	Reference
c-hairy-1 group				
c-hairy-1	chick	posterior	na	35
Hes1	mouse	posterior	N	106
Hey2	chick/mouse	posterior	N	107
deltaC	zebrafish	posterior	Y	108
c-hairy-2 group				
c-hairy-2	chick	anterior	na	106
Lfng	chick/mouse	anterior	Y	109,110
Hes5	mouse	anterior	N	111
Nkd1	Mouse	anterior	nd	112
Hes7	mouse	anterior	Y	113,114
her1	zebrafish	anterior	Y	115,116
her7	zebrafish	anterior	Y	115,116
deltaD	zebrafish	anterior	Y	108
X-Delta-2	Xenopus	anterior	na	13
Other				
hey1	zebrafish	unknown	nd	117
ESR9/10	Xenopus	unknown	na	40
Snail1	Mouse	anterior (cycles out of phase)	Early embryonic lethal	118
Snail2	Chick	anterior	na	118
Axin2	mouse	Posterior (cycles out of phase)	Y	51

na = not applicable; nd = not done.

is proposed to proceed in the opposite direction (rostral-caudal) at a much slower rate (a single wavefront transversing the entire somitic region in the time it takes to make all of the somites). Such cyclical expression in the PSM has now been described in a number of other genes in a wide range of vertebrates (see Table 2).

Because this type of expression is highly dynamic, any particular embryo will only show one frozen moment in time; and to fully appreciate the progression of expression, one must look at a large population of embryos. To assist in the description of such an expression pattern in individual embryos, nomenclature has been developed to describe three standardized intermediate stages in this process[12] (Fig. 3). Phase I involves a broad expression domain throughout the caudal PSM. During phase II, the domain has begun to sweep in a rostral direction and lies in the middle of the PSM. By phase III, the expression domain has refined to a strong band in the forming somite at the rostral limit of the PSM and weaker expression has begun to become apparent again in the caudal PSM. The exact location of the rostral stripe in phase III is completely gene-dependent (Table 2). One group of genes (the *c-hairy-2* group) become restricted to the anterior border of the somite, while the remainder (with a few exceptions) are in the posterior part (the *c-hairy-1* group). There

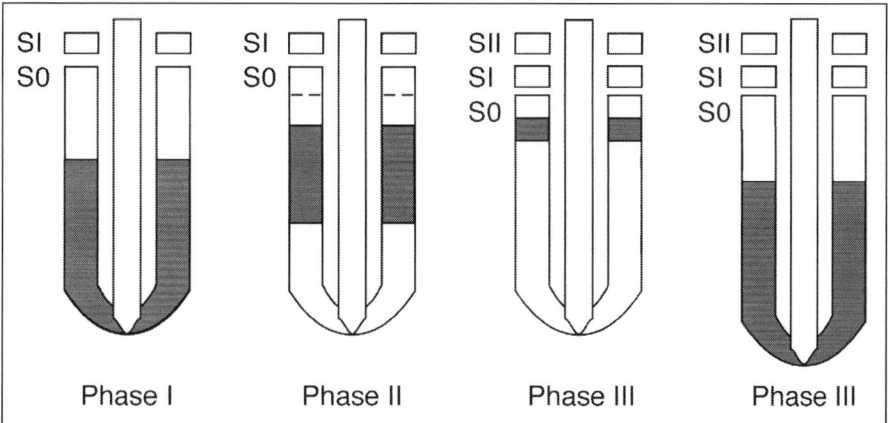

Figure 3. A schematic representation of the standard nomenclature for the three phases of cyclical gene expression in the PSM of mouse and chick, as defined in reference 12.

Table 3. **Summary of cycling gene orthologues between species. Identification of probable orthologues of cycling genes in other vertebrate species. The HES subfamily assignments are based on reference 119. An asterisk indicates genes that are known to be expressed cyclically in the PSM.**

Xenopus	**Mouse**	**Chick**	**Zebrafish**
hairy1	*Hes1**	*c-hairy2**	*her6*
Hairy2a/2b	*Hes4*	*c-hairy1**	*hairy1*
	Hes5*	*Hes5*	*her11**
ESR5	*Hes7**		*her1*/her7**
Hey1	*Hey1*	*Hey1*	*hey1**
	Hey2*	*Hey2**	*gridlock*
Lfng	*Lfng**	*Lfng**	*Lfng*
	Snail1	*Snail2*	
?? X-Delta-1	*Dll1*		*deltac**
?? X-Delta-2*			*deltad**

are representatives of many of these genes in all vertebrates (Table 3). Null mutations in some of these genes have been studied in mouse (*Lfng, Hes1, Hes5, Hes7* and *Hey-2*) and zebrafish (*her1, her7, deltac* and *deltad*). Most of these mutants have segmentation defects, suggesting that cycling genes play critical roles in somitogenesis. Exceptions are the mouse *Hes1, Hes5* and *Hey-2* genes,[36-39] and this may indicate the presence of functional redundancies within the *HES* gene family in the segmentation process.

Do These Genes also Cycle in Xenopus*?*

The anterior bands of expression of cyclical genes in the rostral PSM are very similar to the static stripes observed in *Xenopus*. However, until recently, none of the *Xenopus* orthologues of the cycling genes were reported as showing oscillatory expression in the caudal PSM. However, *ESR9* and *ESR10* have now been demonstrated to cycle,[40] and re-examination of the *X-Delta-2* expression pattern has revealed evidence of transcript cycling.[13] In both cases, cyclical expression is limited to the TZ and TBD. This is a much smaller and more compact sub-domain of the total PSM than the cycling region of mouse or chick and this may make cyclical expression more difficult to detect in *Xenopus*. Thus it may be that careful re-examination of other orthologues of genes known to cycle in other vertebrates (such as *Lfng*) may reveal that they also cycle in the *Xenopus* TBD. As highlighted above, such dynamic patterns are difficult to detect and may only be identified by examination of large numbers of embryos. A classic example of this is the zebrafish *her1* gene. This was initially described to be "expressed in stripes in the PSM that correspond to alternate somite primordia".[41] Careful re-examination of the expression pattern has revealed that *her1* actually oscillates in the PSM[42] in exactly the same manner as its mouse homologue (*Hes7*). Although this may be true for some reportedly noncycling *Xenopus* orthologues of amniote cycling genes, it is clear that some orthologues, notably *Hairy2a* (*c-hairy-1* group), are not expressed at all in the TZ/TBD and thus will be unlikely to oscillate. This suggests that there is not complete evolutionary conservation of function.

Transcriptional Control of the "Clock"

As with the static striped genes discussed previously, the majority of cyclically expressed genes are either components of the Notch signaling pathway or downstream targets. It has become clear that in this case also, control of cyclical gene expression is controlled by Notch signaling. Because few genes have been shown to cycle in *Xenopus*, experiments investigating how such cyclical expression is controlled have been restricted to amniote vertebrates. However, sequence comparisons reveal that the proximal promoters of a large number of *HES* family members are well conserved across all vertebrate species (see above) and thus the conclusions of these experiments are likely to also apply equally to *Xenopus*. Of these studies, the case of the mouse *Hes1* gene is remarkable. This gene naturally oscillates transcriptionally with a two hour period in cultured cells that have been stimulated with either serum or a Delta signal.[43] Furthermore, the period of these oscillations is the same as the inferred period of *Hes1* expression within the PSM of the mouse. The promoter of this gene is very similar to that of the *Hairy2a* promoter described above. A model of how the *Hes1* promoter maintains cyclical gene transcription is as follows:[43] In the absence of a Notch-mediated Delta signal, a CSL protein bound to the SPS sites recruits histone deacetylases (HDAC) which inhibit transcription from the promoter. Once the cell receives a Delta signal, the NICD translocates to the nucleus and converts the CSL protein to an activator of transcription, probably by displacement of the HDACs with histone acetyltransferases (HAT). Thus *Hes1* transcript and protein levels begin to rise, with protein levels lagging transcript levels by about 15 minutes. As Hes1 protein (itself a DNA-binding protein with transcriptional repressor activity) levels rise, it binds to N boxes within the *Hes1* promoter and represses the gene. Both *Hes1* transcript and protein are unstable: as with the *Xenopus Hairy2a/2b* genes, the 3' UTR of the transcript contains a conserved 25bp sequence that targets the transcript for rapid degradation,[32] leading to a half-life of around 25 minutes, with the protein having a similar half-life.[43] As the Hes1 protein is degraded, the promoter is able to be activated again by NICD and the process begins once more. Intriguingly, in vitro cycling has been shown to be temperature sensitive, with a drop from 37^0 to 30^0 resulting in the disruption of both synthesis and degradation rates.[43] This may explain why heat-shock treatment of *Xenopus* embryos causes somitogenesis defects,[17,44] with the increased temperature disturbing the cycling process (see next section). Certainly the heat-shock sensitive zone corresponds to the domain where cycling genes are expressed. This proposal is supported by the observation that the phenotype of the heat-shock induced segmentation disruptions are very similar to those of mouse null mutants for various Notch-pathway genes. Although this would be an easy hypothesis to test, the effect of heat-shock on cycling gene expression in *Xenopus* has not yet been investigated.

However, since in the mouse the Notch receptor and Delta ligands are expressed equally throughout the cycling PSM, this part of the cycle is constant. Thus one might expect a constant rate of NICD production, which should dampen the oscillations of *HES* gene expression until an equilibrium level was attained. However another direct target of Notch signaling, the glycosyltransferase Lfng, helps maintain cyclical activation of the Notch pathway.[45-47] This protein is located in the Golgi apparatus of the cell, where it posttranslationally modifies newly produced Notch receptors. Exactly how Lfng functions to maintain cycling is unknown and is a hotly debated topic. In mouse and chick, the *Lfng* transcript is expressed in phase with *Hes1* and is likewise rapidly degraded. However the transcriptional control mechanism of *Lfng* is thought to be very different from the simple system driving the *HES* family genes.[46,47] Although the Lfng loop synergizes with the *Hes1* loop, in vitro studies clearly demonstrate that *Hes1* is able to cycle with the same two hour period in cultured cells that lack detectable *Lfng*.[43] However in vivo, in embryos lacking functional Lfng expression, oscillatory gene expression is no longer observed.

Embryological Insights into the Nature of the "Wavefront"

The observed passage of an easily visible cell change (segmentation) that occurs from anterior to posterior was for many years the primary focus of investigations into the process of somitogenesis. Initially, two theories were proposed as to how this wavefront might progress.

1. "Propagation", in which cells change one after another in a rostral to caudal progression, based on input from the rostral neighbouring cell.
2. "Kinematic", in which there is no signal propagation, but cells undergo autonomous change at times that are predetermined by the location of the cell in the rostral-caudal axis. This graded programme is set up by a process some time before the change actually occurs.

The first embryological experiments in *Xenopus* designed to test the nature of the wavefront were performed as far back as 1967 by Deuchar and Burgess.[48] In these experiments, sections of mesoderm and neural tube (approximately 2 prospective somites in width) were removed from embryos immediately caudal to the last formed somite, leaving a little PSM as well as the TB domain. When allowed to continue development, the remaining PSM extended and segmented following the same timetable as non-operated control embryos. Likewise, in embryos cut completely in half posterior to the forming somite, there was little or no effect on the rate or number of somites produced in the caudal half. They concluded that "no essential stimulus or instruction is passed caudal during the segmentation of progressively more posterior regions of the somitic mesoderm", thus the wavefront of somite formation must be "kinematic". To explain how the kinematic wave of segmentation might be set up, it was suggested that this wave is preceded by a "prior wave" (or determination front) of somite determination that is also kinematic and is set up as early as gastrulation. This was prompted by the observations of the effects of a 15 minute heat shock (37° instead of the optimal 23°) of *Xenopus* embryos after somitogenesis had begun.[17,44] This treatment causes a single discrete zone of absence or irregular boundary formation whereby a few somites are effected, but then segmentation returns to normal. The temperature sensitive period traverses the embryo from anterior to posterior at about the same rate that somites form, but some hours beforehand. This process is not associated with cell differentiation[17,44] as individual cells in abnormally segmented regions appear to undergo normal cell movements, albeit uncoordinated ones, and still form striated myofibrillar elements. In addition, the size or number of somites formed caudal to the heat-shock zone is completely normal. When the embryos are developing at 23°C, the heat-sensitive zone is approximately four and a half hours prior to segmentation. Since at this temperature each somite takes around 45 minutes to form, the sensitive step can then be mapped to when cells are 5-6 prospective somites from the forming boundary. Thus the heat sensitive step lies at the junction of the TZ and TBD regions of the *Xenopus* PSM, where cycling has been observed to be occurring[13] (Fig. 2). Similarly, treatment of embryos with nocodazole (an anti-microtubule drug) results in similar disruption of somite patterning, with the nocodazole-sensitive stage of development being identical to heat-shock.[14]

In chick, the location of the determination front in the PSM has been mapped by experiments involving anterior-posterior inversions of somitomeres.[49] When somites S0 and S-I are dissected and replaced into the PSM in the opposite anterior-posterior polarity and the manipulated embryo allowed to continue developing for 3-24 hours, they will form somites with properly positioned boundaries and perfectly reversed polarity. This indicates that this region of the PSM is already fully determined and the "determination front" must have already moved caudally. Inversion of somites S-II to -IV results in the formation of ectopic boundaries and abnormal anterior-posterior patterning, suggesting partial determination. Finally, inversion of somites S-V to S-XII produces normal segmentation, an observation also noted in *Xenopus*.[48] This suggests that in chick the caudal two-thirds of the PSM is undetermined and the determination front lies around somite S-IV. It is interesting that this corresponds to the approximate location of the heat-shock and nocodazole sensitive zones in *Xenopus*, as well as to the zone of extension (Fig. 2). However, it has taken a further two decades to elucidate the molecular mechanisms driving this wave of change.

The Molecular Nature of the Wavefront

There is now strong evidence from studies in *Xenopus*, chick and zebrafish that the determination front is generated by two opposing gradients of signaling. These involve retinoic acid (RA) signaling from the anterior PSM[13] and FGF and Wnt signaling from the posterior PSM.[49-51]

FGF and Wnt Signaling

At the posterior end of the PSM, paraxial cell precursors are maintained in an undifferentiated state by high levels of FGF8 signaling.[49,50] Transcription of the FGF8 gene only occurs in these cells at the extreme caudal end of the PSM,[52] and this mRNA is progressively degraded as the cells mature. This sets up a gradient of FGF mRNA and protein in the caudal third of the PSM and is directly translated into a gradient of FGF signaling pathway (MAPK/ERK) response (Fig. 2). Evidence linking this gradient to the determination wavefront is provided by experiments in chick,[49] zebrafish[50] and *Xenopus*[13] in which the levels of FGF8 in the PSM are disturbed. Such alterations in FGF8 levels strongly perturb somitogenesis: increasing FGF8 levels shifts the determination front rostrally, while blocking FGF signaling shifts the wavefront caudally. In zebrafish, the timing of the FGF-inhibition sensitive period of development is in somites S-IV and V, in the same physical location as the heat-shock and nocodazole sensitive zones in *Xenopus* and the location of the determination front indicated by embryological experiments (Fig. 2).

In mouse, Wnt3a has been suggested as controlling expression of FGF8 in the TB.[51] Like FGF8, Wnt3a is transcribed only in the TB and ectopic increase or decrease of levels has the same phenotypic effects as FGF8. In *vestigial tail* mouse mutants that have a hypomorphic allele of Wnt3a and defects in somitogenesis, FGF8 expression in the TB is lost. Likewise, mice homozygous for null mutations of the Wnt signaling effector molecules Dishevelled 1 and 2 also have defects in somite segmentation.[53] However, the relationship between Wnt3a, FGF8 and somitogenesis is likely to be more complex, since in mouse Axin2 (a direct target and inhibitor of Wnt3a signaling) is expressed cyclically in the PSM in the mouse.[51] Remarkably, this is out of phase with *Lfng* and all other known cycling genes (with the exception of *Snail1*). However, the effects of such cyclic repression of Wnt signaling on FGF levels in the PSM are unknown.

Retinoic Acid Signaling

Retinoic acid (RA; a biologically active derivative of Vitamin A) has been implicated as having a role in somitogenesis by studies of Vitamin A deficient quail embryos,[54,55] mouse embryos that are null for components of the RA signaling pathway[56-59] and treatment of *Xenopus* embryos (after stage 18, ~20 hpf) with RA.[13] Intriguingly, in *Xenopus* the defects are observed around four hours after treatment begins, placing the RA sensitive zone in the same region as heat shock and FGF inhibition sensitive zones (Fig. 2), strongly hinting that RA has a role in setting up the determination wavefront.

Two recent papers have revealed that RA mediates an additional level of control on the FGF8 signaling gradient, by inhibition of the FGF signal in the PSM and neuroepithelium.[13,55] In *Xenopus*, the RA synthetic enzyme Raldh2 is expressed in high levels in mature somites and the somitomere and TZ regions, whilst the TBD expresses Cyp26AI (also known as CYP26) which is an enzyme involved in RA catabolism. These two enzymes act to form a gradient of RA activity within the PSM in a rostral (high) to caudal (low) direction (Fig. 2). These two gradients actively antagonise each other. RA, although it does not inhibit the expression of FGF8, directly induces expression of MKP3, a dual specificity phosphatase that dephosphorylates and inactivates ERK, thereby inactivating the MAPK pathway of FGF signaling.[13] Conversely, FGF signaling in *Xenopus* upregulates CYP26 expression in the TBD,[13] and in the chick downregulates Raldh2 in the PSM and mature somites.[55] Further supporting evidence of a role for RA signaling in somitogenesis is provided by germ cell nuclear factor (GCNF). This is an orphan member of the nuclear hormone receptor family, closely related at the protein sequence level to retinoic acid receptor.[60] GCNF is required for RA signaling in *Xenopus*,[61] possibly acting by direct transcriptional repression of CYP26 and mouse knockouts of GCNF have severe defects of somitogenesis.[62]

How Do These Signaling Gradients Trigger a Molecular Switch?

It is likely that FGF8 signaling maintains paraxial cells in the cycling state and inhibits their differentiation, whilst RA signaling stops the "clock", triggering the formation of the segmental prepattern and also activating the maturation process. How might this switch be triggered at the molecular level? The most likely model makes use of observations of the complementary expression patterns of Mesp class bHLH transcription factors within the PSM. *Mesogenin/Mespo* is only expressed in the TBD in *Xenopus* (and in the cycling region of the PSM in other vertebrates);[63-66] *Paraxis* is expressed only within the noncycling domain of the PSM in a mutually exclusive pattern to that of *Mesogenin/Mespo*;[65] and *Thylacine/Mesp2* are expressed very transiently in the region where cells transit from unsegmented to segmented fate, in stripes in the anterior half of prospective somites.[24,67] *Mesogenin* is upregulated by FGF signaling and downregulated by RA signaling; while *Paraxis* behaves in the opposite manner,[13] although in neither case was the directness of this response tested. RA also upregulates *Thylacine* expression directly at the transcription level. Thus *Mesogenin* is exclusively present in cycling PSM and *Paraxis* and *Thylacine* in noncycling PSM. These transcription factors are therefore the only known markers that distinguish the two states of the PSM and are thus candidates for the switching mechanism. The promoter structures of the *HES* class of cycling genes (see above) give a clue to how this switch may operate. These contain conserved hexameric binding sites (E boxes, N boxes and SPS hexamers) which are potential binding sites for Mesp class transcription factors.[68] One model might be that when Mesogenin binds to these promoters, it acts as a permissive factor for transcriptional oscillation, whereas when Paraxis binds to the same promoter it interferes with proteins required for cyclical transcription (such as Mesogenin or HES family members). In this model, the default state of noncycling PSM would be posterior (since Paraxis is expressed throughout the noncycling PSM and is thought to specify posterior fate.[69] Thylacine would then operate to produce the specification of anterior half-segment identity.[67] Thus the pattern of expression of Mesp family genes (the "Mesp code") may specify the identity of cells in the PSM: *Mesogenin* positive, *Paraxis* and *Thylacine* negative cells are cycling PSM; *Mesogenin* and *Thylacine* negative, *Paraxis* positive cells are posterior noncycling; and *Mesogenin* negative, *Paraxis* and *Thylacine* positive cells are anterior noncycling.

What are the Morphomechanical Mechanisms Required for Somite Separation?

Once the PSM is patterned, and the anterior and posterior half-segments have been determined, this information is translated into boundary formation via a series of downstream genes. These include membrane-bound cell adhesion molecules (cadherins and NCAM), actin regulatory proteins (Ena/VASP), extracellular matrix components (SPARC) and cell surface ligands and receptors (Ephs and Ephrins). Although it is not known exactly how the highly co-ordinated cell movements in this process occur, studies of these effector molecules provide some clues for future avenues of research.

Cadherins

Cadherins are a superfamily of integral membrane glycoproteins expressed on the cell surface. These have been shown to have important roles in many different developmental processes (reviewed in ref. 70). They function by promoting cell-cell adhesion in the formation of cellular junctions (such as synaptic junctions and intercalated disks in cardiac muscle), as well as in the establishment of cellular polarity and the sorting of embryonic tissues. Of particular relevance is that, in the embryo, it is thought that the expression of different cadherins plays a role in guiding migrating cells through differential interactions with neighbouring cells. Since somite formation involves highly specific cell movements at the time of segmentation, these proteins are obvious candidates for mediating this process.

Membership of the cadherin superfamily is judged by the presence of one or more calcium-binding structures (cadherin repeats) in the extra-cellular domain (ECD) of a membrane-bound protein. Cadherins cluster on the cell surface and bind to cadherins on adjacent cells through a calcium-dependent interaction, usually homotypic. In *Xenopus*, two such molecules have been investigated in relation to their role in somite segmentation: C-cadherin[71] and paraxial protocadherin.[31,72] C-cadherin belongs to the type I cadherin sub-family which is characterized by having 5 cadherin repeats in the ECD and a conserved β-catenin binding site at the carboxy-terminal part of the intracellular domain (ICD). Through this interaction, these cadherins are linked to the actin cytoskeleton and thus form quite rigid cell-cell contacts. They may also be involved in the triggering of the Wnt signaling pathway (which acts through β-catenin) in response to contact with other cells or the extra-cellular matrix. In *Xenopus*, C-cadherin is the most prevalent type I cadherin in somites and differentitated muscle and is also expressed in the PSM.[73] Clues to its function in somitogenesis have been provided by experiments in which mutant type I cadherins have been overexpressed ectopically in the PSM.[71] Two dominant negative constructs were tested, one inhibiting just cadherin-mediated cell-cell adhesion and the other also potentially blocking Wnt signaling. In both cases the phenotype was the same, resulting in extensive disorganisation of somitic tissue, but with some segmentation still observed. At the segmental boundary, dominant negative expressing cells do not have the same elongated shape and cell-cell contacts and cell protrusions are much less frequent than in the control. This may suggest that somitic cells are still capable of migration, but that co-ordination of cell movements required for somite rotation have been lost. This defect is apparently downstream of the establishment of the segmental prepattern, since in these embryos there was no disturbance of *X-Delta2* expression, suggesting that the segmental prepatterning was intact.

Paraxial protocadherin (PAPC) belongs to the protocadherin subfamily of cadherins, which have six cadherin repeats in the ECD and lack the β-catenin binding site in the ICD. PAPC has homotypic cell adhesion properties and is expressed throughout the PSM, including the TZ and TBD regions and is upregulated in the somitomere region in a static striped pattern corresponding to anterior half-segments of prospective somites S-0 and S-I[72] (see Fig. 2). Transcription of the PAPC gene is controlled by transcription factors of the Mesp family and thus appears to be driven downstream of the segmental prepattern. The TBD and TZ expression is activated by *Mesogenin/ Mespo*, whilst the static stripes are controlled by *Thylacine*.[31] As is the case for the other static stripe genes discussed above, downregulation of PAPC in the posterior presumptive somite halves is controlled by Notch-mediated repression. The effects of two different mutant protocadherins have been tested by overexpression in the PSM. Firstly, a secreted dominant negative PAPC (which interferes with cell-cell binding) produces a disruption of co-ordinated rotation, with myotomal cells being fragmented into small groups in a similar manner to the type I cadherin experiments. In this case, markers of anterior (*X-Delta2*) and posterior half somites (*Hairy2a*) are still expressed in these cells, but their distribution is randomized. A second mutant PAPC (activated PAPC, which causes 10-fold more potent cell-cell interaction) produces different defects, in which there is no rotation at all and the myotomal nuclei remain aligned with those in the PSM. However, the position of cells from anterior and posterior half somites is also highly disorganized. It is thought that PAPC is unlikely to be required for the establishment of half-segment identity, but is instead involved in maintaining that identity. This may occur by a simple mechanism in which cells in the anterior compartment of the forming somite (PAPC expressing) have differential cell adhesion properties to adjacent cells in the posterior half of the previously formed somite (nonPAPC expressing), thereby preventing cell mixing and generating a boundary along which the somitic furrow forms. However this model does not explain why such a boundary does not also form between the anterior and posterior compartments of the same somite. Thus boundary formation is likely to be controlled by additional mechanisms controlling somite cell behaviour, such as the Eph signaling pathway (see below).

In mouse, PAPC is expressed in similar static stripes in the noncycling PSM and these stripes are dependent on the murine homologue of *Thylacine, Mesp2*.[74,75] However, neonatal mice carrying a targeted null mutation for this gene do not have skeletal abnormalities.[74] This may be due to compensation by another member of the protocadherin family, of which there are more than 60. Supporting this hypothesis is the observation that overexpression of a dominant negative secreted protocadherin in mouse PSM leads to segmental defects.[75]

Roles for other members of the cadherin family in somitogenesis have been suggested, notably in mouse, where the type I N-cadherin and cad11 have also been shown to have a role in somitogenesis.[76] However, these are thought to influence the differentiation of mature somites, rather than the segmentation process itself.[76-78] This may also be the case in *Xenopus*, where N-cadherin is expressed in a similar domain in mature somites.[79]

Ena/VASP

In *Xenopus* a member of the Ena/VASP family of actin-regulatory proteins (Xena) has been recently shown to be required for cell rearrangements during somitogenesis.[80] Xena protein co-localizes at intersomitic junctions with components of cell-matrix adhesion complexes including beta1-integrin, vinculin, FAK and tenascin. Inhibition of Xena using ectopic expression of a dominant negative version of the protein leads to abnormal somite rotation and failure of intersomitic boundary formation. It is postulated that Xena regulates integrin-dependent adhesion and migration by providing a link between cell-surface integrins and the actin cytoskeleton in a manner similar to that discussed for the cadherins (see above) and is well positioned to control changes in adhesive strength and cell motility during somitogenesis.

NCAM

Neural cell adhesion molecule (NCAM) is one of the most prevalent cell adhesion molecules in vertebrate development. Amongst its many roles in the embryo, it is involved in myogenesis and myoblast migration.[81] In *Xenopus*, NCAM overexpression throughout the embryo (via RNA injection at the two cell stage) disrupts somite segmentation and organisation.[82] In chick, anti-NCAM antibodies administered to explanted PSM interrupts calcium-independent cell adhesion,[83] but the effects of such disruption are not as dramatic as seen with an anti-N-cadherin antibody, suggesting that NCAM has a lesser role in somitogenesis than N-cadherin. Indeed, null mutant mice lacking all isoforms of NCAM have no reported somitogenesis defects.[84]

SPARC

The extra-cellular matrix (ECM) is composed of proteins that are required for structural maintenance (such as fibronectin and laminin), growth factors and matricellular proteins (which mediate cell-ECM interactions, but do not have a structural role (reviewed in ref. 85). When tissues undergo changes in cell-cell or cell-matrix contact, such as during embryonic development or tissue remodelling in the adult, it is the matricellular proteins that mediate changes in cell properties, including cell adhesiveness and proliferation state. SPARC (also known as osteonectin, BM-40 and 43k protein) in particular appears to function as a counteradhesive protein, as well as an inhibitor of cellular proliferation.[85]

SPARC expression is regulated at the transcriptional and posttranscriptional levels by many different growth factors. Particularly relevant for a role in somitogenesis, it is upregulated by RA signaling[86] and downregulated by FGF signaling.[87] In the *Xenopus* PSM, SPARC expression is limited to the somitic clefts and may play a role in the de-adhesion required when the somitic furrow forms prior to the cell movements required for somite rotation to occur. Indeed, injection of anti-SPARC antibodies into the blastocoel cavity of blastula stage *Xenopus* embryos has a dramatic effect on somitogenesis at tailbud stage (stage 24, ~26 hpf), including a total loss of segmental boundaries and totally disorganized myotome, although myotomal differentiation still occurs.[88]

Eph Receptors and Ephrin Ligands

As mentioned above, cadherin-mediated cell-cell adhesion and signaling is unlikely to be sufficient for correct segmentation to occur. Signaling through the Eph family of receptors is also a key component of this process. Eph receptors are divided into two sub-classes (EphA and EphB) of transmembrane tyrosine kinases involved in a wide range of morphogenetic processes in embryonic development (reviewed in ref. 89). There are two classes of ligands for these receptors. Class A ephrins are tethered to the membrane by a GPI linkage and preferentially bind EphA receptors. Class B Ephrins contain a transmembrane domain and an ICD and preferentially bind EphB receptors. An important exception for the purposes of this discussion is EphA4 (the main Eph receptor expressed in vertebrate PSM) which can bind to both types of ligand. Eph receptors and ligands tend to show complementary patterns of expression and are likely to function in the initial stages of cellular compartmentalisation. Ligand/receptor interactions trigger bidirectional signaling,[90] and these signals are important for inhibiting cell mixing across boundaries. This usually occurs by promoting either attraction or repulsion between the Eph-expressing and Ephrin-expressing cell populations. Various Ephs and Ephrins have been described as being expressed in the PSM and mature somites of mouse,[91-93] chick,[94] fish[95] and *Xenopus*.[96,97] The most illuminating experiments on the role of this signaling pathway in somitogenesis have been done in zebrafish. Here signaling between the EphA4 receptor and the Ephrins B2 and A-L1 required for the formation of the somitic furrow and epithelialization of the forming somite.[95,98] These are the only members of the Eph/Ephrin family that are expressed in the PSM of fish. EphA4 is expressed in the anterior of mature somites and prospective somites S0 and S-I. Ephrins B2 and A-L1 are both expressed in the posterior domain of mature somites and presumptive somites S0, S-I and S-II.[95] As with cadherins, the disruption of Eph signaling disrupts somite formation, but does not effect the establishment of the segmental prepattern in the PSM. Thus it is thought that Eph signaling is required to translate the prepattern into the events of boundary formation and differentiation. Activation of signaling may lead to localized cell de-adhesion or repulsion between cells at the boundary and this may initiate cell migration and formation of the somitic furrow. Interestingly, in *fused somites* mutant zebrafish that lack proper segmentation, somitic furrow formation can be restored by expression of an ICD-truncated EphA4 receptor. This suggests that receptor-mediated signaling is not required and that ligand-mediated reverse signaling is sufficient for furrow formation.[98] Although EphA4 clearly has a role in the segmentation process, and in the mouse EphA4 transcription is directly under the control of Mesp2,[68] targeted deletion of the EphA4 gene has no phenotypic effect on somitogenesis.[99] This could be explained by another Eph family member compensating for its loss. Indeed, in mammals there are at least 14 receptors and 8 ligands, and in chick EphA3 is co-expressed with EphA4 in the PSM.[94] In *Xenopus*, the expression patterns of Eph/Ephrins have not been well characterized. However, limited analysis seems to show EphA4 and EphrinB2 in the PSM, although their precise locations within this tissue cannot be determined from the published data.[97,100] No experiments have yet been reported testing the roles of these molecules in somitogenesis.

Conclusion

Studies of somitogenesis were pioneered in amphibians using classical embryological techniques. However the *Xenopus laevis* system has somewhat fallen out of favour as a tool to dissect the mechanisms underlying somitogenesis. Nonetheless, important discoveries have still been made using this system and, given the remarkably high degree of evolutionary conservation of the molecular mechanisms driving this process, this model organism still has a large part to play in the elucidation of the finer details. In particular, the field as a whole may very well benefit from a re-examination of many of the early *Xenopus* embryological experiments using molecular assays.

Acknowledgements

I wish to thank Mark Solloway for critical comments on the text. DBS is supported by a Westfield-Belconnen Fellowship, the Victor Chang Cardiac Research Institute and National Health and Medical Research Council Project Grants 303705 & 404804.

References

1. Nieuwkoop PD, Faber J. Normal table of Xenopus laevis (Daudin). New York: Garland; 1994.
2. Muntz L. Myogenesis in the trunk and leg during development of the tadpole of Xenopus laevis (Daudin 1802). J Embryol Exp Morphol 1975; 33:757-774.
3. Kielbowna L. The formation of somites and early myotomal myogenesis in Xenopus laevis, Bombina variegata and Pelobates fuscus. J Embryol Exp Morphol 1981; 64:295-304.
4. Hamilton L. The formation of somites in Xenopus. J Embryol Exp Morph 1969; 22:253-264.
5. Youn BW, Malacinski GM. Somitogenesis in the amphibian Xenopus: scanning electron microscope analysis of intrasomitic cellular arrangements during somite rotation. J Embryol Exp Morphol 1981; 64:23-43.
6. Keller R. The origin and morphogenesis of amphibian somites. Curr Top Dev Biol 2000; 47(183-246).
7. Chanoine C, Hardy S. Xenopus muscle development: from primary to secondary myogenesis. Dev Dyn 2003; 226:12-23.
8. Hopwood N, Pluck A, Gurdon J. Xenopus Myf-5 marks early muscle cells and can activate muscle genes ectopically in early embryos. Development 1991; 111:551-560.
9. Hopwood N, Pluck A, Gurdon J. MyoD expression in the forming somites is an early response to mesoderm induction in Xenopus embryos. EMBO J 1989; 8:3409-3417.
10. Niehrs C, Keller R, Cho K et al. The homeobox gene goosecoid controls cell migration in Xenopus embryos. Cell 1993; 72:491-503.
11. Jacobson A. Somitomeres: mesodermal segments of vertebrate embryos. Development 1988; 104 (Suppl):209-220.
12. Pourquie O, Tam P. A nomenclature for prospective somites and phases of cyclic gene expression in the presomitic mesoderm. Dev Cell 2001; 1:619-620.
13. Moreno T, Kintner C. Regulation of Segmental Patterning by Retinoic Acid Signaling during Xenopus Somitogenesis. Dev Cell 2004; 6:205-218.
14. Davidson D. Segmentation in frogs. Development 1988; 104(Suppl):221-229.
15. Tam P, Meier S, Jacobson A. Differentiation of the metameric pattern in the embryonic axis of the mouse. II. Somitomeric organization of the presomitic mesoderm. Differentiation 1982; 21:109-122.
16. Packard D. The influence of axial structures on chick somite formation. Dev Biol 1976; 53:36-48.
17. Pearson M, Elsdale T. Somitogenesis in amphibian embryos. I. Experimental evidence for an interaction between two temporal factors in the specification of somite pattern. J Embryol Exp Morphol 1979; 51:27-50.
18. Youn B, Keller R, Malacinski G. An atlas of notochord and somite morphogenesis in several anuran and urodelean amphibians. J Embryol Exp Morphol 1980; 59:223-247.
19. Wilson P, Oster G, Keller R. Cell rearrangement and segmentation in Xenopus: direct observation of cultured explants. Development 1989; 105:155-166.
20. Kulesa P, Fraser S. Cell dynamics during somite boundary formation revealed by time-lapse analysis. Science 2002; 298:991-995.
21. Cooke J, Zeeman E. A clock and wavefront model for control of the number of repeated structures during animal morphogenesis. J Theor Biol 1976; 58:455-476.
22. Cooke J. A gene that resuscitates a theory—somitogenesis and a molecular oscillator. Trends Genet 1998; 14:85-88.
23. Jen W, Wettstein D, Turner D et al. The Notch ligand, X-Delta-2, mediates segmentation of the paraxial mesoderm in Xenopus embryos. Development 1997; 124:1169-1178.
24. Sparrow DB, Jen WC, Kotecha S et al. Thylacine 1 is expressed segmentally within the paraxial mesoderm of the Xenopus embryo and interacts with the Notch pathway. Development 1998; 125(11):2041-2051.
25. Jen W, Gawantka V, Pollet N et al. Periodic repression of Notch pathway genes governs the segmentation of Xenopus embryos. Genes Dev 1999; 13:1486-1499.
26. Cossins J, Vernon A, Zhang Y et al. Hes6 regulates myogenic differentiation. Development 2002; 129(9):2195-2207.
27. Lai E. Notch signaling: control of cell communication and cell fate. Development 2004; 131:965-973.
28. Coffman C, Harris W, Kintner C. Xotch, the Xenopus homolog of Drosophila notch. Science 1990; 249:1438-1441.
29. Wu J, Wen L, Zhang W et al. The secreted product of Xenopus gene lunatic Fringe, a vertebrate signaling molecule. Science 1996; 273:355-358.
30. Lamar E, Deblandre G, Wettstein D et al. Nrarp is a novel intracellular component of the Notch signaling pathway. Genes Dev 2001; 15:1885-1899.
31. Kim S, Jen W, De Robertis E et al. The protocadherin PAPC establishes segmental boundaries during somitogenesis in xenopus embryos. Curr Biol 2000; 10:821-830.
32. Davis R, Turner D, Evans L et al. Molecular targets of vertebrate segmentation: two mechanisms control segmental expression of Xenopus hairy2 during somite formation. Dev Cell 2001; 1:553-565.

33. Nellesen D, Lai E, Posakony J. Discrete enhancer elements mediate selective responsiveness of enhancer of split complex genes to common transcriptional activators. Dev Biol 1999; 213:33-53.

34. Gajewski M, Voolstra C. Comparative analysis of somitogenesis related genes of the hairy/Enhancer of split class in Fugu and zebrafish. BMC Genomics 2002; 3:21.

35. Palmeirim I, Henrique D, Ish-Horowicz D et al. Avian hairy gene expression identifies a molecular clock linked to vertebrate segmentation and somitogenesis. Cell 1997; 91:639-648.

36. Ishibashi M, Ang S, Shiota K et al. Targeted disruption of mammalian hairy and Enhancer of split homolog-1 (HES-1) leads to up-regulation of neural helix-loop-helix factors, premature neurogenesis and severe neural tube defects. Genes Dev 1995; 9:3136-3148.

37. Cau E, Gradwohl G, Casarosa S et al. Hes genes regulate sequential stages of neurogenesis in the olfactory epithelium. Development 2000; 127:2323-2332.

38. Donovan J, Kordylewska A, Jan Y et al. Tetralogy of fallot and other congenital heart defects in Hey2 mutant mice. Curr Biol 2002; 12:1605-1610.

39. Sakata Y, Kamei C, Nakagami H et al. Ventricular septal defect and cardiomyopathy in mice lacking the transcription factor CHF1/Hey2. Proc Natl Acad Sci USA 2002; 99:16197-16202.

40. Li Y, Fenger U, Niehrs C et al. Cyclic expression of esr9 gene in Xenopus presomitic mesoderm. Differentiation 2003; 71:83-89.

41. Muller M, v Weizsacker E, Campos-Ortega J. Expression domains of a zebrafish homologue of the Drosophila pair-rule gene hairy correspond to primordia of alternating somites. Development 1996; 122:2071-2078.

42. Sawada A, Fritz A, Jiang Y et al. Zebrafish Mesp family genes, mesp-a and mesp-b are segmentally expressed in the presomitic mesoderm and Mesp-b confers the anterior identity to the developing somites. Development 2000; 127:1691-1702.

43. Hirata H, Yoshiura S, Ohtsuka T et al. Oscillatory expression of the bHLH factor Hes1 regulated by a negative feedback loop. Science 2002; 298:840-843.

44. Elsdale T, Pearson M, Whitehead M. Abnormalities in somite segmentation following heat shock to Xenopus embryos. J Embryol Exp Morph 1976; 35:625-635.

45. Dale J, Maroto M, Dequeant M et al. Periodic notch inhibition by lunatic fringe underlies the chick segmentation clock. Nature 2003; 421:275-278.

46. Morales A, Yasuda Y, Ish-Horowicz D. Lunatic fringe expression is controlled during segmentation by a cyclic transcriptional enhancer responsive to notch signaling. Dev Cell 2002; 3:63-74.

47. Cole S, Levorse J, Tilghman S et al. Clock regulatory elements control cyclic expression of Lunatic fringe during somitogenesis. Dev Cell 2002; 3:75-84.

48. Deuchar E, Burgess A. Somite segmentation in amphibian embryos: is there a transmitted control mechanism? J Embryol Exp Morphol 1967; 17:349-358.

49. Dubrulle J, McGrew M, Pourquie O. FGF signaling controls somite boundary position and regulates segmentation clock control of spatiotemporal Hox gene activation. Cell 2001; 106:219-232.

50. Sawada A, Shinya M, Jiang Y et al. Fgf/MAPK signalling is a crucial positional cue in somite boundary formation. Development 2001; 128:4873-4880.

51. Aulehla A, Wehrle C, Brand-Saberi B et al. Wnt3a plays a major role in the segmentation clock controlling somitogenesis. Dev Cell 2003; 4:395-406.

52. Dubrulle J, Pourquie O. fgf8 mRNA decay establishes a gradient that couples axial elongation to patterning in the vertebrate embryo. Nature 2004; 427:419-422.

53. Hamblet N, Lijam N, Ruiz-Lozano P et al. Dishevelled 2 is essential for cardiac outflow tract development, somite segmentation and neural tube closure. Development 2002; 129:5827-5838.

54. Maden M, Graham A, Zile M et al. Abnormalities of somite development in the absence of retinoic acid. Int J Dev Biol 2000; 44:151-159.

55. Diez del Corral R, Olivera-Martinez I, Goriely A et al. Opposing FGF and retinoid pathways control ventral neural pattern, neuronal differentiation and segmentation during body axis extension. Neuron 2003; 40:65-79.

56. Abu-Abed S, Dolle P, Metzger D et al. Developing with lethal RA levels: genetic ablation of Rarg can restore the viability of mice lacking Cyp26a1. Development 2003; 130:1449-1459.

57. Niederreither K, Subbarayan V, Dolle P et al. Embryonic retinoic acid synthesis is essential for early mouse post-implantation development. Nat Genet 1999; 21:444-448.

58. Niederreither K, Vermot J, Schuhbaur B et al. Embryonic retinoic acid synthesis is required for forelimb growth and anteroposterior patterning in the mouse. Development 2002; 129:3563-3574.

59. Sakai Y, Meno C, Fujii H et al. The retinoic acid-inactivating enzyme CYP26 is essential for establishing an uneven distribution of retinoic acid along the anterio-posterior axis within the mouse embryo. Genes Dev 2001; 15:213-225.

60. Chen F, Cooney A, Wang Y et al. Cloning of a novel orphan receptor (GCNF) expressed during germ cell development. Mol Endocrinol 1994; 8:1434-1444.

61. Barreto G, Borgmeyer U, Dreyer C. The germ cell nuclear factor is required for retinoic acid signaling during Xenopus development. Mech Dev 2003; 120:415-428.

62. Chung A, Katz D, Pereira F et al. Loss of orphan receptor germ cell nuclear factor function results in ectopic development of the tail bud and a novel posterior truncation. Mol Cell Biol 2001; 21(2):663-77.

63. Joseph E, Cassetta L. Mespo: a novel basic helix-loop-helix gene expressed in the presomitic mesoderm and posterior tailbud of Xenopus embryos. Mech Dev 1999; 82:191-194.

64. Buchberger A, Bonneick S, Arnold H. Expression of the novel basic-helix-loop-helix transcription factor cMespo in presomitic mesoderm of chicken embryos. Mech Dev 2000; 97:223-226.

65. Yoon J, Moon R, Wold B. The bHLH class protein pMesogenin1 can specify paraxial mesoderm phenotypes. Dev Biol 2000; 222:376-391.

66. Yoo K, Kim C, Park H et al. Characterization and expression of a presomitic mesoderm-specific mespo gene in zebrafish. Dev Genes Evol 2003; 213:203-206.

67. Saga Y, Hata N, Koseki H et al. Mesp2: a novel mouse gene expressed in the presegmented mesoderm and essential for segmentation initiation. Genes Dev 1997; 11:1827-1839.

68. Nakajima Y, Morimoto M, Takahashi Y et al. Identification of Epha4 enhancer required for segmental expression and the regulation by Mesp2. Development 2006; 133:2517-2525.

69. Johnson J, Rhee J, Parsons S et al. The anterior/posterior polarity of somites is disrupted in paraxis-deficient mice. Dev Biol 2001; 229:176-187.

70. Wheelock M, Johnson K. Cadherins as modulators of cellular phenotype. Annu Rev Cell Dev Biol 2003; 19(207-35).

71. Giacomello E, Vallin J, Morali O et al. Type I cadherins are required for differentiation and coordinated rotation in Xenopus laevis somitogenesis. Int J Dev Biol 2002; 46:785-792.

72. Kim S, Yamamoto A, Bouwmeester T et al. The role of paraxial protocadherin in selective adhesion and cell movements of the mesoderm during Xenopus gastrulation. Development 1998; 125:4681-4690.

73. Levi G, Ginsberg D, Girault J et al. EP-cadherin in muscles and epithelia of Xenopus laevis embryos. Development 1991; 113:1335-1344.

74. Yamamoto A, Kemp C, Bachiller D et al. Mouse paraxial protocadherin is expressed in trunk mesoderm and is not essential for mouse development. Genesis 2000; 27:49-57.

75. Rhee J, Takahashi Y, Saga Y et al. The protocadherin papc is involved in the organization of the epithelium along the segmental border during mouse somitogenesis. Dev Biol 2003; 254:248-261.

76. Horikawa K, Radice G, Takeichi M et al. Adhesive subdivisions intrinsic to the epithelial somites. Dev Biol 1999; 215:182-189.

77. Radice G, Rayburn H, Matsunami H et al. Developmental defects in mouse embryos lacking N-cadherin. Dev Biol 1997; 181:64-78.

78. Linask K, Ludwig C, Han M et al. N-cadherin/catenin-mediated morphoregulation of somite formation. Dev Biol 1998; 202:85-102.

79. Simonneau L, Broders F, Thiery J. N-cadherin transcripts in Xenopus laevis from early tailbud to tadpole. Dev Dyn 1992; 194:247-260.

80. Kragtorp KA, Miller JR. Regulation of somitogenesis by Ena/VASP proteins and FAK during Xenopus development. Development 2006; 133:685-695.

81. Martin B, Harland R. Hypaxial muscle migration during primary myogenesis in Xenopus laevis. Dev Biol 2001; 239:270-280.

82. Kintner C. Effects of altered expression of the neural cell adhesion molecule, N-CAM, on early neural development in Xenopus embryos. Neuron 1988; 1:545-555.

83. Duband J, Dufour S, Hatta K et al. Adhesion molecules during somitogenesis in the avian embryo. J Cell Biol 1987; 104:1361-1374.

84. Cremer H, Chazal G, Goridis C et al. NCAM is essential for axonal growth and fasciculation in the hippocampus. Mol Cell Neurosci 1997; 8:323-335.

85. Brekken R, Sage E. SPARC, a matricellular protein: at the crossroads of cell-matrix communication. Matrix Biol 2001; 19:816-827.

86. Nomura S, Hashmi S, McVey J et al. Evidence for positive and negative regulatory elements in the 5'-flanking sequence of the mouse sparc (osteonectin) gene. J Biol Chem 1989; 264:12201-12207.

87. Delany A, Canalis E. Basic fibroblast growth factor destabilizes osteonectin mRNA in osteoblasts. Am J Physiol 1998; 274:C734-740.

88. Purcell L, Gruia-Gray J, Scanga S et al. Developmental anomalies of Xenopus embryos following microinjection of SPARC antibodies. J Exp Zool 1993; 265:153-164.

89. Murai K, Pasquale E. 'Eph'ective signaling: forward, reverse and crosstalk. J Cell Sci 2003; 116:2823-2832.

90. Cowan C, Henkemeyer M. Ephrins in reverse, park and drive. Trends Cell Biol 2002; 12:339-346.

91. Krull C, Lansford R, Gale N et al. Interactions of Eph-related receptors and ligands confer rostrocaudal pattern to trunk neural crest migration. Curr Biol 1997; 7:571-580.

92. Nieto M, Gilardi-Hebenstreit P, Charnay P et al. A receptor protein tyrosine kinase implicated in the segmental patterning of the hindbrain and mesoderm. Development 1992; 116:1137-1150.

93. Wang H, Anderson D. Eph family transmembrane ligands can mediate repulsive guidance of trunk neural crest migration and motor axon outgrowth. Neuron 1997; 18:383-396.

94. Baker R, Antin P. Ephs and ephrins during early stages of chick embryogenesis. Dev Dyn. 2003; 228: 128-142.

95. Durbin L, Brennan C, Shiomi K et al. Eph signaling is required for segmentation and differentiation of the somites. Genes Dev 1998; 12:3096-3109.

96. Scales J, Winning R, Renaud C et al. Novel members of the eph receptor tyrosine kinase subfamily expressed during Xenopus development. Oncogene 1995; 11:1745-1752.

97. Helbling P, Saulnier D, Robinson V et al. Comparative analysis of embryonic gene expression defines potential interaction sites for Xenopus EphB4 receptors with ephrin-B ligands. Dev Dyn 1999; 216: 361-373.

98. Barrios A, Poole R, Durbin L et al. Eph/Ephrin signaling regulates the mesenchymal-to-epithelial transition of the paraxial mesoderm during somite morphogenesis. Curr Biol 2003; 13:1571-1582.

99. Dottori M, Hartley L, Galea M et al. EphA4 (Sek1) receptor tyrosine kinase is required for the development of the corticospinal tract. Proc Natl Acad Sci USA 1998; 95:13248-13253.

100. Winning R, Sargent T. Pagliaccio, a member of the Eph family of receptor tyrosine kinase genes, has localized expression in a subset of neural crest and neural tissues in Xenopus laevis embryos. Mech Dev 1994; 46:219-229.

101. Koyano-Nakagawa N, Kim J, Anderson D et al. Hes6 acts in a positive feedback loop with the neurogenins to promote neuronal differentiation. Development 2000; 127:4203-4216.

102. Kondow A, Hitachi K, Ikegame T et al. Bowline, a novel protein localized to the presomitic mesoderm, interacts with Groucho/TLE in Xenopus. Int J Dev Biol 2006; 50:473-479.

103. Rones M, Woda J, Mercola M et al. Isolation and characterization of Xenopus Hey-1: a downstream mediator of Notch signaling. Dev Dyn 2002; 225:554-560.

104. Glinka A, Wu W, Delius H et al. Dickkopf-1 is a member of a new family of secreted proteins and functions in head induction. Nature 1998; 391:357-362.

105. de Souza F, Gawantka V, Gomez A et al. The zinc finger gene Xblimp1 controls anterior endomesodermal cell fate in Spemann's organizer. EMBO J 1999; 18:6062-6072.

106. Jouve C, Palmeirim I, Henrique D et al. Notch signalling is required for cyclic expression of the hairy-like gene HES1 in the presomitic mesoderm. Development 2000; 127:1421-1429.

107. Leimeister C, Dale K, Fischer A et al. Oscillating expression of c-Hey2 in the presomitic mesoderm suggests that the segmentation clock may use combinatorial signaling through multiple interacting bHLH factors. Dev Biol 2000; 227:91-103.

108. Jiang Y, Aerne B, Smithers L et al. Notch signalling and the synchronization of the somite segmentation clock. Nature 2000; 408:475-479.

109. McGrew M, Dale J, Fraboulet S et al. The lunatic fringe gene is a target of the molecular clock linked to somite segmentation in avian embryos. Curr Biol 1998; 8:979-982.

110. Forsberg H, Crozet F, Brown N. Waves of mouse Lunatic fringe expression, in four-hour cycles at two-hour intervals, precede somite boundary formation. Curr Biol 1998; 8:1027-1030.

111. Dunwoodie S, Clements M, Sparrow D et al. Axial skeletal defects caused by mutation in the spondylocostal dysplasia/pudgy gene Dll3 are associated with disruption of the segmentation clock within the presomitic mesoderm. Development 2002; 129:1795-1806.

112. Ishikawa A, Kitajima S, Takahashi Y et al. Mouse Nkd1, a Wnt antagonist, exhibits oscillatory gene expression in the PSM under the control of Notch signaling. Mech Dev 2004; 121:1443-1453.

113. Bessho Y, Miyoshi G, Sakata R et al. Hes7: a bHLH-type repressor gene regulated by Notch and expressed in the presomitic mesoderm. Genes Cells 2001; 6:175-185.

114. Bessho Y, Sakata R, Komatsu S et al. Dynamic expression and essential functions of Hes7 in somite segmentation. Genes Dev 2001; 15:2642-2647.

115. Holley S, Geisler R, Nusslein-Volhard C. Control of her1 expression during zebrafish somitogenesis by a delta-dependent oscillator and an independent wave-front activity. Genes Dev 2000; 14:1678-1690.

116. Oates A, Ho R. Hairy/E(spl)-related (Her) genes are central components of the segmentation oscillator and display redundancy with the Delta/Notch signaling pathway in the formation of anterior segmental boundaries in the zebrafish. Development 2000; 129:2929-2946.

117. Winkler C, Elmasri H, Klamt B et al. Characterization of hey bHLH genes in teleost fish. Dev Genes Evol 2003; 213:541-553.

118. Dale JK, Malapert P, Chal J et al. Oscillations of the snail genes in the presomitic mesoderm coordinate segmental patterning and morphogenesis in vertebrate somitogenesis. Dev Cell 2006; 10:355-366.

119. Davis R, Turner D. Vertebrate hairy and Enhancer of split related proteins: transcriptional repressors regulating cellular differentiation and embryonic patterning. Oncogene 2001; 20:8342-8357.

CHAPTER 5

Role of Delta-Like-3 in Mammalian Somitogenesis and Vertebral Column Formation

Gavin Chapman and Sally L. Dunwoodie*

Abstract

Somitogenesis is a term that encompasses somite formation, patterning and differentiation and it is a process that is fundamental to the formation of the axial skeleton in vertebrates. Notch signalling is a mechanism used to specify cell fate in many different contexts, with signalling occurring between cells in contact. Notch signalling is fundamental to the formation and patterning of somites and importantly a ligand of Notch, is mutated in the abnormal vertebral segmentation syndrome spondylocostal dysostosis. Here we discuss what is known about the expression and function of this ligand, Delta-like-3, during somitogenesis and vertebral column formation in mouse and humans.

Introduction

Somites are the progenitors of skeletal muscle, tendons, dermis and skeleton of the trunk. The ordered arrangement of these structures is inherent in the regimented and reiterative way in which somites are formed and patterned. The formation of somites is one of the most intriguing processes to occur during vertebrate development. This is perhaps because of the inherent beauty of what we currently understand of the mechanism; an oscillation of gene expression in cells within spatial groups provides periodicity and synchrony; and a wavefront that interacts with this clock to specify competence to form a somite. Oscillatory gene expression is correlated to a "segmentation clock" and it is in the presomitic mesoderm (somite progenitor tissue) that the clock and wavefront intersect. The Notch signalling pathway, which is evolutionarily conserved and widely utilised, controls the expression of these oscillatory genes. In mammals, the intrinsic role that Notch signalling plays in somitogenesis has been revealed through genetic studies in both humans and mouse. In particular two ligands of Notch, Delta-like 1 (Dll1) and Delta-like-3 (Dll3), are central to somite formation, patterning and differentiation. It is currently unclear why two ligands are required and specifically what role each ligand plays during somitogenesis. *Dll3* is of particular interest since this gene encodes the most divergent Delta-like ligand and it carries mutations in the abnormal vertebral segmentation syndrome spondylocostal dysostosis.

Somitogenesis

Somites are paired structures that form on either side of the midline of the embryo and constitute the paraxial mesoderm of the trunk. In the mouse embryo, somite formation begins at 8 days post

*Corresponding Author: Sally L Dunwoodie—Associate Professor Faculties of Medicine and Science, University of New South Wales, Developmental Biology Program, Victor Chang Cardiac Research Institute, 384 Victoria Street, Darlinghurst, Sydney, NSW Australia. Email: s.dunwoodie@victorchang.edu.au

Somitogenesis, edited by Miguel Maroto and Neil V. Whittock. ©2008 Landes Bioscience and Springer Science+Business Media.

coitum (dpc) following germ layer formation and ceases by 14.5 dpc when some 60 somites have formed.[1] Somites form in a rostral to caudal progression as the trunk elongates, in mammals this occurs every 90-120 minutes[2] producing approximately 65 somites in mouse and 42 in humans. Somites are derived from unsegmented paraxial mesoderm termed presomitic mesoderm; this process is referred to as initial segmentation. Newly formed or nascent somites consist of an epithelium arranged to produce a tightly packed sphere. These nascent somites begin to differentiate about 20 hours after formation as cells in the ventral half undergo a transition to mesenchyme, while the dorsal portion (dermomyotome) remains epithelial. This mesenchymal transition represents the first morphological sign of somite maturation and demonstrates that the somite is compartmentalised along the dorsoventral axis; these compartments are further subdivided to give rise to distinct cell types.[3, 4] Somite maturation comes to fruition when differentiated cell types arise: the dorsal epithelium (dermomyotome) gives rise to progenitors of striated muscle and skin, with the ventral mesenchyme (sclerotome) producing progenitors of bone and tendon. The somite's position along the rostro-caudal axis affects the fate of the somitic derivatives. For example, rostral to the forelimbs and caudal to the hind limbs, somite derivatives constitute the bones and tendons of the vertebral column and the dorsal or epaxial muscle. Between the limbs, somites are also the progenitors of the bones and tendons of the ribs and ventral or hypaxial muscle. At the level of the limbs themselves, muscle progenitors migrate into the limb buds from the dermomyotome and establish the dorsal and ventral muscle masses. Grafting experiments show that some of these rostrocaudal distinctions between somites are established in the presomitic mesoderm.[5] When presomitic mesoderm from the prospective thoracic region is transplanted to the prospective cervical region, ectopic rib-bearing vertebra form. In contrast, similar transplantation experiments show that the myogenic population escapes this early segmental specification and develops according to its new environment.[6,7] The contribution therefore that somitic tissue makes to the trunk and limbs, is significant. For this reason and the fact that somite formation is a reiterative process, many have been drawn to their study. In addition to dorsoventral patterning, somites are also polarised with clear antero-posterior identity. The importance of this is highlighted by the fact that somite polarity underpins initial segmentation, resegmentation and pattering the peripheral nervous system of the trunk. As mentioned above, initial segmentation refers to the formation of somites, this occurs only in rostral presomitic mesoderm when tissue with anterior and posterior identity is juxtaposed.[9] Resegmentation refers to the condensation of ventral mesenchymal sclerotome, to form the vertebrae; the anterior segment of one somite and posterior segment of its neighbour combine to form a single vertebra. Somites also impact on other systems of the trunk; the anterior sclerotome is permissive to peripheral nerve components such as neural crest cells and axons of motorneurons while the posterior sclerotome is refractory resulting in the segmented arrangement of the peripheral nervous system of the trunk.[9,10]

A number of theoretical models have been proposed to explain somite formation and its antero-posterior compartmentalisation.[11] One model that is widely recognised as the most feasible is the "clock and wavefront" model, which proposed that an oscillator or clock acts in an intracellular manner to synchronise groups of cells within the presomitic mesoderm.[12] This clock is considered to oscillate in conjunction with a wavefront travelling in a rostro-caudal direction along the embryonic axis. When a group of cells, synchronised by the clock, is reached by the wavefront, then they undergo the same fate, which is to form a somite.

The advent of molecular biology revolutionised the way we study somitogenesis since it resulted in the identification of genetic components of this process, which have been functionally examined. Mutational analysis in mouse gave the first clues as to which signalling pathways were required for somite formation in mammals. Fibroblast growth factor (FGF) signalling was implicated, since a null mutation in the *FGF receptor1 (fgfr1)* gene results in embryos that lack somites despite having generated paraxial mesoderm.[13] Wnt signalling was also shown to be involved in somite formation since both the *Wnt3a* null and hypomorphic *(vestigial tail)* alleles lead to aberrant somite formation.[14,15] The final member of the triumvirate required for somite formation was the receptor *Notch1*.[16] Today, the anterior boundary of the FGF8 gradient in the presomitic mesoderm is considered to represent the determination or wavefront of competence proposed by Cooke and Zeeman (1976), with Wnt3a acting upstream of FGF8 in the presomitic mesoderm and Notch signalling as the perpetuator of the segmentation clock.[17,18] As for how segmental specification occurs, clear insight has come from studying the action of FGF8, which tightly coordinates the segmentation process as well as the spatiotemporal expression of Hox genes. Therefore, the position of the segment boundary and the future regional identity of the somites is linked.[17,19]

The Notch Signalling Pathway

Notch and Signal Transduction

Notch is a type I membrane spanning receptor that exists on the cell surface as a heterodimer.[20,21] In mammals, the extracellular portion of the receptor contains between 29 and 36 EGF-like repeats and three Lin-12/Notch (Lin) repeats. EGF-like repeats 11 and 12 in *Drosophila* Notch are responsible for ligand binding and are conserved in vertebrate Notch proteins.[22] The remaining portion of the receptor spans the membrane and contains the intracellular domain that mediates Notch signaling upon ligand binding. Notch receptors interact with ligands of the DSL (Delta, Serrate and LAG2), Contactin or MAGP families (reviewed in refs. 23,24-26). The Notch receptor exists on the cell surface as a heterodimer because it is cleaved at site 1 (S1 cleavage) by a Furin-like convertase during receptor trafficking through the Golgi.[20,21] Interaction between DSL Ligand and the Notch receptor renders Notch susceptible to proteolytic cleavage at Site 2 (S2 cleavage) by TNF-α converting enzyme (TACE), releasing the extracellular domain.[27,28] Activation of TACE is poorly understood despite the fact that ligand binding and consequent S2 cleavage of Notch represents the critical event in receptor activation. Subsequent proteolytic cleavage events (S3 and S4 cleavage) of the remaining membrane bound Notch receptor are carried out by the γ-secretase complex in a constitutive manner [29,30] and result in the release of the Notch intracellular domain (Notch IC). γ-secretase represents one enzyme activity responsible for generation of the amyloidogenic peptide $A\beta_{42}$ from amyloid precursor protein (APP) and hence the pathology associated with Alzheimer's disease[31] but γ-secretase does not appear to cleave Notch heterodimers because it prefers transmembrane proteins with short extracellular regions as substrates.[32] γ-secretase activity requires Presenilin Nicastrin, Pen-2 and Aph1.[33] Nicastrin recognises free amino terminals stubs of type1 transmembrane substrates such as that of S2-cleaved Notch1 while Presenilin is thought to be the catalytic subunit of the γ-secretase complex.[34-36] Once released, Notch IC translocates to the nucleus by virtue of two nuclear localisation sequences where it forms a complex with the DNA-binding protein CSL (CBF1/Su(H)/Lag-1; also known as RBP-jκ).[37,38] The interaction between Notch IC and CSL causes the displacement of histone deactylases previously complexed with CSL,[39] recruitment of the histone acetyltransferase p300 with Notch IC and subsequent activation of Notch target genes.[40,41] Known direct targets of Notch include members of the hairy/enhancer-of-split (HES), HES-related (HERP) family of bHLH transcription factors, Lunatic fringe, Cyclin D1, p21[WAF1], Nodal and GFAP.[42-48] HES and HERP-family transcription factors act to repress transcription of Neurogenin and Mash1 by recruiting TLE, the mammalian homologue of *Drosophila* Groucho.[49-51] In addition, sequestration of the E47 binding partner renders MyoD and Mash1 inactive.[52,53] Alternative CSL-independent signalling through Notch has also been described.[54,55]

The DSL Ligands

DSL proteins are type I transmembrane ligands of the Notch family of receptors. In mammals five DSL ligands can be grouped into two classes (Delta-like and Serrate-like) based on homology to *Drosophila* Delta or Serrate. Delta-like ligands (Dll1, Dll3 and Dll4) are more related to *Drosophila* Delta and lack the cysteine rich domain found in Serrate-related ligands (Serrate1/Jagged1 and Serrate2/Jagged2) (Fig. 1). The extracellular portion of the protein is characterised by the presence of a signal peptide which is followed by an N-terminal domain that displays limited homology among DSL ligands, a DSL domain[56] and a number of EGF-like repeats. The family also contains a C-terminal intracellular domain following the transmembrane domain. The DSL

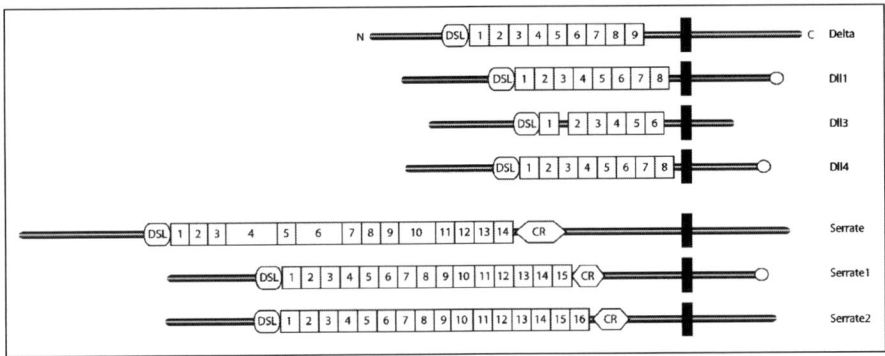

Figure 1. A schematic representation of mammalian DSL ligands compared with *Drosophila* Delta and Serrate. DSL domain (DSL, oval), EGF-like repeats (numbered rectangles), cysteine-rich domain (CR, hexagon), transmembrane domain (black box) and PDZ-binding motif (circle) are indicated.

domain is related to an EGF-like repeat and is required for receptor binding. Maximal receptor binding occurs if the neighbouring EGF-like repeats are present.[57] While relative binding affinities of DSL ligands for Notch receptors have not been exhaustively determined, it is known that Jagged1 can bind to Notch1, Notch2 and Notch3[57,58] and that, Jagged1, Jagged2 and Dll1 can bind to Notch2 to activate Notch signalling.[59]

Binding studies where soluble ligands that lack the transmembrane region and intracellular domains have been generated, indicate that membrane anchorage is not required for receptor binding.[57] While the extracellular domain is responsible for ligand binding, productive signalling however, through Notch cleavage and intracellular domain release requires that the soluble ligand be clustered[60] or immobilised on a substrate.[61] However activation of Notch signalling via anchored soluble ligands are orders of magnitude less than when signalling is activated through cell-cell contact with ligand-expressing cells.[60] Indeed ligand-induced Notch signalling achieved by coculture can be inhibited by addition of free or clustered soluble ligand.[62,63] Thus the membrane tether does not merely act as an anchor allowing clustering but is important for maximal Notch signalling. Consistent with this is the observed importance of endocytosis of ligand into the ligand-expressing cells for proper cell fate outcomes of Notch signalling.[64,65] Such a mechanism is believed to remove the extracellular domain of the receptor allowing further proteolysis. The requirement for endocytosis for effective Notch signalling has highlighted the importance of the intracellular domain of DSL ligands. Further evidence exists in support of this as *Drosophila* Delta and Serrate that lack the intracellular domain act to antagonise the function of full-length ligands.[66,67] In addition, *Shibire* mutants that affect *Drosophila* Dynamin protein cause a defect in endocytosis,[68,69] and *shibire* mutations inhibit Notch signalling when present in the receiving or ligand-expressing (sending) cells.[32,70] Recent advances also implicate two E3-ubiquitin ligases in the endocytosis of DSL ligands. Mindbomb and Neuralized have each been shown to ubiquitylate the intracellular domain of Delta.[71-73] Mindbomb and Neuralized act in the ligand-expressing (or signal sending cell) to cause ubiquitin-dependent internalisation of Delta.[71-74] Mouse Neur2 and Mib1 appear to act at distinct steps of Dll1 endocyotosis.[75]

The intracellular domain of the receptor may have functions additional to allowing regulated endocytosis of the ligand. Like Notch receptors, mouse Dll1 undergoes successive ADAM protease- and γ-secretase- cleavages that result in release of its intracellular domain.[76,77] The intracellular domain of Delta-like and Serrate-like ligands can enter the nucleus following γ-secretase cleavage[78] and, in the case Jagged1, can activate AP1-dependent reporters.[76] Putative nuclear localisation sequences (NLS) are found in DSL ligands except Dll3 and Dll4. They appear to be important for signalling because mutation of the NLS in *Xenopus* Serrate-1 abrogates the intracellular domain's inhibitory function on neurogenesis.[79]

For some of the DSL ligands, protein-protein interactions have been examined; Dll1 and Jagged1 via their C-terminal PDZ-binding motifs, bind to the PDZ domain-containing proteins Activin receptor interacting protein 1 (Acvrinp1) and AF6 (homologue of *Drosophila* Canoe), respectively.[80-82] Although not required for Notch receptor signalling in neighbouring cells, the PDZ-binding motif of Jagged1 is essential for its ability to cause loss of contact inhibition in RKE cells.[81] Interestingly, loss of contact inhibition requires full-length Jagged1 but is independent of Notch receptor signalling, implying the existence of a separate signal in ligand-expressing cells.

Apart from their positive functions, DSL ligands have also been shown to inhibit Notch signalling. Impaired Notch signalling has been observed when DSL ligands are expressed in the same cell (in *cis*) as Notch.[83,84] When expressed in the same cells, Notch1 interacts with Dll1 or Jagged1 inside the cell but not on the cell surface.[85] Co-expression of ligands with Notch1 inhibits activity of a Hes-5 promoter reporter and yet does not impair cell surface presentation of Notch1. Although the mechanism through which this occurs is not known, co-expression of Lunatic fringe inhibits cell autonomous interaction between the DSL ligand and Notch.[85]

Glycosylation of Notch and the DSL Ligands

Ligand-induced Notch signalling is altered by the glycosylation state of the receptor and currently this is known to be a four-step process. In *Drosophila*, EGF-like repeats are modified by addition of *O*-fucose to serine and threonine residues by *O*-fucosyltransferase (OFUT-1).[86,87] This modification is required for ligand-receptor interaction and consistent with this OFUT-1 null mouse embryos undergo abnormal somitogenesis.[88] The Fringe family proteins (Lunatic, Manic, Radical) catalyse the addition of *N*-acetylglucosamine to this fucose and further modification occurs through the action of β4galactosyltransferase 1.[89] Fringe modification in *Drosophila* potentiates Delta-dependent Notch signalling whilst Serrate-dependent Notch signalling is inhibited by the action of Fringe.[90,91] The same is true for Notch1, Dll1 and Jagged1, whilst Lunatic fringe potentiates both Dll1 and Jagged1-mediated Notch2 signalling.[92] Fringe activity increases the capacity of Notch to bind to Delta but not to Serrate/Jagged (see ref. 93). A limited analysis of the glycosylation state of DSL ligands has been performed; this shows that Dll1 and Jagged 1 are substrates for *O*-fucose glycosylation by Fringe proteins.[94]

Lunatic fringe is the only mammalian Fringe protein required for normal somitogenesis.[95-97] In *Lunatic fringe* null mouse embryos, somite formation is abnormal, antero-posterior somite identity is disorganised and severe axial skeletal defects result, which include a reduction in the number of caudal vertebrae. *Lunatic fringe* mRNA levels oscillate in the presomitic mesoderm, each oscillation corresponding to the formation of a somite.[98-100] It was not appreciated that there were distinct rostral and caudal expression domains implicit in the cyclical expression of the *Lunatic fringe* gene, until separate elements were identified that regulated rostral and caudal presomitic mesoderm expression.[42,43] Cyclical *Lunatic fringe* expression is altered in the presomitic mesoderm of Notch pathway mutants, indicating that *Lunatic fringe* expression is driven by Notch signalling.[97,101-103] The identification of functional CSL binding sites that affect *Lunatic fringe* expression in the presomitic mesoderm demonstrates that this gene is indeed a direct target of Notch signalling.[43] While mutant analysis in mouse demonstrates that the lack of Lunatic fringe results in abnormal somite formation, it has recently been shown that it is the oscillation of Lunatic fringe that is required; noncycling overexpression of Lunatic fringe results in the same somitic defect as the complete absence of expression.[104,105]

Notch Signalling and Somite Formation

Multiple ligands (Dll1, Dll3, Dll4, Jagged1, Jagged2) and Notch receptors (Notch1, Notch2, Notch3, Notch4) exist in mammals (Fig. 1). More than a single receptor and ligand are expressed during somitogenesis; however, differential expression indicates that they may perform unique functions (see Table 1). Transcripts of *Notch1, Notch2, Dll1, Dll3* and *Jagged1* are each localised

Table 1. Expression of Notch receptors and ligands during somitogenesis in mouse

Receptor/Ligand	Expression			Phenotype			References
	PSM	I-III	IV-	lethality	somite/vertebra	anterior/posterior somite identity	
Notch1	regionalised	+	medial/dorsal	10-11dpc	irregular	present and organised	101, 142, 143
Notch2	regionalised	+	medial/dorsal	11dpc, postnatal[a]	none		142, 144-146
Notch3	-	somitocoel	dorsal	viable	none		47, 142
Notch4	-	-	-	viable	none		147, 148
Dll1	regionalised	posterior	posterior	~12dpc	irregular	absent	101, 106, 117, 124
Dll3	regionalised	anterior	-	postnatal	irregular	present and disorganised	102, 106
Dll4	-	-	-	-	-		148, 149
Jagged1	regionalised	posterior (I)	-	11dpc	none		96, 150
Jagged2	-	-	-	postnatal	none		96, 151

PSM: presomitic mesoderm; I-III: somites I to III; IV-: somites rostral to and including IV; dpc: days post coitum; a: hypomorphic Notch2 allele.

in the presomitic mesoderm as well as in somites to differing extents, while *Notch3* expression has only been detected in somites. Mutant analysis in mouse demonstrates that only *Notch1*, *Dll1* and *Dll3* are required for normal somite formation but interestingly they each display differences with respect to antero-posterior somite patterning. In Notch1 mutants antero-posterior somite patterning appears normal. In Dll3 mutants, anterior and posterior tissue is present but disorganised and Dll1 mutants lack antero-posterior identity all together.

Dll3

Dll3 was originally isolated on the basis of its expression in mesoderm and the primitive streak during gastrulation.[106] Of the three Delta-like ligands in mammals, Dll3 is the most divergent, showing only 30% homology overall to *Drosophila* Delta, compared to 47% for Dll1 and 43% for Dll4. This is supported by the observation that Dll1 and Dll4 are more closely related to each other than either is to Dll3 (Fig. 2). Structurally, Dll3 also differs from other Delta-like ligands in that it lacks EGF-like repeat 2 that is otherwise highly conserved across mammalian DSL ligands.[107] Moreover, a number of highly conserved residues present in the DSL domains of all other DSL ligands are absent from Dll3. Currently, Dll3 has only been identified in mammals and it represents the most divergent of all DSL family proteins.

A number of processes that act on the DSL ligands have been identified; these include ADAM protease cleavage, γ-secretase cleavage, ubiquitylation and internalisation and glycosylation. Many of these modifications focus on the intracellular domain, which in Dll3 bears no homology to other DSL ligands and is at least half their size. DSL ligands can undergo juxtamembrane cleavage by TACE or Kuzbanian ADAM proteases and subsequent γ-secretase cleavage.[77-79,108-110] The released intracellular domain (ICD) localises to the nucleus as mentioned above. Given the lack of conservation in protease cleavage sites it is impossible to predict if Dll3 undergoes cleavage. As

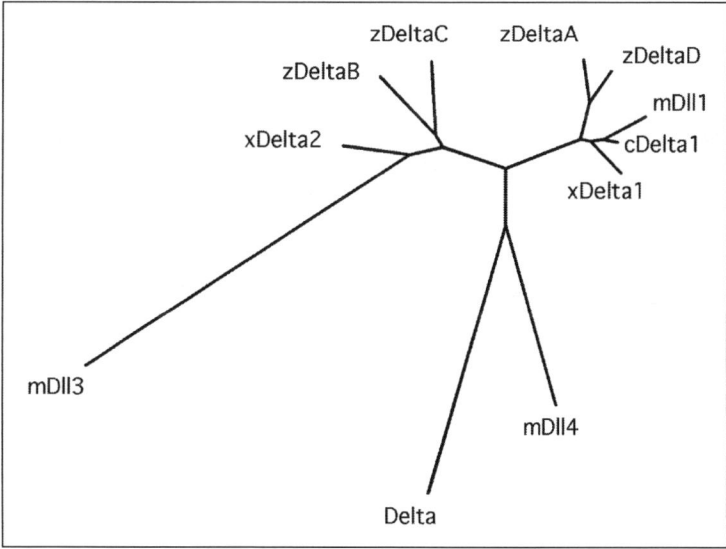

Figure 2. Phylogenetic tree of vertebrate DSL ligands. The tree was created by quartet puzzling using the maximum-likelihood method with the JTT (Jones, Taylor, Thornton) model of substitution.[139,140] Amino-acid sequences were aligned with CLUSTALW and the resulting tree drawn with Phylip.[141] Branch lengths are proportional to the number of accumulated substitutions. Mouse Delta-like1 (mDll1), Mouse Delta-like3 (mDll3), Mouse Delta-like4 (mDll4), chicken Delta1 (cDelta1), *Xenopus* Delta1 (xDelta1), *Xenopus* Delta2 (xDelta2), zebrafish DeltaA (zDeltaA), zebrafish DeltaB (zDeltaB), zebrafish DeltaC (zDeltaC), zebrafish DeltaD (zDeltaD), *Drosophila* Delta *Delta).

to whether the ICD of Dll3 would localise to the nucleus is also difficult to predict, since there is no obvious NLS. Mindbomb and Neuralized-dependent ubiquitylation of lysine residues in the intracellular domain results in endocytosis of Delta-like ligands.[71-73] The lack of lysine residues in the intracellular domain of Dll3 indicates that it is unlikely to be ubiquitylated and thus undergo endocytosis. Surprisingly, all Notch ligands (Dll1, Dll3, Dll4, Jagged1, Jagged2) co-immunopre-cipitate in HEK-293A cells with mouse Mindbomb1 (Mib1) and when co-expressed the ligands accumulate in cytoplasmic vesicles having previously been localised on the plasma membrane or in mesh-like patterns in the cytoplasm.[111] Fringe proteins glycosylate DSL ligands, the functional significance of which is not known.[94] Two EGF-like repeats of Dll3 have the *O*-fucosylation con-sensus site ($C^2XXGG(S/T)C^3$)[112] and thus it is likely that Dll3 also is modified in this way. Finally, the C-terminal PDZ-ligand binding site found in Jagged1, Dll1 and Dll4 is absent in Dll3[81,82] thus, the protein-protein interactions of Dll3 are likely to differ from those of other DSL ligands.

Recently, some features concerning the function of Dll3 have been reported and show that, as predicted from its divergent amino acid sequence, Dll3 has some unique properties.[113] Dll3 and Dll1 are present on the cell surface to the same extent; despite this and unlike Dll1, Dll3 can neither bind Notch1 in *trans* nor activate signalling. Instead Dll3, like Dll1,[85] can bind Notch1 in *cis* and inhibit ligand-induced *trans*-activation. In addition the ability of Dll3 to function in *trans* or in *cis*, unlike Dll1[85] does not seem to be modified by Lunatic fringe.[113] Although the mechanism of Dll3-mediated *cis*-inhibition of Notch1 remains to be established, it is clear that Dll3 is unique amongst DSL ligands in its inability to *trans*-activate Notch signalling.

Dll1 and Dll3 Perform Different Functions during Somitogenesis in Mammals

Dll1 and Dll3 are the only Notch ligands shown to be required for somite formation in mam-mals and they function in a nonredundant manner during somitogenesis.[102,114-116]

Dll1 and Dll3 Are Differentially Expressed during Somitogenesis

During somitogenesis gene expression of Notch receptors and ligands has been determined at the transcript level and consequently there is no information concerning the expression of the proteins.[106,117] Dll1 and Dll3 are both expressed in the presomitic mesoderm but not always in the same regions or to the same extent (Fig. 3A,B). Dll3 transcripts are localised along the length of the presomitic mesoderm, the rostral extent corresponds to the most recently formed somite border (B0).[118] Dll1 transcripts accumulate approximately half a somite's width caudal to this and from here expression extends caudally along the length of the presomitic mesoderm. Thus in the somite that is next to form (S0), Dll3 is expressed in the anterior half and Dll1 in the posterior half. Once the somite has formed, Dll3 transcripts are detected at a low level in the anterior of the most recently formed somites (SI-II), while Dll1 expression continues in the posterior somite halves (Fig. 3A,B).

Anterior and Posterior Somitic Tissue Is Randomly Distributed in Dll3 Mutants

Three *Dll3* mutant alleles exist: *Dll3,pu Dll3neo* and *Dll3.oma* The *Dll3pu* allele was generated during an X-ray mutagenesis screen at the Oak Ridge National Laboratory and the phenotype originally described by Gruneberg.[119] Subsequently, the mutation was identified and characterised by Kusumi and colleagues; a full length *Dll3* transcript is generated from the *Dll3pu* allele however, a 4-nucleotide deletion in the third exon results in a frameshift and truncation of the Dll3 protein N-terminal to the DSL domain.[115] Gene targeting in embryonic stem cells generated the *Dll3neo* null allele, in which the DSL, EGF-like repeats and transmembrane domain were deleted.[102] *Dll3oma* is a newly described, spontaneously arising allele, whereby a single nucleotide substitution in EGF-like repeat 5 results in a glycine to cysteine conversion at amino acid 409.[120] Compound *Dll3neo/Dll3pu* mutants indicate that each of these alleles affects somitogenesis and formation of the axial skeleton to an equal extent, thus *Dll3pu* is also considered to be a null allele.[102] It also appears

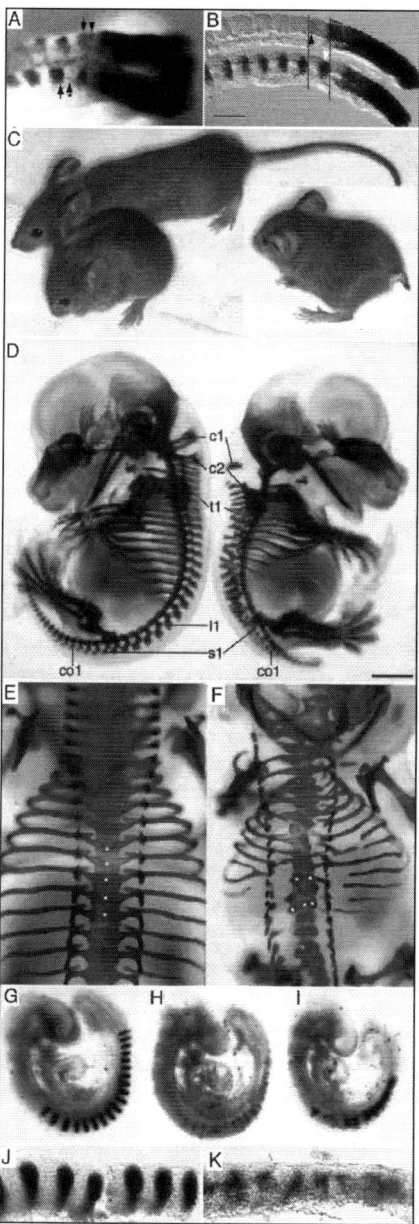

Figure 3. Dll3 is required for during somitogenesis for formation of the axial skeleton. A,B) RNA in situ hybridisation localises Dll3 transcripts to the presomitic mesoderm and somites. Dorsal caudal view of 10-somite embryo *Dll3* (brown) and *Dll1* (blue) transcripts are colocalised within the presomitic mesoderm but not the somites. *Dll3* transcripts are localised to the anterior boundary of the forming somite (black arrowhead) and the nascent somite (black arrowhead*). *Dll1* transcripts are localised to the posterior somite boundaries (black arrow). B) Lateral view of caudal part of 9 dpc embryo split along the midline (fine vertical lines align somites, posterior to right). The embryo was dissected in half, the *Dll3* riboprobe hybridised to the top half and the *Dll1* probe hybridised to the bottom half. *Dll3* is expressed in a broad band of cells at the anterior of the forming somite (short thick black line), this expression is refined to a faint narrow band at the anterior of the nascent somite (black arrowhead). *Dll1* is expressed in a broad band presumably at the posterior of the forming somite (white line), this domain narrows to the posterior boundary of the somites (white arrowhead). C-F) *Dll3^{neo}/Dll3^{neo}* mutants have a truncated body axis and skeletal dysplasia. C) *Dll3^{neo}/Dll3^{neo}* mutants have a shortened body and tail compared with *Dll3/Dll3^{neo}* mice. D) Lateral view of Alcian Blue-stained embryos (14.5 dpc). The positions of vertebrae: cervical (c1 and c2), thoracic (t1), lumbar (l1), sacral (s1) and coccygeal (co1) are indicated. E,F) Dorsal view of developing skeleton. E) *Dll3/Dll3^{neo}* embryo from left in D. F) *Dll3^{neo}/Dll3^{neo}* embryo from right in D. Red dots indicate centrum corresponding to the position of t1, white dots indicate centrum of thoracic vertebrae. Note that in *Dll3^{neo}/Dll3^{neo}* embryos, ossification centres lie two and three in a row instead of lying in column as seen in *Dll3/Dll3^{neo}*. G-K) Uncx4.1 expression shows that anteroposterior somite polarity is disrupted in *Dll3^{neo}/Dll3^{neo}* mutants at 9.5 dpc. G) RNA in situ hybridization shows that *Uncx4.1* is localized to the posterior of the somite in *Dll3/Dll3^{neo}* embryos. J) Expression is clearly restricted to the posterior of epithelial somites dissected from G. H,I) *Uncx4.1* expression is reduced in *Dll3^{neo}/Dll3^{neo}* mutants. Periodic expression is partially retained in some embryos (H) or lost (I). K) Trunk paraxial mesoderm dissected from H; expression appears periodic but is not restricted to the posterior somite (note the lack of epithelial structure). This figure is reproduced from Dunwoodie et al 1997[106], 2002[102] with the permission of The Company of Biologists Ltd.

that *Dll3^oma* represents a null allele given the similarity in phenotype between *Dll3^oma/Dll3^oma* and *Dll3^neo/Dll3^neo* individuals. *Dll3* null mice are easily identifiable as they have a shortened trunk and a small tail (Fig. 3C). This is due to a disorganised axial skeleton along its entire length (Fig. 3D-F). Hemivertebrae are associated with disorganised neural arches, rib bifurcations and absences and the loss of some 25 coccygeal (tail) vertebrae. In *Dll3* null mutant embryos, somite boundary formation is delayed and irregular and condensation of the paraxial mesenchyme into epithelial somites is reduced.[102,115] The somitic tissue has anterior and posterior identity since *Cerberus1* (anterior) and *Uncx4.1* (posterior) markers are expressed.[102,115] However, unlike normal somites, which have clearly defined anterior and posterior domains, the somitic tissue is not polarised as *Uncx4.1* expression is scattered along its length rather than being restricted to the posterior half of the somite (Fig. 3G-K). This random distribution of somitic tissue with anterior or posterior identity has morphological ramifications affecting both the peripheral nervous system and the vertebral column. Components of the peripheral nervous system of the trunk, dorsal root ganglia and spinal nerve axons, can only pass through somitic tissue with anterior identity; this results in their segmented distribution along the length of the trunk.[9] In *Dll3* null mutants, the random arrangement of somitic tissue with anterior and posterior identity results in an equally disorganised arrangement of dorsal root ganglia and spinal nerve axons.[102,121] This loss of somite polarity has a significant impact of vertebra and rib formation in *Dll3* null mutants since the anterior part of one somite gives rise to the caudal half of the vertebral body and the intervertebral disc, whereas the caudal compartment of the neighbouring somite generates the rostral half of the vertebral body and the pedicle of the neural arch.[122] Thus, since *Dll3* null mutant somites lack ordered antero-posterior polarity, a vertebral column develops which shows a randomised arrangement of vertebral components.[123,124]

Somitic Tissue Lacks Anterior and Posterior Identity in Dll1 Mutants

In *Dll1* null mutant embryos, somites are not fully epithelialised and segment borders are not maintained, despite this, segmentation of the dermomyotome is apparent.[114] The sclerotome however remains loosely packed, with the absence of a condensed posterior half affecting the arrangement of the spinal root ganglia and axons. The ganglia are fused rather than being segmented and axons of the spinal nerves, now no longer restricted to pass through the anterior sclerotome, are not segmented. The expression of genes that mark anterior *(EphA4, Cerberus1)* and posterior *(Uncx4.1)* somite identity are significantly downregulated *(Cerberus1)* or not detected *(EphA4, Uncx4.1)* in *Dll1* null mutants.[101,124] This suggests that somites of *Dll1* null mutants largely lack anterior and posterior identity, exactly what effect this has on the rostral and caudal components of the vertebrae is not known, as *Dll1* null mutant embryos die at ~12 dpc prior to vertebral formation.

Mesp2 Is Central to the Establishment of Anterior-Posterior Somite Polarity

Essentially, the difference in somitogenesis between the *Dll1* and *Dll3* null mutants is centred on the specification of anterior and posterior somite identity; *Dll1* mutant somites lack anterior and posterior identity, while in *Dll3* mutants anterior and posterior tissue is specified but not polarised within a somite. Saga and colleagues present evidence that this occurs largely through the action of Dll1, Dll3 and Mesp2.[124,125] Mesp2 is a bHLH transcription factor, which is expressed in the anterior of the presomitic mesoderm just prior to somite formation and is essential for the formation of somite boundaries as well as the specification of anterior somite identity.[126,127] Using single and compound mutants of *Dll1, Dll3, Mesp2* and *Presenilin1,* Saga and colleagues have concluded that Dll1, and Dll3 and Mesp2 operate a signalling network that is required to establish anterior-posterior polarity in the rostral presomitic mesoderm.[124,125] The signalling network relies on feedback loops of Dll1 and Mesp2 for establishment of anterior-posterior polarity and Dll3 for localisation of Dll1 and Mesp2 expression in the rostral presomitic mesoderm. Mutation in *Dll1* and *Dll3* also affects gene expression in the caudal presomitic mesoderm where genes are expressed in an oscillatory manner.

Table 2. *Differential effects of Dll1 and Dll3 null mutations on gene expression in presomitic mesoderm*

Gene	Dll1[tm1Gos]	Dll3[neo]/Dll3[pu]
Dynamic expression		
Lunatic fringe	SDR; not dynamic[101]	DR; not dynamic, caudal not detected[102]
	DR; dynamic[43]	DR; not dynamic, caudal not detected[103]
		DR; not dynamic, caudal detected[97]
		DR; dynamic[43]
Hes1	SDR[128]	Not dynamic[102]
Hes3	-	-
Hes5	SDR[101]	Not detected[102]
Hes7	Not dynamic[105]	Normal[103]
Stage specific expression		
Mesp2	SDR; diffuse[97]	DR; diffuse[102]
		DR; diffuse[103]
		DR; diffuse[97]
Cerberus1	DR[101]	Diffuse[102]
Eph4	SDR[101]	-
Hey1	-	Diffuse[102]
HeyL/3	Not detected[131]	SDR[103]

DR: down regulated; SDR: severely down regulated.

Cyclical Gene Expression in Disrupted in the Presomitic Mesoderm in Dll1 and Dll3 Mutants

The reiterative nature of somite formation is driven from the presomitic mesoderm by the "segmentation clock" which, is observed at the molecular level by oscillatory gene expression. In mouse, the cycling genes identified to date include *Lunatic Fringe, Hes1, Hes5, Hes7, Hey2* and *Axin2*.[18,98,99,102,128-131] Apart from *Axin2* (inhibitor of Wnt signalling) which appears to act upstream of Notch, the other genes are direct targets of Notch signalling, with Lunatic Fringe (glycosyltransferase) acting back on the signalling pathway by modifying Notch and DSL ligands. The expression of these cyclical genes is differentially affected in the presomitic mesoderm of *Dll1* and *Dll3* mutant mice. Generally speaking, expression is severely down regulated in *Dll1* null mutants, while in *Dll3* mutants, expression is differentially disrupted (Table 2). In some cases, for example with *Lunatic fringe* and *Hes1* in *Dll3* null embryos, a band of expression persists in the rostral presomitic mesoderm despite the severe reduction in caudal gene expression. This may reflect the fact that for *Lunatic fringe* at least, distinct regulatory regions control expression in the presomitic mesoderm; one directs rostral expression, another controls cyclical expression in the caudal region.[43]

Mutation in Dll3 Causes Spondylocostal Dysostosis

Spondylocostal dysostosis (SCD; OMIM 277300) are a group of disorders characterised by vertebral segmentation defects and rib anomalies that result in short stature and abdominal protrusion. The axial skeletal defects include hemivertebrae, rib fusions and deletions with a nonprogressive kyphoscoliosis. Autosomal dominant and autosomal recessive (AR) modes of inheritance have been observed. Linkage analysis of several SCD pedigrees with AR inheritance mapped the affected region to 7.8 cM at 19q13.1-q13.3.[132] The Dll3 gene was located within this interval and represented a good candidate for causing SCD because mutation in mouse *Dll3* exhibited similar vertebral defects.[115] Sequence analysis identified mutations in *DLL3* in each of

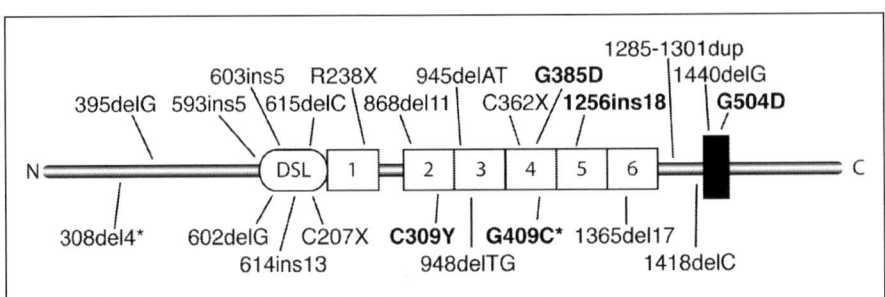

Figure 4. SCD mutations mapped to a schematic representation of Dll3. Mutations causing premature termination of the protein are shown in normal text while missense mutations are shown in bold. Mutations in Dll3 that cause a SCD-like phenotype in mouse are indicated with an asterisk. Mutations in Dll3 shown were originally reported in references 115,116,120, 121,133-135.

the three SCD pedigrees.[116] The vertebral disruptions caused by mutation in *DLL3* are distinct from those in *DLL3* -independent cases of SCD; thus *DLL3* is responsible for causing ARSCD type 1, which results in "pepple beach" vertebral morphology.[133] Subsequent sequence analysis of *DLL3* has revealed a number of genetic lesions that cause either premature termination of the protein, or point mutations within the N-terminal domain, DSL, EGF-repeats or transmembrane domain.[116,121,133-135] (Fig. 4). These mutations have not yet been functionally analysed however, one mutation results in an unpaired cysteine that is likely to disrupt folding of EGF-like repeat 2 and/or intracellular transport of the protein. Mutations in NOTCH3, responsible for CADASIL syndrome, that result in unpaired cysteines in EGF-like repeats were shown in vitro to prevent the receptor from localising to the cell surface.[136] More recently, *DLL3*-independent cases of SCD have been identified. In the first case, SCD linkage to 15q21.3-15q26.1 was demonstrated and a 4-bp duplication was identified in the human *MESP2* gene.[137] Here, the radiological features of the vertebral column are distinct from the *DLL3*-dependent ARSCD type I, thus *MESP2* mutation causes ARSCD type 2. In the second case, a candidate gene approach was used to show that a point mutation in LUNATIC FRINGE caused ARSCD type 3, which also has distinct radiological features of the vertebral column.[138] Functional analysis showed that the amino acid change, phe-188-leu, rendered the LUNATIC FRINGE enzymatically inactive and consequently it was unable to modulate Notch1 signalling in an in vitro coculture assay.[138]

Dll3 Conclusions and the Future

Dll3 is clearly defined as a gene with paramount importance to the process of somitogenesis and thus vertebral column formation in mouse and humans. Despite this, our understanding of its function in Notch signalling is rudimentary. Importantly, Dll3 has recently been shown to interact with Notch, but only as an inhibitor of signalling in *cis*; the ability of Dll3 to *trans*-activate Notch signalling has not been demonstrated. It is not yet clear where in the cell Dll3 interacts with Notch, what affects their ability to interact and how interaction inhibits Notch signalling. It is also not known how the function of the Dll3 protein is affected by posttranslational modifications; this represents a key area of investigation for the future. In addition, how the functions of both Dll3 and Dll1 are coordinated during somitogenesis needs to be clarified. Information concerning these issues is clearly important for an enhanced understanding of somitogenesis. However, since Notch signalling is involved in cell fate specification in a number of contexts during embryonic and adult life and due to the fact that Dll1 and Dll3 are expressed widely, these issues have a broader relevance than to somitogenesis alone.

Acknowledgements

The authors thank Merridee Wouters of the Computational Biology and Bioinformatics Program at the Victor Chang Cardiac Research Institute for preparing Figure 2. This work is supported by NHMRC grant 404804. SLD is a Pfizer Foundation Australia Senior Research Fellow. G.C. is supported by a National Health and Medical Research Council (Australia) C.J. Martin Fellowship (No. 158043) and a Cancer Institute of NSW fellowship.

References

1. Kaufman MH, Bard JBL. Somites and their derivatives (muscles, dermis and vertebrae). In: Kaufman MH, Bard, J B L, ed. The Anatomical Basis of Mouse Development. Edinburgh: Academic Press; 1999:51-59.
2. Tam PP. The control of somitogenesis in mouse embryos. J Embryol Exp Morphol 1981; 65 Suppl:103-128.
3. Brent AE, Tabin CJ. Developmental regulation of somite derivatives: muscle, cartilage and tendon. Curr Opin Genet Dev 2002; 12(5):548-557.
4. Brent AE, Schweitzer R, Tabin CJ. A somitic compartment of tendon progenitors. Cell 2003; 113(2):235-248.
5. Kieny M, Mauger A, Sengel P. Early regionalization of somitic mesoderm as studied by the development of axial skeleton of the chick embryo. Dev Biol 1972; 28(1):142-161.
6. Cauwenbergs P, Butler J, Cosmos E. Intraspecific chick/chick chimaeras: dystrophic somitic mesoderm transplanted to a normal host forms muscles with a dystrophic phenotype. Neurosci Lett 1986; 68(2):149-154.
7. Butler J, Cauwenbergs P, Cosmos E. Fate of brachial muscles of the chick embryo innervated by inappropriate nerves: structural, functional and histochemical analyses. J Embryol Exp Morphol 1986; 95:147-168.
8. Stern CD, Keynes RJ. Interactions between somite cells: the formation and maintenance of segment boundaries in the chick embryo. Development 1987; 99(2):261-272.
9. Keynes RJ, Stern CD. Segmentation in the vertebrate nervous system. Nature 1984; 310(5980):786-789.
10. Rickmann M, Fawcett JW, Keynes RJ. The migration of neural crest cells and the growth of motor axons through the rostral half of the chick somite. J Embryol Exp Morphol 1985; 90:437-455.
11. Maroto M, Pourquie O. A molecular clock involved in somite segmentation. Curr Top Dev Biol 2001; 51:221-248.
12. Cooke J, Zeeman EC. A clock and wavefront model for control of the number of repeated structures during animal morphogenesis. J Theor Biol 1976; 58(2):455-476.
13. Yamaguchi TP, Harpal K, Henkemeyer M et al. fgfr-1 is required for embryonic growth and mesodermal patterning during mouse gastrulation. Genes Dev 1994; 8(24):3032-3044.
14. Takada S, Stark KL, Shea MJ et al. Wnt-3a regulates somite and tailbud formation in the mouse embryo. Genes Dev 1994; 8(2):174-189.
15. Greco TL, Takada S, Newhouse MM et al. Analysis of the vestigial tail mutation demonstrates that Wnt-3a gene dosage regulates mouse axial development. Genes Dev 1996; 10(3):313-324.
16. Conlon RA, Reaume AG, Rossant J. Notch1 is required for the coordinate segmentation of somites. Development 1995; 121(5):1533-1545.
17. Dubrulle J, McGrew MJ, Pourquie O. FGF signaling controls somite boundary position and regulates segmentation clock control of spatiotemporal Hox gene activation. Cell 2001; 106(2):219-232.
18. Aulehla A, Wehrle C, Brand-Saberi B et al. Wnt3a plays a major role in the segmentation clock controlling somitogenesis. Dev Cell 2003; 4(3):395-406.
19. Dubrulle J, Pourquie O. From head to tail: links between the segmentation clock and antero-posterior patterning of the embryo. Curr Opin Genet Dev 2002; 12(5):519-523.
20. Blaumueller CM, Qi H, Zagouras P et al. Intracellular cleavage of Notch leads to a heterodimeric receptor on the plasma membrane. Cell 1997; 90(2):281-291.
21. Logeat F, Bessia C, Brou C et al. The Notch1 receptor is cleaved constitutively by a furin-like convertase. Proc Natl Acad Sci USA 1998; 95(14):8108-8112.
22. Rebay I, Fleming RJ, Fehon RG et al. Specific EGF repeats of Notch mediate interactions with Delta and Serrate: implications for Notch as a multifunctional receptor. Cell 1991; 67(4):687-699.
23. Fleming RJ. Structural conservation of Notch receptors and ligands. Semin Cell Dev Biol 1998; 9(6):599-607.
24. Hu QD, Ang BT, Karsak M et al. F3/contactin acts as a functional ligand for Notch during oligodendrocyte maturation. Cell 2003; 115(2):163-175.
25. Cui XY, Hu QD, Tekaya M et al. NB-3/Notch1 Pathway via Deltex1 Promotes Neural Progenitor Cell Differentiation into Oligodendrocytes. J Biol Chem 2004; 279(24):25858-25865.

26. Miyamoto A, Lau R, Hein PW et al. Microfibrillar proteins MAGP-1 and MAGP-2 induce Notch1 extracellular domain dissociation and receptor activation. J Biol Chem 2006; 281(15):10089-10097.

27. Brou C, Logeat F, Gupta N et al. A novel proteolytic cleavage involved in Notch signaling: the role of the disintegrin-metalloprotease TACE. Mol Cell 2000; 5(2):207-216.

28. Mumm JS, Schroeter EH, Saxena MT et al. A ligand-induced extracellular cleavage regulates gamma-secretase-like proteolytic activation of Notch1. Mol Cell 2000; 5(2):197-206.

29. Schroeter EH, Kisslinger JA, Kopan R. Notch-1 signalling requires ligand-induced proteolytic release of intracellular domain. Nature 1998; 393(6683):382-386.

30. Okochi M, Steiner H, Fukumori A et al. Presenilins mediate a dual intramembranous gamma-secretase cleavage of Notch-1. EMBO J 2002; 21(20):5408-5416.

31. De Strooper B, Saftig P, Craessarerts K et al. Deficiency of presenilin-1 inhibits the normal cleavage of amyloid precursor protein. Nature 1998; 391:387-390.

32. Struhl G, Greenwald I. Presenilin-mediated transmembrane cleavage is required for Notch signal transduction in Drosophila. Proc Natl Acad Sci USA 2001; 98(1):229-234.

33. Edbauer D, Winkler E, Regula JT et al. Reconstitution of gamma-secretase activity. Nat Cell Biol 2003; 5(5):486-488.

34. Li Y-M, Xu M, Lai M-T et al. Photoactivated γ-secretase inhibitors directed to the active site covalently label presenilin 1. Nature 2000; 405:689-694.

35. Esler WP, Kimberly WT, Ostaszewski BL et al. Transition-state analogue inhibitors of gamma-secretase bind directly to presenilin-1. Nat Cell Biol 2000; 2(7):428-434.

36. Shah S, Lee SF, Tabuchi K et al. Nicastrin functions as a gamma-secretase-substrate receptor. Cell 2005; 122(3):435-447.

37. Jarriault S, Brou C, Logeat F et al. Signalling downstream of activated mammalian Notch. Nature 1995; 377(6547):355-358.

38. Hsieh JJ, Henkel T, Salmon P et al. Truncated mammalian Notch1 activates CBF1/RBPJk-repressed genes by a mechanism resembling that of Epstein-Barr virus EBNA2. Mol Cell Biol 1996; 16(3):952-959.

39. Kao HY, Ordentlich P, Koyano-Nakagawa N et al. A histone deacetylase corepressor complex regulates the Notch signal transduction pathway. Genes Dev 1998; 12(15):2269-2277.

40. Oswald F, Tauber B, Dobner T et al. p300 acts as a transcriptional coactivator for mammalian Notch-1. Mol Cell Biol 2001; 21(22):7761-7774.

41. Wallberg AE, Pedersen K, Lendahl U et al. p300 and PCAF act cooperatively to mediate transcriptional activation from chromatin templates by notch intracellular domains in vitro. Mol Cell Biol 2002; 22(22):7812-7819.

42. Cole SE, Levorse JM, Tilghman SM et al. Clock regulatory elements control cyclic expression of Lunatic fringe during somitogenesis. Dev Cell 2002; 3(1):75-84.

43. Morales AV, Yasuda Y, Ish-Horowicz D. Periodic Lunatic fringe expression is controlled during segmentation by a cyclic transcriptional enhancer responsive to notch signaling. Dev Cell 2002; 3(1):63-74.

44. Iso T, Kedes L, Hamamori Y. HES and HERP families: multiple effectors of the Notch signaling pathway. J Cell Physiol 2003; 194(3):237-255.

45. Ronchini C, Capobianco AJ. Induction of cyclin D1 transcription and CDK2 activity by Notch(ic): implication for cell cycle disruption in transformation by Notch(ic). Mol Cell Biol 2001; 21(17):5925-5934.

46. Rangarajan A, Talora C, Okuyama R et al. Notch signaling is a direct determinant of keratinocyte growth arrest and entry into differentiation. EMBO J 2001; 20(13):3427-3436.

47. Krebs LT, Xue Y, Norton CR et al. Characterization of Notch3-deficient mice: normal embryonic development and absence of genetic interactions with a Notch1 mutation. Genesis 2003; 37(3):139-143.

48. Ge W, Martinowich K, Wu X et al. Notch signaling promotes astrogliogenesis via direct CSL-mediated glial gene activation. J Neurosci Res 2002; 69(6):848-860.

49. Paroush Z, Finley RL, Jr., Kidd T et al. Groucho is required for Drosophila neurogenesis, segmentation and sex determination and interacts directly with hairy-related bHLH proteins. Cell 1994; 79(5):805-815.

50. Fisher A, Caudy M. The function of hairy-related bHLH repressor proteins in cell fate decisions. Bioessays 1998; 20(4):298-306.

51. Grbavec D, Stifani S. Molecular interaction between TLE1 and the carboxyl-terminal domain of HES-1 containing the WRPW motif. Biochem Biophys Res Commun 1996; 223(3):701-705.

52. Sasai Y, Kageyama R, Tagawa Y et al. Two mammalian helix-loop-helix factors structurally related to Drosophila hairy and Enhancer of split. Genes Dev 1992; 6(12B):2620-2634.

53. Hirata H, Ohtsuka T, Bessho Y et al. Generation of structurally and functionally distinct factors from the basic helix-loop-helix gene Hes3 by alternative first exons. J Biol Chem 2000; 275(25):19083-19089.

54. Shawber C, Nofziger D, Hsieh JJ-D et al. Notch signaling inhibits muscle cell differentiation through a CBF-independent pathway. Development 1996; 122:3765-3773.

55. Bush G, diSibio G, Miyamoto A et al. Ligand-induced signaling in the absence of furin processing of Notch1. Dev Biol 2001; 229(2):494-502.
56. Nye JS, Kopan R. Vertebrate ligands for Notch. Curr Biol 1995; 5:966-969.
57. Shimizu K, Chiba S, Kumano K et al. Mouse jagged1 physically interacts with notch2 and other notch receptors. Assessment by quantitative methods. J Biol Chem 1999; 274(46):32961-32969.
58. Shimizu K, Chiba S, Saito T et al. Physical interaction of Delta1, Jagged1 and Jagged2 with Notch1 and Notch3 receptors. Biochem Biophys Res Commun 2000; 276(1):385-389.
59. Shimizu K, Chiba S, Hosoya N et al. Binding of Delta1, Jagged1 and Jagged2 to Notch2 rapidly induces cleavage, nuclear translocation and hyperphosphorylation of Notch2. Mol Cell Biol 2000; 20(18):6913-6922.
60. Shimizu K, Chiba S, Saito T et al. Integrity of intracellular domain of Notch ligand is indispensable for cleavage required for release of the Notch2 intracellular domain. EMBO J 2002; 21(3):294-302.
61. Varnum-Finney B, Wu L, Yu M et al. Immobilization of Notch ligand, Delta-1, is required for induction of notch signaling. J Cell Sci 2000; 113 Pt 23:4313-4318.
62. Sun X, Artavanis-Tsakonas S. Secreted forms of DELTA and SERRATE define antagonists of Notch signaling in Drosophila. Development 1997; 124(17):3439-3448.
63. Hicks C, Ladi E, Lindsell C et al. A secreted Delta1-Fc fusion protein functions both as an activator and inhibitor of Notch1 signaling. J Neurosci Res 2002; 68(6):655-667.
64. Klueg KM, Muskavitch MA. Ligand-receptor interactions and trans-endocytosis of Delta, Serrate and Notch: members of the Notch signalling pathway in Drosophila. J Cell Sci 1999; 112(Pt 19):3289-3297.
65. Parks AL, Klueg KM, Stout JR et al. Ligand endocytosis drives receptor dissociation and activation in the Notch pathway. Development 2000; 127(7):1373-1385.
66. Sun X, Artavanis-Tsakonas S. The intracellular deletions of Delta and Serrate define dominant negative forms of the Drosophila Notch ligands. Development 1996; 122(8):2465-2474.
67. Hukriede NA, Gu Y, Fleming RJ. A dominant-negative form of Serrate acts as a general antagonist of Notch activation. Development 1997; 124(17):3427-3437.
68. Chen MS, Obar RA, Schroeder CC et al. Multiple forms of dynamin are encoded by shibire, a Drosophila gene involved in endocytosis. Nature 1991; 351(6327):583-586.
69. van der Bliek AM, Meyerowitz EM. Dynamin-like protein encoded by the Drosophila shibire gene associated with vesicular traffic. Nature 1991; 351(6325):411-414.
70. Seugnet L, Simpson P, Haenlin M. Requirement for dynamin during Notch signaling in Drosophila neurogenesis. Dev Biol 1997; 192(2):585-598.
71. Deblandre GA, Lai EC, Kintner C. Xenopus neuralized is a ubiquitin ligase that interacts with XDelta1 and regulates Notch signaling. Dev Cell 2001; 1(6):795-806.
72. Lai EC, Deblandre GA, Kintner C et al. Drosophila neuralized is a ubiquitin ligase that promotes the internalization and degradation of delta. Dev Cell 2001; 1(6):783-794.
73. Itoh M, Kim CH, Palardy G et al. Mind bomb is a ubiquitin ligase that is essential for efficient activation of Notch signaling by Delta. Dev Cell 2003; 4(1):67-82.
74. Li Y, Baker NE. The roles of cis-inactivation by Notch ligands and of neuralized during eye and bristle patterning in Drosophila. BMC Dev Biol 2004; 4:5.
75. Song R, Koo BK, Yoon KJ et al. Neuralized-2 regulates notch ligand in cooperation with mind bomb-1. J Biol Chem 2006.
76. LaVoie MJ, Selkoe DJ. The Notch ligands, Jagged and Delta, are sequentially processed by alpha-secretase and presenilin/gamma-secretase and release signaling fragments. J Biol Chem 2003; 278(36):34427-34437.
77. Six E, Ndiaye D, Laabi Y et al. The Notch ligand Delta1 is sequentially cleaved by an ADAM protease and gamma-secretase. Proc Natl Acad Sci USA 2003; 100(13):7638-7643.
78. Ikeuchi T, Sisodia SS. The Notch ligands, Delta1 and Jagged2, are substrates for presenilin-dependent "gamma-secretase" cleavage. J Biol Chem 2003; 278(10):7751-7754.
79. Kiyota T, Kinoshita T. The intracellular domain of X-Serrate-1 is cleaved and suppresses primary neurogenesis in Xenopus laevis. Mech Dev 2004; 121(6):573-585.
80. Pfister S, Przemeck GK, Gerber JK et al. Interaction of the MAGUK family member Acvrinp1 and the cytoplasmic domain of the Notch ligand Delta1. J Mol Biol 2003; 333(2):229-235.
81. Ascano JM, Beverly LJ, Capobianco AJ. The C-terminal PDZ-ligand of JAGGED1 is essential for cellular transformation. J Biol Chem 2003; 278(10):8771-8779.
82. Hock B, Bohme B, Karn T et al. PDZ-domain-mediated interaction of the Eph-related receptor tyrosine kinase EphB3 and the ras-binding protein AF6 depends on the kinase activity of the receptor. Proc Natl Acad Sci USA 1998; 95(17):9779-9784.
83. Henrique D, Hirsinger E, Adam J et al. Maintenance of neuroepithelial progenitor cells by Delta-Notch signalling in the embryonic chick retina. Curr Biol 1997; 7(9):661-670.

84. de Celis JF, Bray SJ. The Abruptex domain of Notch regulates negative interactions between Notch, its ligands and Fringe. Development 2000; 127(6):1291-1302.

85. Sakamoto K, Ohara O, Takagi M et al. Intracellular cell-autonomous association of Notch and its ligands: a novel mechanism of Notch signal modification. Dev Biol 2002; 241(2):313-326.

86. Okajima T, Irvine KD. Regulation of notch signaling by o-linked fucose. Cell 2002; 111(6):893-904.

87. Okajima T, Xu A, Irvine KD. Modulation of notch-ligand binding by protein O-fucosyltransferase 1 and fringe. J Biol Chem 2003; 278(43):42340-42345.

88. Shi S, Stanley P. Protein O-fucosyltransferase 1 is an essential component of Notch signaling pathways. Proc Natl Acad Sci USA 2003; 100(9):5234-5239.

89. Chen J, Moloney DJ, Stanley P. Fringe modulation of Jagged1-induced Notch signaling requires the action of beta 4galactosyltransferase-1. Proc Natl Acad Sci USA 2001; 98(24):13716-13721.

90. Bruckner K, Perez L, Clausen H et al. Glycosyltransferase activity of Fringe modulates Notch-Delta interactions. Nature 2000; 406(6794):411-415.

91. Moloney DJ, Panin VM, Johnston SH et al. Fringe is a glycosyltransferase that modifies Notch. Nature 2000; 406(6794):369-375.

92. Hicks C, Johnston SH, diSibio G et al. Fringe differentially modulates Jagged1 and Delta1 signalling through Notch1 and Notch2. Nat Cell Biol 2000; 2(8):515-520.

93. Weinmaster, Kintner. Modulation of notch signaling during somitogenesis. Annu Rev Cell Dev Biol 2003; 19:367-395.

94. Panin VM, Shao L, Lei L et al. Notch ligands are substrates for protein O-fucosyltransferase-1 and Fringe. J Biol Chem 2002; 277(33):29945-29952.

95. Evrard YA, Lun Y, Aulehla A et al. lunatic fringe is an essential mediator of somite segmentation and patterning. Nature 1998; 394(6691):377-381.

96. Zhang N, Gridley T. Defects in somite formation in lunatic fringe-deficient mice. Nature 1998; 394(6691):374-377.

97. Zhang N, Norton CR, Gridley T. Segmentation defects of Notch pathway mutants and absence of a synergistic phenotype in lunatic fringe/radical fringe double mutant mice. Genesis 2002; 33(1):21-28.

98. Aulehla A, Johnson RL. Dynamic expression of lunatic fringe suggests a link between notch signaling and an autonomous cellular oscillator driving somite segmentation. Dev Biol 1999; 207(1):49-61.

99. Forsberg H, Crozet F, Brown NA. Waves of mouse Lunatic fringe expression, in four-hour cycles at two-hour intervals, precede somite boundary formation. Curr Biol 1998; 8(18):1027-1030.

100. McGrew MJ, Dale JK, Fraboulet S et al. The lunatic fringe gene is a target of the molecular clock linked to somite segmentation in avian embryos. Curr Biol 1998; 8(17):979-982.

101. Barrantes IB, Elia AJ, Wunsch K et al. Interaction between Notch signalling and Lunatic fringe during somite boundary formation in the mouse. Curr Biol 1999; 9(9):470-480.

102. Dunwoodie SL, Clements M, Sparrow DB et al. Axial skeletal defects caused by mutation in the spondylocostal dysplasia/pudgy gene Dll3 are associated with disruption of the segmentation clock within the presomitic mesoderm. Development 2002; 129(7):1795-1806.

103. Kusumi K, Mimoto MS, Covello KL et al. Dll3 pudgy mutation differentially disrupts dynamic expression of somite genes. Genesis 2004; 39(2):115-121.

104. Dale JK, Maroto M, Dequeant ML et al. Periodic notch inhibition by lunatic fringe underlies the chick segmentation clock. Nature 2003; 421(6920):275-278.

105. Serth K, Schuster-Gossler K, Cordes R et al. Transcriptional oscillation of lunatic fringe is essential for somitogenesis. Genes Dev 2003; 17(7):912-925.

106. Dunwoodie SL, Henrique D, Harrison SM et al. Mouse Dll3: a novel divergent Delta gene which may complement the function of other Delta homologues during early pattern formation in the mouse embryo. Development 1997; 124(16):3065-3076.

107. Lissemore JL, Starmer WT. Phylogenetic analysis of vertebrate and invertebrate Delta/Serrate/LAG-2 (DSL) proteins. Mol Phylogenet Evol 1999; 11(2):308-319.

108. Qi H, Rand MD, Wu X et al. Processing of the notch ligand delta by the metalloprotease Kuzbanian. Science 1999; 283(5398):91-94.

109. Bland CE, Kimberly P, Rand MD. Notch-induced proteolysis and nuclear localization of the Delta ligand. J Biol Chem 2003; 278(16):13607-13610.

110. Ikeuchi T, Sisodia SS. Cell-free generation of the notch1 intracellular domain (NICD) and APP-CT-fgamma: evidence for distinct intramembranous "gamma-secretase" activities. Neuromolecular Med 2002; 1(1):43-54.

111. Koo BK, Lim HS, Song R et al. Mind bomb 1 is essential for generating functional Notch ligands to activate Notch. Development 2005; 132(15):3459-3470.

112. Harris RJ, Spellman MW. O-linked fucose and other posttranslational modifications unique to EGF modules. Glycobiology 1993; 3(3):219-224.

113. Ladi E, Nichols JT, Ge W et al. The divergent DSL ligand Dll3 does not activate Notch signaling but cell autonomously attenuates signaling induced by other DSL ligands. J Cell Biol 2005; 170(6):983-992.
114. Hrabe de Angelis M, McIntyre J 2nd, Gossler A. Maintenance of somite borders in mice requires the Delta homologue DII1. Nature 1997; 386(6626):717-721.
115. Kusumi K, Sun ES, Kerrebrock AW et al. The mouse pudgy mutation disrupts Delta homologue Dll3 and initiation of early somite boundaries. Nat Genet 1998; 19(3):274-278.
116. Bulman MP, Kusumi K, Frayling TM et al. Mutations in the human delta homologue, DLL3, cause axial skeletal defects in spondylocostal dysostosis. Nat Genet 2000; 24(4):438-441.
117. Bettenhausen B, Hrabe de Angelis M, Simon D et al. Transient and restricted expression during mouse embryogenesis of Dll1, a murine gene closely related to Drosophila Delta. Development 1995; 121(8):2407-2418.
118. Pourquie O, Tam PP. A nomenclature for prospective somites and phases of cyclic gene expression in the presomitic mesoderm. Dev Cell 2001; 1(5):619-620.
119. Gruneberg H. Genetical studies on the skeleton of the mouse XXIX Pudgy. Genet Res 1961; 2:384-393.
120. Shinkai Y, Tsuji T, Kawamoto Y et al. New mutant mouse with skeletal deformities caused by mutation in delta like 3 (dll3) gene. Exp Anim 2004; 53(2):129-136.
121. Sparrow DB, Clements M, Withington SL et al. Diverse requirements for Notch signalling in mammals. Int J Dev Biol 2002; 46(4):365-374.
122. Aoyama H, Asamoto K. The developmental fate of the rostral/caudal half of a somite for vertebra and rib formation: experimental confirmation of the resegmentation theory using chick-quail chimeras. Mech Dev 2000; 99(1-2):71-82.
123. Saga Y, Takeda H. The making of the somite: molecular events in vertebrate segmentation. Nat Rev Genet 2001; 2(11):835-845.
124. Takahashi Y, Inoue T, Gossler A et al. Feedback loops comprising Dll1, Dll3 and Mesp2 and differential involvement of Psen1 are essential for rostrocaudal patterning of somites. Development 2003; 130(18):4259-4268.
125. Takahashi Y, Koizumi K, Takagi A et al. Mesp2 initiates somite segmentation through the Notch signalling pathway. Nat Genet 2000; 25(4):390-396.
126. Saga Y, Hata N, Koseki H et al. Mesp2: a novel mouse gene expressed in the presegmented mesoderm and essential for segmentation initiation. Genes Dev 1997; 11(14):1827-1839.
127. Nomura-Kitabayashi A, Takahashi Y, Kitajima S et al. Hypomorphic Mesp allele distinguishes establishment of rostrocaudal polarity and segment border formation in somitogenesis. Development 2002; 129(10):2473-2481.
128. Jouve C, Palmeirim I, Henrique D et al. Notch signalling is required for cyclic expression of the hairy-like gene HES1 in the presomitic mesoderm. Development 2000; 127(7):1421-1429.
129. Bessho Y, Sakata R, Komatsu S et al. Dynamic expression and essential functions of Hes7 in somite segmentation. Genes Dev 2001; 15(20):2642-2647.
130. Bessho Y, Hirata H, Masamizu Y et al. Periodic repression by the bHLH factor Hes7 is an essential mechanism for the somite segmentation clock. Genes Dev 2003; 17(12):1451-1456.
131. Leimeister C, Dale K, Fischer A et al. Oscillating expression of c-Hey2 in the presomitic mesoderm suggests that the segmentation clock may use combinatorial signaling through multiple interacting bHLH factors. Dev Biol 2000; 227(1):91-103.
132. Turnpenny PD, Bulman MP, Frayling TM et al. A gene for autosomal recessive spondylocostal dysostosis maps to 19q13.1-q13.3. Am J Hum Genet 1999; 65(1):175-182.
133. Turnpenny PD, Whittock N, Duncan J et al. Novel mutations in DLL3, a somitogenesis gene encoding a ligand for the Notch signalling pathway, cause a consistent pattern of abnormal vertebral segmentation in spondylocostal dysostosis. J Med Genet 2003; 40(5):333-339.
134. Whittock NV, Ellard S, Duncan J et al. Pseudodominant inheritance of spondylocostal dysostosis type 1 caused by two familial delta-like 3 mutations. Clin Genet 2004; 66(1):67-72.
135. Bonafe L, Giunta C, Gassner M et al. A cluster of autosomal recessive spondylocostal dysostosis caused by three newly identified DLL3 mutations segregating in a small village. Clin Genet 2003; 64(1):28-35.
136. Karlstrom H, Beatus P, Dannaeus K et al. A CADASIL-mutated Notch 3 receptor exhibits impaired intracellular trafficking and maturation but normal ligand-induced signaling. Proc Natl Acad Sci USA 2002; 99(26):17119-17124.
137. Whittock NV, Sparrow DB, Wouters MA et al. Mutated MESP2 Causes Spondylocostal Dysostosis in Humans. Am J Hum Genet 2004; 74(6):1249-1254.
138. Sparrow DB, Chapman G, Wouters MA et al. Mutation of the LUNATIC FRINGE gene in humans causes spondylocostal dysostosis with a severe vertebral phenotype. Am J Hum Genet 2006; 78(1):28-37.
139. Strimmer K, von Haeseler A. Quartet puzzling: A quartet maximum-likelihood method for reconstructing tree topologies. Mol Biol Evol 1996; 13:964-969.

140. Jones DT, Taylor WR, Thornton JM. The rapid generation of mutation data matrices from protein sequences. Comput Appl Biosci 1992; 8(3):275-282.
141. Thompson JD, Higgins DG, Gibson TJ. CLUSTAL W: improving the sensitivity of progressive multiple sequence alignment through sequence weighting, position-specific gap penalties and weight matrix choice. Nucleic Acids Res 1994; 22(22):4673-4680.
142. Williams R, Lendahl U, Lardelli M. Complementary and combinatorial patterns of Notch gene family expression during early mouse development. Mech Dev 1995; 53(3):357-368.
143. Reaume AG, Conlon RA, Zirngibl R et al. Expression analysis of a Notch homologue in the mouse embryo. Dev Biol 1992; 154(2):377-387.
144. Swiatek PJ, Lindsell CE, del Amo FF et al. Notch1 is essential for postimplantation development in mice. Genes Dev 1994; 8(6):707-719.
145. Hamada Y, Kadokawa Y, Okabe M et al. Mutation in ankyrin repeats of the mouse Notch2 gene induces early embryonic lethality. Development 1999; 126(15):3415-3424.
146. McCright B, Gao X, Shen L et al. Defects in development of the kidney, heart and eye vasculature in mice homozygous for a hypomorphic Notch2 mutation. Development 2001; 128(4):491-502.
147. Shirayoshi Y, Yuasa Y, Suzuki T et al. Proto-oncogene of int-3, a mouse Notch homologue, is expressed in endothelial cells during early embryogenesis. Genes Cells 1997; 2(3):213-224.
148. Krebs LT, Xue Y, Norton CR et al. Notch signaling is essential for vascular morphogenesis in mice. Genes Dev 2000; 14(11):1343-1352.
149. Shutter JR, Scully S, Fan W et al. Dll4, a novel Notch ligand expressed in arterial endothelium. Genes Dev 2000; 14(11):1313-1318.
150. Xue Y, Gao X, Lindsell CE et al. Embryonic lethality and vascular defects in mice lacking the Notch ligand Jagged1. Hum Mol Genet 1999; 8(5):723-730.
151. Jiang R, Lan Y, Chapman HD et al. Defects in limb, craniofacial and thymic development in Jagged2 mutant mice. Genes Dev 1998; 12(7):1046-1057.
152. Aoyama H, Asamoto K. Determination of somite cells: independence of cell differentiation and morphogenesis. Development 1988; 104(1):15-28.
153. Baron M. An overview of the Notch signalling pathway. Semin Cell Dev Biol 2003; 14(2):113-119.

Chapter 6

Mesp-Family Genes Are Required for Segmental Patterning and Segmental Border Formation

Yumiko Saga* and Yu Takahashi

Abstract

Elaborate somite patterning is based on the dynamic gene regulation within the presomitic mesoderm (PSM) derived from the primitive streak and tailbud in the later stage embryo. Notch signaling and the regulators are major players involved in the all events required for the temporally and spatially coordinated somite formation. PSM can be subdivided at least two domains based on the regulation and maybe the function of genes expressed. In the posterior PSM, a basic-HLH protein Hes7 plays a central role to generate traveling wave of gene expression by negatively regulating the transcription of the target genes, which may lead defining soimte spacing and future segmental unit. In the anterior PSM, cells start to prepare segmental patterning by acquiring rostral or caudal identity of somite primordia and defining segmental border. In this process, Mesp2, another basic HLH protein plays a critical role. Genetic evidence is provided how Mesp2 regulates Notch signaling to establish segmental identity in the anterior PSM.

Introduction

Loss of function studies revealed unexpected fact that Notch signaling pathway is involved in the somitogenesis, the vertebrate segmentation event, which is a different mechanism used for segmentation in the fruit fly, Drosophila.[1] In addition, a finding of oscillation gene that reflect segmentation clock was fascinating enough to put forward the somitogenesis as a most sophisticated developmental system.[2] The metameric structure is generated through the clock mechanism, which accompanies periodic activation of Notch signaling. The traveling wave generated in the posterior PSM slows down and finally stops in the anterior PSM and the expression become segmental by localizing either in rostral or caudal compartment in somite primordia. Here, segmental identity, rostor-caudal polarity is established and segmental border will be defined. Several years ago, we cloned a unique gene Mesp2, encoding a basic HLH-type transcription factor, because of the unique expression pattern in the anterior PSM just before segmental border formation.[3] Obviously this gene is not involved in the events in the posterior PSM, so it has independent function of segmentation clock. But genetic analyses have revealed that it affects expression of almost all clock genes and it works within the Notch signaling cascade. The reciprocal interaction implicated by the genetic study leads us to generate so complicated scheme, which is still be subject to change. In this article, I like to review the events in the anterior PSM by focusing on the function of Mesp-family genes.

*Corresponding Author: Yumiko Saga—Division of Mammalian Development National Institute of Genetics, Yata 1111, Mishima, 411-8540, Japan. Email: ysaga@lab.nig.ac.jp

Somitogenesis, edited by Miguel Maroto and Neil V. Whittock. ©2008 Landes Bioscience and Springer Science+Business Media.

Background of Mesp1 and Mesp2

Initially *Mesp1* is isolated as a posterior specific cDNA by a subtraction between cDNAs expressed in posterior and anterior region of 7.5 dpc mouse embryos.[4] The expression of *Mesp1* is observed in the nascent mesoderm at the onset of gastrulation. That is why it was named *Mesp1* (mesoderm posterior 1). *Mesp2* was identified by a cross hybridization during screening of *Mesp1* genome.[3] These two genes locate in the #7 chromosome head to head orientation and the proteins share almost identical bHLH motif although outside of the domain shows considerable diversity (Fig. 1A). *Mesp2* also shows the expression in the early mesoderm. In addition, both genes are expressed in the paraxial mesoderm just before segment border formation. Therefore these two genes are redundant in their domain structure and the expression pattern, indicating redundant function during embryogenesis. Unexpectedly, however, single knockout mice for Mesp1 or Mesp2 exhibit distinct phenotype (Fig. 1B-E). Mesp1-null mouse exhibits cardia bifida, due to the delayed migration of cardiac precursor cells, which is composed of descendants of *Mesp1*-expressing cells.[5] However, Mesp1-null mouse forms morphologically normal somites and shows normal rostro-caudal patterning within each somite, indicating that Mesp1 is not essential for somitogenesis. Mesp2-null mouse exhibits normal heart morphogenesis, instead it shows defective somitogenesis.[3] However, Mesp1 and Mesp2 double-null mouse exhibits very sever phenotype, featured by a complete lack of embryonic mesoderm,[6] indicating a redundant function of Mesp1 and Mesp2 for mesoderm formation (Fig. 1F).

Function of Mesp2 during Somitogenesis

Expression Pattern of Mesp2

The initial expression of *Mesp2* is observed in the nascent mesoderm as similar to that of *Mesp1*, however, the expression is weaker than that of *Mesp1* and disappears before 8.0 dpc. The second expression domain appears at 8.5 dpc as a single pair of bands in the paraxial mesoderm just lateral to the node. During somite formation, the expression is observed as sharp bands at S-1 or S-2 region in the PSM. Thus the transcriptional on-off is strictly regulated during somitogenesis. The expression continues until 13.5 dpc when somitogenesis ends. The temporal change of *Mesp2* expression pattern is revealed by explant cultures of bisected PSM fragments, with the half being fixed immediately and the other half fixed after cultivation for a certain times (Fig. 2A-F).[7]

Mesp2 expression appears as a broad single band occupying about single somite length and the caudal half expression is gradually downregulated leaving the rostral half band intact and finally the band disappears (Fig. 2G). The time required for a single cycle is perfectly matched with the time for single somite formation. This expression pattern is intimately involved in the function of Mesp2 during somitogenesis (see below).

Phenotype of the Mesp2 Knockout Mouse

The Mesp2-null embryo lacks initial segment border formation, which is a commonly observed phenotype among other segmentation mutants. However in addition, the embryo exhibits a distinct phenotype for the vertebral morphology. Mesp2-null fetus has vertebrae with completely fused pedicles of neural arches (Fig. 3A-B) and shows proximal fusion of ribs (Fig. 3C-D).[3,7,8] Since the pedicle of neural arch is derived from the caudal half of a somite, Mesp2-null embryo must have caudalized somites. Actually, molecular analysis revealed that this is a case. Caudal somite marker *Uncx4.1*, which is only expressed in the caudal half of somite, are expressed in entire somitic region of Mesp2-null embryo (Fig. 3E-F) and reversely rostral somite marker such as *Tbx18* is completely absent (Fig. 3G-H). Therefore, Mesp2 is required to have a rostral property of the somite. In contrast, a mutant that lacks a function of presenilin1, a Notch signal mediator involved in nuclear translocation of the Notch intracellular domain,[8,9] exhibit completely reverse phenotype, which lacks pedicle of neural arch and the somite has no expression of *Uncx4.1*. The finding of those mutant mice leads us to further investigate the mechanism to establish rostro-caudal polarity.

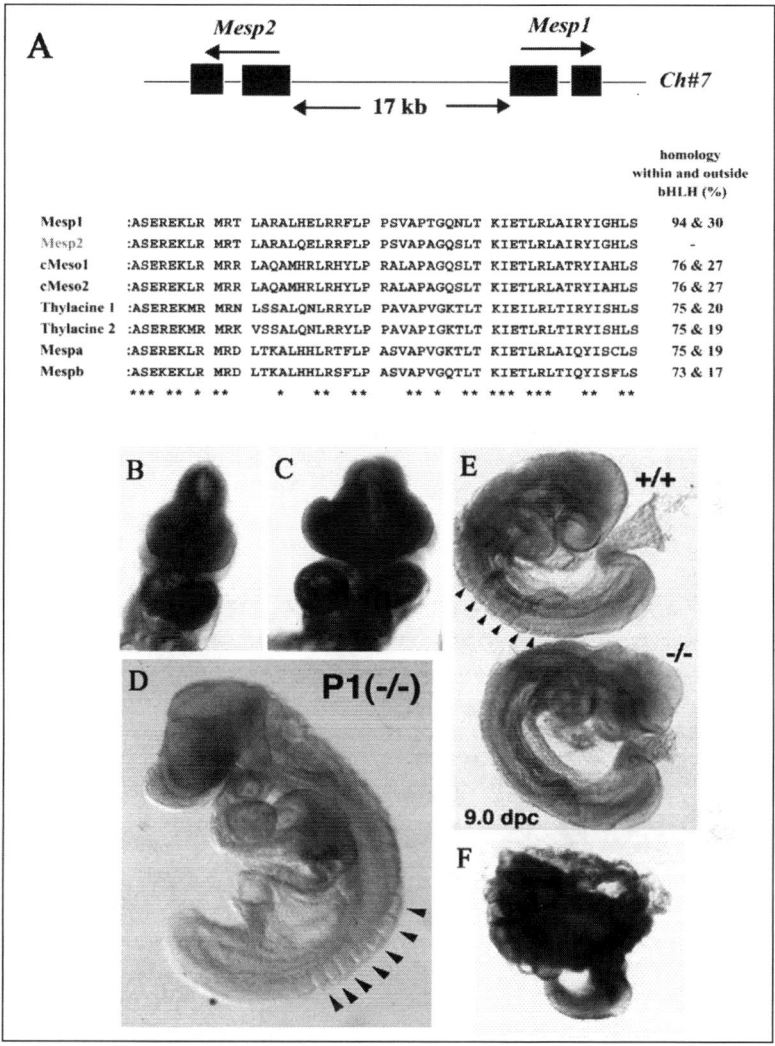

Figure 1. Genomic organization of the *Mesp1* and *Mesp2*, the comparison of amino acid sequences in the bHLH motives among vertebrates (A) and comparison of phenotypes among Mesp1-null (B-D), Mesp2-null (E) and Mesp,1 Mesp2-double-null (F) embryos. Mesp1-null embryo shows cardia bifida (C) but the somites are normally segmented (D, arrowheads). Mesp2-null embryo shows defective segmentation (E). Double-null embryo lacks embryonic mesoderm and the posterior structure (F). B-D) Reprinted from Mechanism of Development 75, Yumiko Saga, Genetic rescue of segmentation defect in Mesp2-deficient mice by *Mesp1* gene replacement, 53-66, ©1998, with permission from Elsevier.

Molecular Mechanism Leading to the Rostro-Caudal Polarity of Somite

Expression analyses of segmentation genes in normal and mutant embryos clearly show that the PSM can be divided into at least two distinct regions.[10] In the posterior PSM (P-PSM), a clock mechanism centered by Hes7 transcription factor and Notch signaling generates oscillating transcriptioal waves of certain genes.[11] In the anterior PSM (A-PSM), accumulating evidence indicates that there is a separate transcriptional regulation from that in P-PSM. Expression

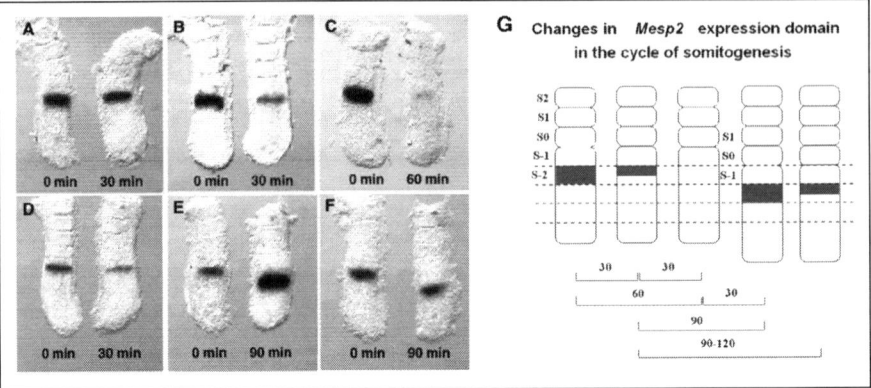

Figure 2. *Mesp2* expression pattern changes during somitogenesis. The results obtained by explant culture are shown (A-F) and the time course of the change in the expression pattern is schematically drawn (G).

pattern of several genes in the A-PSM is resembled to that of *Mesp2*; wider expression domain become narrower and finally localizes to either rostral or caudal somite primordia. The major event in the A-PSM is establishing rostro-caudal polarity of somites and it precedes segment border formation.

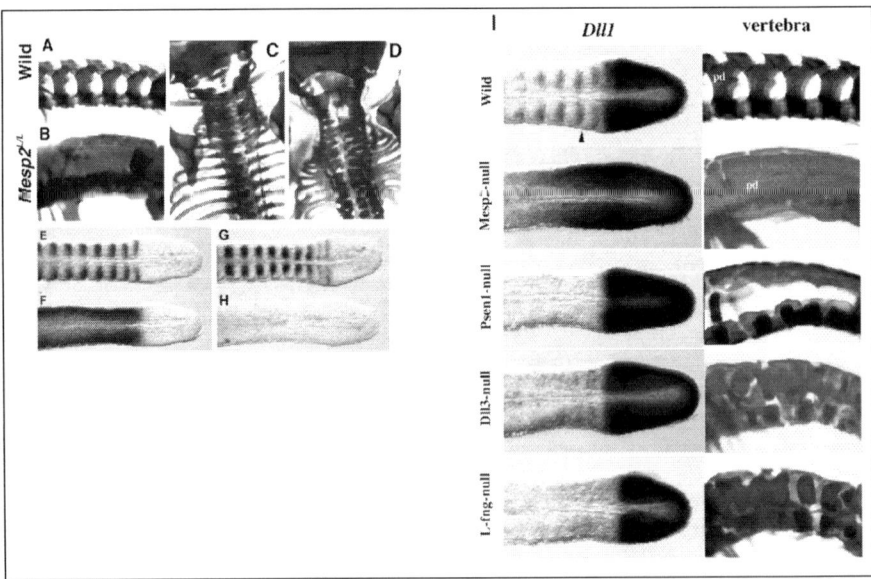

Figure 3. Skeletal morphology and gene expression of Mesp2-null (A-H) and other mutant embryos (I). The pedicle of neural arch in lumber vertebra (B) and the proximal part of the rib (D) in the Mesp2-null embryo is completely fused compared with clearly segmented vertebra (A) and rib (C) in the wild-type. The expression of a caudal marker gene *Uncx4.1* (E) is observed in the entire somitic region (F) and reversely that of a rostral marker gene *Tbx18* (G) is absent in the Mesp2-null embryo (H). I) The expression pattern of *Dll1* prefigures the vertebral morphology of wild-type and other mutant mice. Details are in the text. Pd: pedicle.

In mice, the expression pattern and the level of *Dll1* in the A-PSM appear to prefigure the segmental features of vertebrae. As shown in Figure 3I, *Dll1* expression in the A-PSM in the wild-type embryo is restricted to the caudal-half somite primordia and the caudal expression is maintained after somite border have formed. By contrast, in Psen1-null embryos, which show no caudal expression of *Dll1* in A-PSM with normal *Dll1* expression in P-PSM, the vertebrae are rostralized. In Mesp2-null embryos, in which *Dll1* expression in A-PSM is expanded the vertebrae are caudalized. In addition, in Dll3 or Lunatic Fringe (*L-fng*)-null embryos, randomized *Dll1* expression is observed. The resulting vertebrae show mixed and randomized patterns with respect to their rostrocaudal identity. Consistently, inactivation of *Uncx4.1* (paired-type homeobox transcription factor), which appears to be a downstream target of Dll1, results in the loss of the caudal component of vertebrae,[12,13] strengthening the idea that the expression of *Dll1* in A-PSM and not P-PSM determines rostro-caudal patterning of the vertebrae. In the absence of Mesp2, expression of L-fng, *Dll1* and *Dll3*, components of Notch signaling pathway, fail to be localized in either rostral or caudal region,[14] indicating that Mesp2 is required for the generation of normal patterning of these gene expression. However, the reverse is also true. In the absence of above genes, *Mesp2* expression domain fails to make a clear rostral band. Therefore, reciprocal regulation of Mesp2 and Notch signaling is important to generate rostro-caudal polarity.

Genetics Involved in the Rostro-Caudal Patterning

In mouse embryos, at least two Notch ligands, Dll1 and Dll3 are co-expressed in the PSM and their expression domains are finally segregated into the caudal and rostral halves of formed somites.[15,16] Therefore, either ligand must be involved in the two Notch signaling pathways implicated in the establishing rostro-caudal polarity. However, despite a large number of studies, possible functional differences between Dll1 and Dll3 signals are not clear. Thus, we conducted genetic studies of the roles in rostro-caudal patterning of Dll1- and Dll3-mediated Notch signaling, the relationships between Notch signaling and Mesp2 function and the involvement of Psen1 in Dll1- and Dll3-mediated Notch pathways. Our analysis of these genetic interactions revealed several novel findings.[14] Dll1- and Dll3-Notch signaling and Mesp2 constitute a complex signaling network for stripe formation in the anterior PSM. Feedback loops of *Dll1* and *Mesp2*, in which Dll1-Notch signal induces *Dll1* and *Mesp2* expression and Mesp2 suppresses *Dll1* expression, are essential for establishment of the rostro-caudal polarity, while *Dll3* is necessary for localization and integration of expression of *Dll1* and *Mesp2*.

An intriguing finding is obtained when we have analyzed an interaction between Dll3 and Psen1 on the vertebral morphology. Psen1-null embryo completely lacks *Dll1* and *Uncx4.1* expression and the vertebra is rostralized. However, if we remove Dll3 gene from Pse1-null mouse, we see the partial recovery of caudal component of the vertebra, indicating that Psen1-independent Dll3-Notch signaling might counteract Psen1-dependent Dll1-Notch signaling. Therefore, *Dll1* is activated by the Psen1-dependent Dll1-Notch signaling pathway and suppressed by the Psen1-independent Dll3-Notch pathway. However, this suppressive Dll3 pathway is not sufficiently active in the absence of Mesp2. Genetic analysis of Dll1 and Mesp2 or Dll3 and Mesp2 double-null embryos revealed that Mesp2 plays a major role in suppression of the caudal genes including *Uncx4.1,* more directly than Dll1 or Dll3. As shown in Figure 4, however, both Dll1 and Dll3 influence the expression of *Mesp2*. Collectively, these genes constitute a complex network and interactions among these genes might result in the simultaneous localization of Dll1, Dll3 and Mesp2.

In addition, Dll1-Notch signal is required for both rostral and caudal properties, since it induces *Dll1* itself and *Mesp2*. In contrast to Dll1, Dll3 upregulates *Mesp2* and suppresses *Dll1* and *Uncx4.1*, resulting in the suppression of caudal half properties. It should be noted, however, that the scheme in Figure 4 does not immediately represent signaling cascades within single cells, but instead represents results from complex intercellular interactions among mesodermal cells in A-PSM.

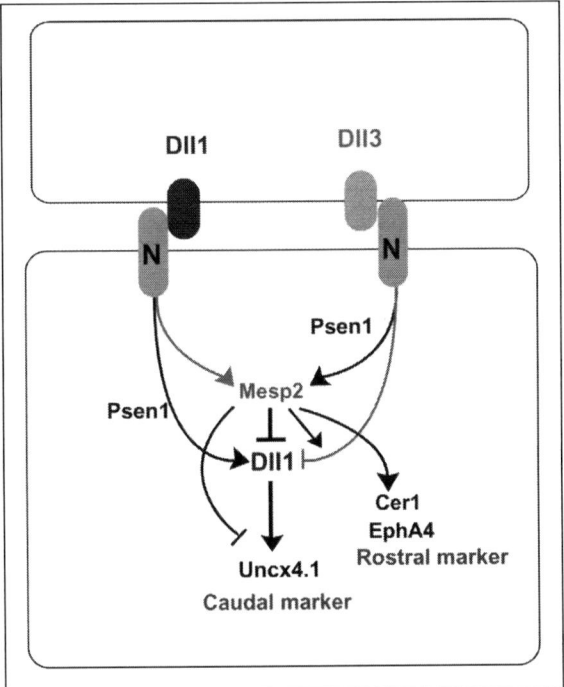

Figure 4. A summary of genetic interactions working in the anterior PSM and required for the establishing rostra-caudal patterning in the anterior PSM. Details are in the text. From: Takahashi Y et al. Development 2003; 130(18):4259-68.[14]

Mesp2 Is also Involved in the Segment Border Formation

The expression of Notch components, such as *Hes7* and *L-fng*, oscillates in the PSM, whereas *Notch1* is expressed throughout the entire PSM and is higher in the anterior PSM. We attempted to visualize the activation of Notch signaling by detecting a processed-NICD (Notch intracellular domain) using a specific antibody (Fig. 5A). Unlike its RNA expression pattern, the activity of Notch1 exhibits a dynamic pattern in the PSM during somitogenesis.[17] Intriguingly, in most of the embryos, Notch activation is in a similar phase to the transcript of *L-fng*, which encodes a glycosyltransferase that can modify Notch activity. In the Dll1-null embryo, no Notch1 activity was found anywhere in the PSM, whereas in the L-fng-null embryo, Notch1 activity was detected throughout the entire PSM but does not seem to oscillate. These observations indicate that the Dll1 ligand is required for Notch1 activation and that L-fng functions as a suppressor of Notch activity, thus generating the oscillation of Notch1 activation.

In the anterior PSM, the oscillation of Notch activity is arrested and localizes in the caudal portion of the S0 somite. By double-staining with anti-active-NICD and anti-Mesp2 antibodies, the spatial relationship between the localization of these two factors was revealed (Fig. 5B). At the initial phase of Mesp2 expression, the two domains partially overlap in such a way that cells in the posterior domain of Notch activation co-express Mesp2. The overlapping domain gradually reduces as Mesp2 expression levels increase. Ultimately, the two expression domains are completely separated from each other and form a clear boundary. Intriguingly, the strongest detectable signals of both Notch activity and Mesp2 are found in close proximity to the boundary on both sides.

The involvement of Mesp2 in boundary formation was further studied by *Mesp2-venus* knock-in mouse in which Mesp2 localization is detected in live embryos using confocal microscopy.[17] In the 8.75 dpc embryo, two to three segmental stripes are detectable, showing an anterior to posterior

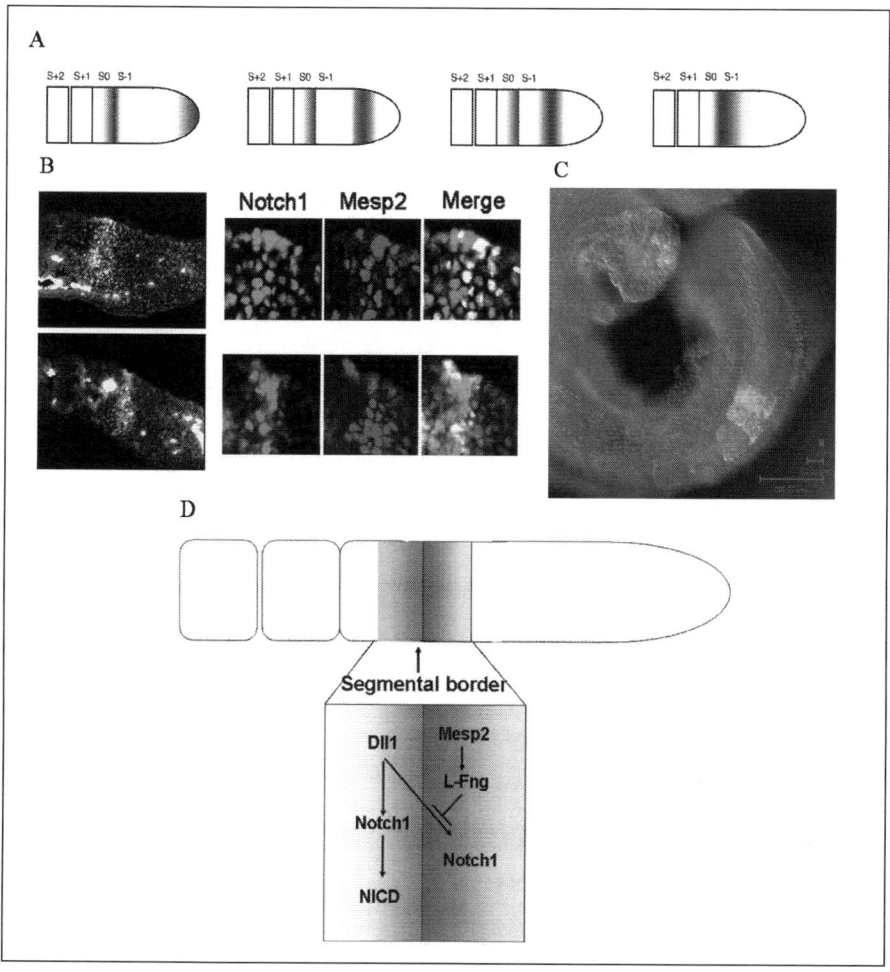

Figure 5. Notch activity during somitogenesis and regulation of Notch signal by Mesp2. A) Schematic representation of Notch1 activity during somitogenesis. Notch1 activity oscillates in the posterior PSM and stabilizes in the anterior PSM. B) The Notch activity in the anterior PSM is repressed by Mesp2. Notch1 activity detected by anti-active Notch.1 C) Visualized in vivo localization of Mesp2-venus fusion protein. D) A model for segmental border formation. Details are in the text. From: Morimoto M et al. Nature 2005; 435(7040):354-9.[17]

gradient for each segment (Fig. 5C). The signal in the most anterior segment is very weak but a second stripe has a clear anterior boundary that matches perfectly to the segmental border and in fact lines the entire border. The strongest third stripe also has a clear anterior border in the PSM in which no sign of a morphological boundary can be detected. These observations strongly suggest that Mesp2 expression demarcate the next segmental border. Thus Notch activity constitutes a core of the segmentation clock and that Mesp2 arrests the oscillation of Notch activity and generate the segmental border by suppressing Notch signaling. The suppression appears to be mediated by the function of L-fng (Fig. 5D). The mechanism to create the segmental segregation is still unknown, but it is noted that EphA4 appear to be a direct transcriptional target of Mesp2 and has been implicated in the cellular segregation.[18]

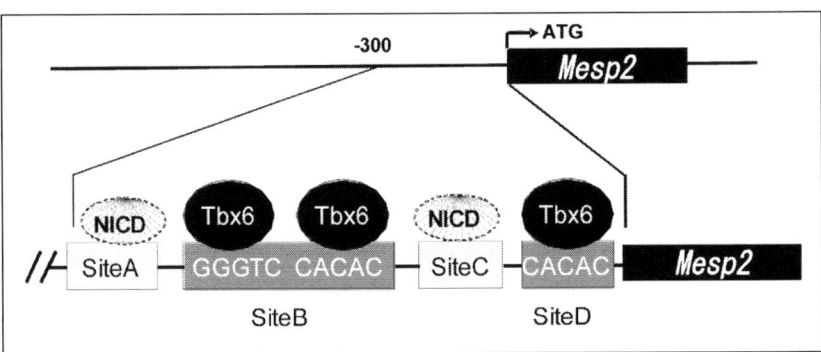

Figure 6. A proposed mechanism underlying the control of *Mesp2* expression. Tbx6 and NICD (ovals) interact with the conserved upstream sites in the *Mesp2* gene, site A to site D (represented by boxes). Tbx6 binds to Site B (two molecules) and site D (single molecule). Site A and site C interact with RBPJk to achieve a significant increase in *Mesp2* expression levels in the presence of Notch signals. This activation fully depends on the binding of Tbx6 to sites B or D. From: Yasuhiko Y et al. Proc Natl Acad Sci USA 2006; 103(10):3651-6.[19]

Regulation of Mesp2 Expression during Somitogenesis

The next crucial question is the mechanism by which Mesp2 is turned on in the anterior PSM. A transgenic approach has revealed that a 300bp portion of the 5'-adjoining sequence of the *Mesp2* ORF includes cis-elements that regulate PSM-specific *Mesp2* expression (Fig. 6). A T-box transcription factor, Tbx6 that binds to the cis-regulatory elements of the *Mesp2* gene was then identified by yeast one-hybrid screening. In somite-stage embryos, however, *Tbx6* is expressed throughout the PSM and also in the tailbud region, whereas *Mesp2* expression is restricted to the anterior PSM, just prior to somite formation and the expression overlaps only in the anterior limit of the *Tbx6* expression domain. This indicates that other unknown factor(s) participate in the pathways that restrict the *Mesp2* expression domain to the anterior PSM. By luciferase reporter assay it was shown that Notch ICD can activate *Mesp2* enhancer in the Tbx6-dependent manner. The current findings thus provide the first evidence that *Mesp2* is a direct target of Notch signaling and is regulated by a novel Notch signal mechanism which is based upon the binding of Tbx6 to transcriptional regulatory sequences.[19]

Functional Redundancy between Mesp1 and Mesp2

We already mentioned that these two genes function redundantly for early mesoderm formation. However, Mesp1, Mesp2-double-null mouse die early without any mesoderm formation, which hampers functional study of these genes during somitogenesis. Since Mesp1-null embryo has segmented somites with normal rostro-caudal polarity, Mesp2, but not Mesp1, is essential for somite formation and the rostro-caudal patterning of somites.[5] However, following methods has revealed functional redundancy between Mesp1 and Mesp2 in somitogenesis.

Gene Replacement

To directly ask whether Mesp1 can rescue the Mesp2 function, we introduced *Mesp1* cDNA into the *Mesp2* allele. The introduced *Mesp1* cDNA could rescue the defects caused by Mesp2 deficiency in a dosage-dependent manner (Fig. 7). Mesp2 [p1/p1] mice that lacked *Mesp2* expression but had four copies of the *Mesp1* gene survived into the adult hood and were fertile.

The skeletal defects and the reduction in expression *of Notch,1 Notch2* and *FGFR-1* observed in Mesp2-null mice were almost completely rescued by the introduced *Mesp1*. Whereas, Mesp2 [p1/-] mice that has three copies of *Mesp1* exhibits defects that are much more severe than in Mesp2 [p1/p1] mice but milder than in Mesp2-null mice. These observation strongly suggest that Mesp1 has similar function to that of Mesp2 with lower efficiency.

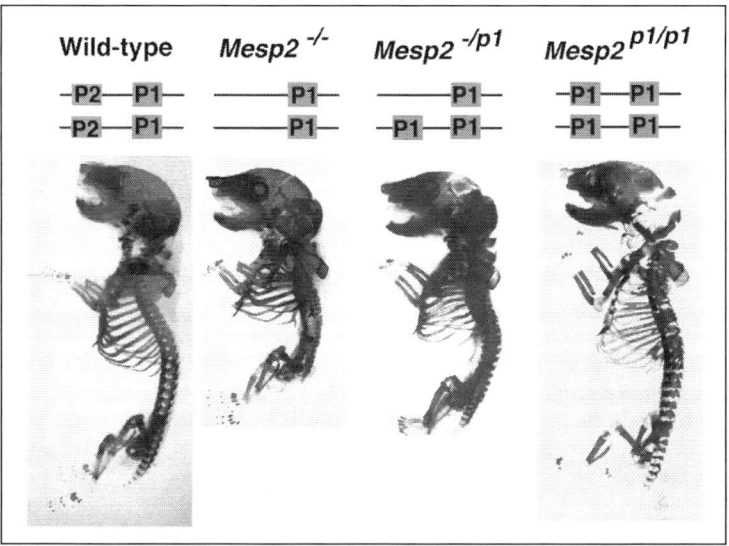

Figure 7. Mesp1 could rescue the function of Mesp2 by dosage dependent manner. *Mesp2* gene is replaced by *Mesp1* cDNA. Mesp2 deficiency is partially rescued by the three copies of *Mesp1* and almost completely rescued by four copies of *Mesp1*. Reprinted from Mechanism of Development 75, Yumiko Saga, Genetic rescue of segmentation defect in Mesp2-deficient mice by *Mesp1* gene replacement, 53-66, ©1998, with permission from Elsevier.

Chimera Study

In order to make much clearer the functional redundancy and the relative contributions of both Mesp1 and Mesp2 in somitogenesis and the cell autonomy of Mesp functions in each aspect of somitogenesis, we performed chimera analysis using either Mesp1/Mesp2 double–null cells or Mesp2-null cells.[20] Both Mesp1/Mesp2 double–null and Mesp2-null cells fail to form initial segment borders and exhibit completely caudalized properties, confirming that the contribution of Mesp1 is minor during both segment border formation and acquisition of rostral properties. In contrast, Mesp1/Mesp2 double–null cells contribute to neither epithelial somite nor dermo-myotome formation and Mesp2-null cells partially contribute to incomplete somitogenesis and dermomyotome development. This indicates that Mesp1 has a significant role in the epithelialization of somitic mesoderm. We also found that the roles of the Mesp genes in epithelialization and that the role of Mesp2 in establishing rostral properties is cell autonomous. However, we also found that epithelial somite formation with normal rostro-caudal patterning by wild-type cells was severely disrupted by non-cell autonomous effects of Mesp mutant cells, supporting that Mesp2 is responsible for the rostro-caudal patterning process itself in the anterior PSM.

Mesp Genes in the Other Vertebrates

As expected by an indispensable role of Mesp genes for somitogenesis, homologous genes are identified in all vertebrates examined, which includes chick (*c-Meso1* and *2*),[21] *Xenopus* (*Thylacine1* and *2*)[22] and zebrafish (*mespa* and *b*).[23] Those contain conserved bHLH domain but outside of the domain shows considerable variation just like between Mesp1 and Mesp2 (Fig. 1A). The expression patterns during somitogenesis are similar and the function also appears to be conserved among the vertebrates. The zebrafish mespb could rescue the function of Mesp2 when it was introduced in the *Mesp2* locus by gene knockin method.[24] Recent study in *Xenopus* embryo provided evidence that *Thylacine* might be directly regulated by retinoic acid in the A-PSM.[25] They also have shown that the concentration of retinoic acid is higher in the A-PSM and might be involved in the maturation of the paraxial mesoderm and could be antagonized with FGF signaling which is involved in the maintenance of immature state of paraxial mesoderm cells.[10] This is an interesting speculation to be evaluated in mice system.

Recently, *MESP2* mutations are found in human patients with Spondylocostal dysostosis (SCD).[26] The recessive mutation results in a formation of truncated protein with intact bHLH motif but lacking the C-terminal region, indicating the presence of a functional domain in the C-terminal region.

Perspective

A most important unsolved problem is where we should place Mesp2 in Notch signal cascade. We showed Mesp2 must be regulated by Notch signal and Mesp2 regulates expression of members in the Notch signaling pathway. The reciprocal regulation is most important to establish rostro-caudal polarity. So far *L-fng* and *EphA4* are shown as possible direct targets of Mesp2. However, microarray analysis reveals that many other genes in the presumptive rostral compartment are affected in Mesp2-null embryo. Interestingly, although some genes are activated by Mesp2, some other genes appear to be suppressed by Mesp2 function. The further study is required for understanding of Mesp2 function in the anterior PSM, which is critically important to clarify the molecular mechanism leading to events occurring somite segmentation.

Mesp family genes are exclusively expressed in the A-PSM, thus the function is restricted in the events occurring in the A-PSM. However, the function must be based on the events in the P-PSM. Most of other genes involved in somitogenesis (components of Notch signaling) are expressed in both anterior and posterior PSM and the function might be different in each step. We have used genetics to investigate the relationship between Mesp2 and Notch signal pathway. However, the phenotype obtained by the loss of function study is the combined results of each step. Therefore it is very difficult to evaluate the phenotype whether it reflect the event in anterior, posterior or both. There is few data indicating a direct link between events in anterior and posterior PSM. This would be a central question and important issue to be investigated in a next several years.

References

1. Howard K, Ingham. Regulatory interactions between the segmentation genes fushi tarazu, hairy and engrailed in the Drosophila blastoderm. Cell 1986; 44(6):949-57.
2. Palmeirim I, Henrique D, Ish-Horowicz D, Pourquié O. Avian hairy gene expression identifies a molecular clock linked to vertebrate segmentation and somitogenesis. Cell 1997; 91(5):639-48.
3. Saga Y, Hata N, Koseki H, Taketo MM. Mesp2: a novel mouse gene expressed in the presegmented mesoderm and essential for segmentation initiation. Genes Dev 1997; 11(14):1827-39.
4. Saga Y, Kitajima S, Miyagawa-Tomita S. Mesp1 expression is the earliest sign of cardiovascular development. Trends Cardiovasc Med 2000; 10(8):345-52.
5. Saga Y. Genetic rescue of segmentation defect in MesP2-deficient mice by MesP1 gene replacement. Mech Dev 1998; 75(1-2):53-66.
6. Kitajima S, Takagi A, Inoue T, Saga Y. MesP1 and MesP2 are essential for the development of cardiac mesoderm. Development 2000; 127(15):3215-26.
7. Takahashi Y, Koizumi K, Takagi A et al. Mesp2 initiates somite segmentation through the Notch signalling pathway. Nat Genet 2000; 25(4):390-6.
8. Struhl G, Greenwald I. Presenilin is required for activity and nuclear access of Notch in Drosophila. Nature 1999; 398(6727):522-5.
9. De Strooper B et al. A presenilin-1-dependent gamma-secretase-like protease mediates release of Notch intracellular domain [see comments]. Nature 1999; 398(6727):518-22.
10. Saga Y, Takeda H. The making of the somite: molecular events in vertebrate segmentation. Nat Rev Genet 2001; 2(11):835-45.
11. Bessho Y, Sakata R, Komatsu S et al. Dynamic expression and essential functions of Hes7 in somite segmentation. Genes Dev, 2001; 15(20):2642-7.
12. Leitges M, Neidhardt L, Haenig B et al. The paired homeobox gene Uncx4.1 specifies pedicles, transverse processes and proximal ribs of the vertebral column. Development 2000; 127(11):2259-67.

13. Mansouri A, Voss AK, Thomas T et al. Uncx4.1 is required for the formation of the pedicles and proximal ribs and acts upstream of Pax9. Development 2000; 127(11):2251-8.

14. Takahashi Y, Inoue T, Gossler A, Saga Y. Feedback loops comprising Dll1, Dll3 and Mesp2 and differential involvement of Psen1 are essential for rostrocaudal patterning of somites. Development 2003; 130(18):4259-68.

15. Hrabe de Angelis M, McIntyre J 2nd, Gossler A. Maintenance of somite borders in mice requires the Delta homologue DII1. Nature 1997; 386(6626):717-21.

16. Dunwoodie SL, Henrique D, Harrison SM, Beddington RS. Mouse Dll:3 a novel divergent Delta gene which may complement the function of other Delta homologues during early pattern formation in the mouse embryo. Development 1997; 124(16):3065-76.

17. Morimoto M, Takahashi Y, Endo M, Saga Y. The Mesp2 transcription factor establishes segmental borders by suppressing Notch activity. Nature 2005; 435(7040):354-9.

18. Nakajima Y, Morimoto M, Takahashi Y et al. Identification of Epha4 enhancer required for segmental expression and the regulation by Mesp2. Development 2006; 133(13):2517-25.

19. Yasuhiko Y, Haraguchi S, Kitajima S et al. Tbx6-mediated Notch signaling controls somite-specific Mesp2 expression. Proc Natl Acad Sci USA 2006; 103(10):3651-6.

20. Takahashi Y, Kitajima S, Inoue T et al. Differential contributions of Mesp1 and Mesp2 to the epithelialization and rostro-caudal patterning of somites. Development 2005; 132(4):787-96.

21. Buchberger A, Seidl K, Klein C et al. cMeso-1, a novel bHLH transcription factor, is involved in somite formation in chicken embryos. Dev Biol 1998; 199(2):201-15.

22. Sparrow DB, Jen WC, Kotecha S et al. Thylacine 1 is expressed segmentally within the paraxial mesoderm of the Xenopus embryo and interacts with the Notch pathway. Development 1998; 125(11):2041-51.

23. Sawada A, Fritz A, Jiang YJ et al. Zebrafish Mesp family genes, mesp-a and mesp-b are segmentally expressed in the presomitic mesoderm and Mesp-b confers the anterior identity to the developing somites. Development 2000; 127(8):1691-702.

24. Nomura-Kitabayashi A, Takahashi Y, Kitajima S et al. Hypomorphic Mesp allele distinguishes establishment of rostrocaudal polarity and segment border formation in somitogenesis. Development 2002; 129:2473-81.

25. Moreno TA, Kintner C. Regulation of segmental patterning by retinoic acid signaling during Xenopus somitogenesis. Dev Cell 2004; 6(2):205-18.

26. Whittock NV, Sparrow DB, Wouters MA et al. Mutated MESP2 causes spondylocostal dysostosis in humans. Am J Hum Genet 2004; 74(6):1249-54.

bHLH Proteins and Their Role in Somitogenesis

Miguel Maroto,* Tadahiro Iimura, J. Kim Dale and Yasumasa Bessho

Abstract

The most obvious manifestation of the existence of a segmented, or metameric, body plan in vertebrate embryos is seen during the formation of the somites. Somites are transient embryonic structures formed in a progressive manner from a nonsegmented mesoderm in a highly regulated process called somitogenesis. As development proceeds different compartments are formed within each somite and these progressively follow a variety of differentiation programs to form segmented organs, such as the different bones that make the axial skeleton, body skeletal muscles and part of the dermis. Transcription factors from the basic helix-loop-helix (bHLH) protein family have been described to be implicated in each of the processes involved in somite formation. bHLH proteins are a family of transcription factors characterized by the presence of a DNA binding domain and a dimerization motif that consists of a basic region adjacent to an amphipathic helix, a loop and a second amphipathic helix. In this chapter we will review a number of bHLH proteins known to play a role in somitogenesis.

Introduction

The first indication of a segmented body plan is seen during somitogenesis in the vertebrate embryo. This process is under tight temporal control and generates the somites, which are transient embryonic structures formed from the nonsegmented paraxial mesoderm in a progressive manner until a specific number of somites are formed along the length of the body axis.[1-4] The tissue precursor of the somites, the presomitic mesoderm (PSM), is organized in two bilateral rods of mesenchymal cells that flank the caudal neural tube, which mature gradually in an anterior to posterior direction. During the formation of somites the most rostral or most mature PSM cells bud off as an epithelial sphere of cells to form the new somite. The generation of somites occurs simultaneously with the recruitment of new progenitor mesenchymal cells from the primitive streak/tail bud as they become part of the caudal region of the PSM.[5,6] The pace of somitogenesis and the total number of somites formed are species-specific phenomena and these parameters vary widely between vertebrate species.[7]

Several theoretical models have been proposed to explain the periodicity and bilateral synchrony of the process of somite formation, the most popular being the "clock and wavefront" model[8] the Meinhardt model[9,10] and the cell cycle model.[11] These theoretical models are only able to partially account for different aspects of somitogenesis and thus, a more complete explanation for somitogenesis would require the global integration of each of them. One common aspect of the models is that they predict the existence of an oscillator or clock mechanism acting in the cells of the PSM prior to segmentation. The first molecular evidence for the existence of such a clock or oscillatory behaviour was provided by the discovery of the dynamic mRNA expression

*Corresponding Author: Miguel Maroto—College of Life Sciences University of Dundee, Dow Street, Dundee DD1 5EH, Scotland, UK. Email: m.maroto@dundee.ac.uk

Somitogenesis, edited by Miguel Maroto and Neil V. Whittock. ©2008 Landes Bioscience and Springer Science+Business Media.

of *c-hairy1* in the chick PSM.[12] After the initial description of the bHLH factor c-hairy1, other related bHLH proteins have been described to be implicated not only in the machinery of the segmentation clock but also in later aspects of somite formation.

The first critical role of bHLH proteins in development was demonstrated by experiments in which the expression of the myogenic bHLH protein, MyoD, was shown to convert nonmuscle cells into myocytes.[13,14] Subsequent studies have demonstrated the importance of this family of transcription factors in multiple aspects of development, cell growth and differentiation, apoptosis and hypoxic responses.[15] bHLH proteins are a family of transcription factors characterized by the presence of a DNA binding and dimerization motif that consists of a basic region adjacent to an amphipathic helix, a loop and a second amphipathic helix. While dimerization between bHLH proteins is mediated through the HLH domains, the basic regions immediately adjacent to the HLH domains contact DNA and have been shown to be critical for DNA binding.[16] After dimerization most bHLH proteins can bind to DNA sequences known as "E boxes" that contain the canonical motif CANNTG.[17-21]

The Hairy/E(Spl) Related Factors

In the past decade many bHLH-type transcription factors structurally related to *Drosophila* hairy have been identified in vertebrates.[22,23] These proteins are closely related to a second group of proteins that includes the bHLH factors of the *Drosophila* Enhancer of Split complex (E(Spl)) and deadpan. The hairy/E(Spl) factors constitute a group of repressors that are evolutionarily conserved in structure among different species. Their sequence includes a bHLH domain, an Orange domain and a conserved C-terminal tetrapeptide WRPW motif. One defining feature of this group of factors is the existence of a conserved proline (hairy subgroup) or glycine (hey subgroup) residue in the basic domain. The hairy/E(Spl) repressor factors can bind to the N-box (CACNAG) DNA sequence, as well as the E-box and they have the capability to repress transcription by several mechanisms. The first mechanism implicates an interaction between the hairy/E(Spl) proteins with the nonHLH corepressor Groucho, which binds to the C-terminal tetrapeptide motif of the hairy/E(Spl) protein.[24,25] This complex is then able to bind to the N- or E-box in the promoter region of the target genes. Although the function of the Orange domain is not clear, it might be involved in the transcriptional repression of hairy related factors.[26,27] In the second mechanism of repression the hairy/E(Spl) factors may dimerize with other bHLH activators and prevent them from activating transcription by forming nonfunctional heterodimers. Lastly since hairy related factors can bind to the E-box, which is the target sequence of bHLH activators, they might inhibit these activators by competing with their binding to the E-box.[22,28]

Palmeirim et al (1997) demonstrated that *c-hairy1* is expressed in the chick PSM in a cyclic manner with a periodicity that corresponds to that of somite formation.[12] *c-hairy1* is initially expressed in a broad domain in the posterior PSM and tail bud. The expression domain then moves towards the rostral PSM and finally becomes restricted to a sharp band in the prospective posterior compartment of the forming somite. This wave-like propagation of gene expression is not due to movement of *c-hairy1* expressing cells but is due to cyclic changes in gene expression in each PSM cell. Since ablation of the caudal PSM or all surrounding tissues does not affect the cyclic PSM expression of *c-hairy1*, it is unlikely that the signals from tail bud or surrounding tissues induce the expression of *c-hairy1*. Rather, each PSM cell seems to possess machinery that generates oscillatory gene expression.[12] This finding was the first molecular evidence of the segmentation clock, whose existence had been predicted by theoretical models.[29,30] A similar cyclic expression has since been reported for a number of *hairy/E(Spl)* related genes in the PSM of different vertebrate species. In addition to *c-hairy1*, *c-hairy2* and *c-Hey2* show cyclic expression in the chick PSM.[31,32] The murine *Hes1*, *Hes5*, *Hes7* and *Hey2* genes, the zebrafish *her1* and *her7* genes and the *Xenopus esr9* and *esr10* also oscillate in the PSM of the respective species.[31-39]

Cycling behaviour is not general for all *hairy/E(Spl)* related genes expressed in the vertebrate PSM. For example in zebrafish *her6* and *her4* are only expressed in two pairs of stripes in the most rostral PSM, which later will become the posterior compartment of the newly formed somite.[40] In

Xenopus hairy2 is expressed as a chevron-like stripe in the rostral PSM and *ESR-4* and *ESR-5* are expressed in the caudal PSM and the anterior compartments of prospective somites.[41] Interestingly, whereas *Hes6* and its ortholog *her13.2* are permanently expressed in the caudal PSM in *Xenopus* and zebrafish, respectively, the expression of *Hes6* is not detectable in the mouse PSM.[42,43] Among hairy related genes *Hes6* has a unique molecular property; it has a six amino acid deletion in the loop region compared to the other Hes family members. Owing to this deletion Hes6 can not act as a transcriptional repressor but rather inhibits Hes1-mediated repression by heterodimer formation in mouse.[42] Moreover, unlike the majority of Hes family members *Hes6* is not an immediate target of Notch signalling.[43] In zebrafish, *her13.2* functions downstream of FGF signalling and is required for the oscillation of *her1* and *her7*, as a binding partner for them.[44]

Most *hairy/E(Spl)* related genes expressed in the PSM act downstream of Notch signalling, a pathway that appears crucial for segmentation.[31,32,38,45-47] The only documented exception to this norm seems to be the murine *Hey2* gene whose expression is maintained in the caudal PSM of several Notch mutant embryos, such as *Delta-like-1*(or *Dll1*)(-/-) and *Notch1*(-/-), suggesting that its expression could be independent of Notch activity.[32] In addition to the group of *hairy/E(Spl)* related genes, a number of other genes have also been shown to display cyclic expression in the PSM and these are also members of the Notch pathway. This group includes the glycosyltransferase Lunatic fringe in chick and mouse.[48-50] While the mRNA for both the receptor Notch and its ligands Delta and Serrate are also expressed in different compartments of the PSM in chick and mouse their expression is not dynamic.[34,45,51] However, the zebrafish Notch ligand *DeltaC* gene does appear to cycle in the PSM while zebrafish *Lunatic fringe* (*Lfng*) is not cyclically expressed in the PSM.[36,52] Interestingly, Notch activity oscillates in the mouse PSM.[53] Moreover, Notch signalling coordinates synchronized oscillations of *her1* and *her7* between neighbouring cells in the zebrafish PSM.[52,54,55] Thus, Notch signalling plays a crucial role in regulating oscillatory gene expression, which in turn controls somitogenesis. To date, several genes, which are not members of Notch signalling, are described to be expressed with oscillatory patterns in the PSM. *Snail1* and *Snail2* show oscillatory expression in the mouse and chick PSM[56,57] and *Axin2* and *Nkd1* displays cyclic expression in the mouse PSM.[58]

Loss-of-function experiments in zebrafish and mouse have indicated that periodic Notch signalling and, in particular, oscillating expression of *hairy* related genes is required for oscillatory gene expression in the PSM and subsequent evenly spaced somite segmentation. In zebrafish inhibition of *her1* and *her7* function by genetic mutations or knockdowns with morpholino oligonucleotides eliminates both somite segmentation and oscillatory expression of *her1*, *her7* and *deltaC* in the caudal PSM.[35,38,59,60] In mouse *Hes1* and *Hes5* expression is lost in the *Dll1*(-/-) mutant mice and the expression of *Lunatic fringe* is impaired in the mouse mutants for *Dll1* and *RBP-Jκ*.[31,45] The *Hes7*(-/-) mutant mice have severe defects in somite segmentation and in cyclic expression of *Lfng*, which becomes uniformly expressed throughout the PSM.[33] Interestingly *Hes7* is similar to zebrafish *her1* in primary structure and a proline residue in the C-terminus following the WRPW motif is conserved in Hes7 and her7,[22,35,61,62] suggesting *Hes7* is both functionally and structurally conserved with both *her1* and *her7*.

The use of intron-derived probes have shown that the dynamic mRNA expression of cyclic genes in the PSM is due to periodic activation and repression of transcription rather than a result of posttranscriptional regulation such as periodic changes in the stability of mRNA.[35,36,63] Moreover, the hairy-related proteins bind to and repress their own promoters through a negative feedback mechanism. Likewise the murine *Hes1* gene encodes a bHLH transcriptional repressor which is able to bind and repress its own promoter.[64] This auto-repression by Hes1 is modulated by phosphorylation of the DNA-binding domain.[65] Promoter analyses of two cyclic genes, namely *Hes7* and *Lfng*, are coherent with this model, since they contain CBF1/RBP-J binding sites, implicated in Notch signalling activation and E-boxes, which are potential binding sites for the hairy/E(Spl) repressors.[61,63,66] Consistent with this idea is the fact that in transfected cells the Notch intracellular domain, which functions as a constitutively activated form of the receptor, activates the *Hes7* promoter, whereas cotransfection with *Hes7* abolishes this activation.[62] In the PSM, both *Hes7* and *Lfng* are exclusively transcribed in the region of the PSM where Hes7 protein is absent and stabilization of Hes7 by proteasome inhibitors enlarges the expression domain of Hes7 protein and decreases *Hes7* and *Lfng* transcription, which are also diffusely upregulated in the *Hes7*(-/-) mouse PSM.[62] These findings strongly support the model that Notch signalling activates both the *Hes7* and *Lfng* promoters and that Hes7 protein periodically represses both promoters and thus these data establish a negative feedback loop as a central mechanism driving oscillatory gene expression in the PSM.

Instability of the repressor seems to be crucial for the proper function of the segmentation clock. In fact the expression of Hes7 protein in mouse oscillates in a similar fashion to that of the mRNA (Fig. 1).[62] Hes1 and Hes7, whose half-lives are around 20 minutes, are ubiquitinated and rapidly degraded by the proteasome pathway.[62,67] This instability of *hairy* related proteins provides rapid flux at the functional level. Extending the half life of Hes7 from 22 min. to 30 min. produces a severe disorganization in both the expression of oscillatory genes and segmentation.[68] Quite surprisingly, in several cell lines serum-shock stimulation can elicit oscillations of *Hes1* expression by forming a negative feedback loop regulated by Hes1, suggesting that the machinery of the segmentation clock, at least part of it, seems to be present in a variety of cell types other than the PSM.[67,69]

Several theoretical models have been proposed to explain oscillatory gene expression based on a negative feedback loop of *hairy* related genes in both the zebrafish and mouse PSM and in cultured cells.[68,70,71] These models are, however, still not able to integrate the complexity observed in the PSM of different vertebrates. While the cyclic hairy/E(Spl) factors are clearly critical components of the machinery of the segmentation clock, the expression of *Snail/Slug* and *Lfng* in mouse and chick, *DeltaC* in zebrafish and *Axin2* in mouse is also cyclic[48,49,54,56,72] and interfering with their expression also severely affects the segmentation clock.[54,72] To date the models proposed have not taken their expression and function into consideration. Future models should aim to integrate all the information available concerning the different cyclic genes in specific species and these should then be tested and substantiated by biological experiments.

pMesogenin1 Family: pMesogenin1, Mespo and cMespo

pMesogenin1, also called Mespo in frog and zebrafish and cMespo in chicken, defines a novel subclass of bHLH proteins whose members play significant roles in paraxial mesoderm development.[73-77] *pMesogenin1* and *cMespo* were cloned in a two-hybrid interaction screening using the E47 bHLH domain as the protein interaction bait. These genes are expressed specifically in the caudal portion of the PSM from gastrulation until the stage when the tailbud loses its potential to provide paraxial mesoderm. The rostral boundary of the *pMesogenin1* expression domain is about two or three somite lengths distant from the posterior boundary of the most newly formed somite in mouse and chicken respectively. Gain-of-function experiments in *Xenopus* embryos show that ectopic pMesogenin1 can induce ectopic presomitic paraxial mesoderm, as evidenced by the upregulation of *ESR-4* and *ESR-5*; which are *hairy/E(Spl)* related genes that are normally coexpressed with *pMesoginl* in the unsegmented paraxial mesoderm of the frog.[41,74] Ectopic expression of pMesogenin1 also suppressed expression of axial mesodermal –notochord- markers and, at high doses, disrupted normal notochord development.[74] The *pMesogenin1* homozygous mutant embryos show a complete failure of somite formation and segmentation of the body trunk and tail. At the molecular level, the absence of pMesogenin1 disrupts the expression of components of the Notch signalling pathway, such as *Notch1*, *Notch2* and *Dll3*, including that of the cyclic genes *Lunatic fringe* and *Hes1*, indicating that pMesogenin1 is an essential upstream regulator of trunk paraxial mesoderm development and segmentation.[77]

The Mesp Family: Mesp, Meso and Thylacine

Timing of the determination of antero-posterior compartment identity has been addressed by experiments of in vitro culture of isolated PSM explants or by PSM inversion in the chick embryo.[78-81] These experiments demonstrated that somitic antero-posterior (A/P) determination takes place at the level of the rostral region of the PSM, in contrast to dorso-ventral or medio-lateral

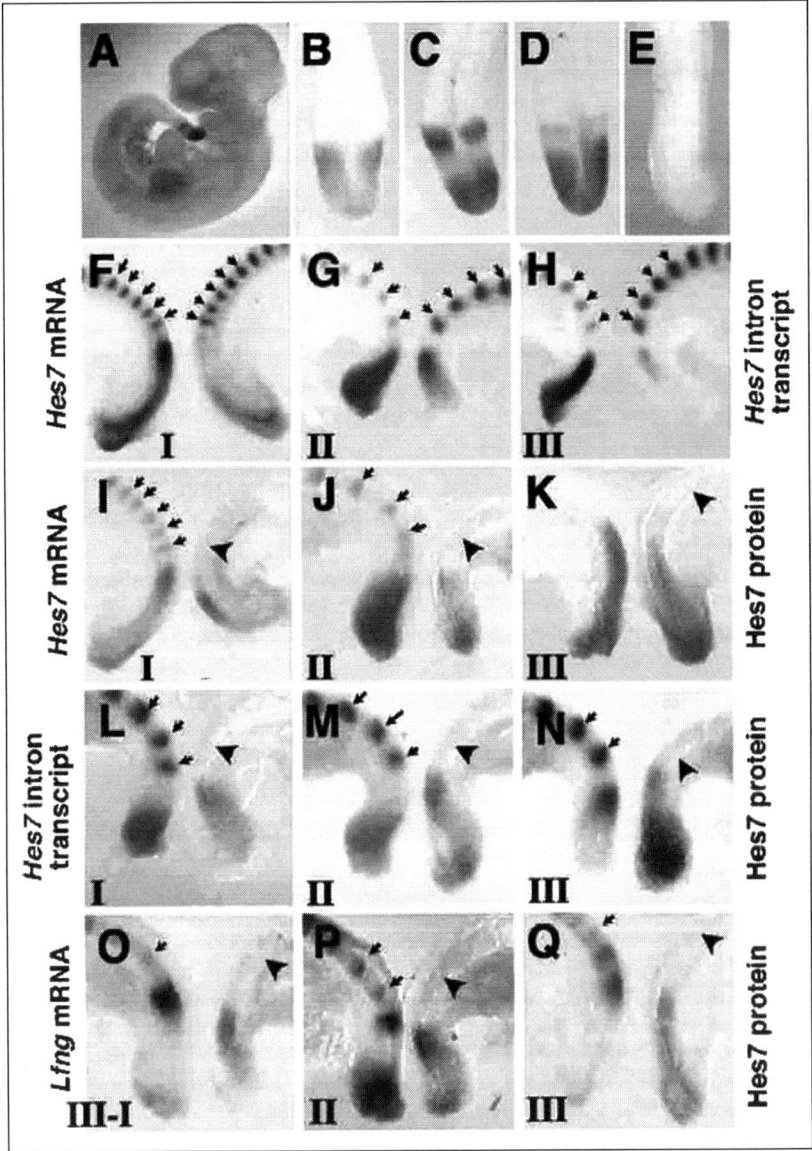

Figure 1. Spatial relationship of Hes7 protein, *Hes7* mRNA, *Hes7* nascent transcript and *Lfng* mRNA in the PSM. A–D) At E10.5, Hes7 immunoreactivity is specifically observed in the PSM of wild-type displaying various expression patterns. E) No signal is detected in *Hes7*-null mice. F–H) Comparison of the expression of *Hes7* mRNA (left) and *Hes7* nascent transcript detected by the *Hes7* intron probe (right) in the bisected caudal portions of E9.5 embryos. I–Q) The expression of Hes7 protein (right) was compared with the expression of *Hes7* mRNA (I–K, left), *Hes7* nascent transcript (L–N, left) *and Lfng* mRNA (O–Q, left). Hes7 protein-positive domains and the regions for *Hes7* nascent transcript and *Lfng* mRNA are mutually exclusive. The established somites are stained with the *Uncx4.1* probe for spatial alignment (arrows). The boundary between S_0 and S_1 is indicated by an arrowhead. Reprinted from Bessho et al (2003)[62] with permission from Cold Spring Harbor Laboratory Press.

somite compartmentalization, which occurs only after the somite has formed.[82] Several genes are expressed in a striped fashion in the rostral-most aspect of the PSM, suggesting the existence of a prepattern of segmentation prior to somite formation. This is the case with the Mesp family of proteins, which are implicated in this somitic A/P determination.

Mespl and Mesp2 are two murine bHLH transcription factors that share an almost identical bHLH motif but their amino acid compositions outside these domains are quite different. They are coexpressed in a striped pattern in the rostral PSM at the level of prospective somite S_{-1}, with a one-somite-wide space separation from the newly formed somite.[83,84] Their expression is dynamic and changes to become restricted to the prospective anterior half somite at S_0 and is finally down-regulated in the newly formed somite. The Mesp family of transcription factors is involved in both segment border formation and establishment of somitic A/P polarity. Genetic analysis of the *Mesp2*-null mutant embryos has led to a model in which Mesp2 is implicated in the specification of the anterior compartment of prospective somites. The *Mesp2* homozygous null mutant mice exhibit severe skeletal malformations attributable to abnormal segmentation. The mutant embryos produce epithelial somites at the cervical region, but the properties of A/P somitic compartments are altered as evidenced at the sclerotomal level by widespread expression of posterior sclerotomal markers, such as *Mox-1, Pax-1* and *Dll1,* throughout the somite.[84] In the prospective thoraco-lumbar region of the *Mesp2(-/-)* embryos, not only the A/P polarity of the sclerotome, but also the somitic boundary is disturbed or absent. These somitic defects are reminiscent of those described for some members of the Notch pathway. In the homozygous *Mesp2(-/-)* mice the expression of *Notch1* and *Notch2* in the PSM is drastically down-regulated, whereas *Dll1* is not severely affected. Moreover, in the *Dll1* and *RBP-J* homozygous null mutant mice the expression of *Mesp2* is drastically decreased,[45] indicating that Mesp2 could interact with the Notch pathway via a feedback loop. Interestingly the lack of expression of *FGFR-1* in the rostral portion of the PSM and the newly formed somite in these *Mesp2(-/-)* embryos suggests that Mesp2 could also interact with the FGF signalling pathway. Based on their results Takahashi et al (2003) proposed a model for A/P patterning involving Mesp2, in which two different Notch pathways are activated in the A/P somitic compartments with mutually exclusive effects upon compartment specification.[85] One is the classic presenilin-dependent Notch pathway responsible for the induction of *Dll1* expression in the posterior half of the somite, whereupon Dll1 acts to promote posterior character. The other pathway is a presenilin1-independent Notch pathway responsible for suppressing *Dll1* expression in the anterior half of the somite. Localized expression of Mesp2 participates in this control; when Mesp2 expression in the rostral PSM becomes restricted to the prospective anterior half somite, it suppresses the *Dll1*-inducing or presenilin-dependent Notch pathway and potentiates the anterior compartment character; and presenilin-independent *Dll3*-inducing Notch pathway.[85,86]

Mesp2 also plays an essential role in somite border formation in mouse.[87] Somite boundaries form at the interface between Notch1-active and -negative domains, as demonstrated by immunostaining with an antibody against active-NICD. This antibody revealed Notch activity oscillates throughout the whole PSM in a narrow domain which is initiated in the caudal PSM and sweeps from the tailbud to the border between S_0 and S_{-1} which is a domain that demarcates the future somite boundary. After reaching this level, the stripe of NICD activity becomes restricted to the posterior half of S_0. This stripe of Notch1 activity is in a domain which is mutually exclusive from the *Mesp2* and *Lnfg* expression domains in the anterior half of S_{-1}. Genetic analysis indicates that the Notch1 positive/negative interface is generated by suppression of *Notch1* expression by Mesp2 through induction of *Lnfg*, a suppressor of Notch activity. Consistent with this observation, biochemical analysis shows that Mesp2 could potentially activate *lunatic fringe* transcription by directly binding to its promoter. Mesp2 also directly binds to the enhancer sequence of *EphA4* which is involved in somite epithelialization and whose expression resembles that of *Mesp2* and is absent in *Mesp2(-/-)* embryos.[88] Taken together, these data imply Mesp2 induces the formation of segmental borders by generating genetic expression boundaries and activating genes involved in the regulation of somite epithelialization. Although the expression of *Mesp1* appears to be normal in the *Mesp2(-/-)* mice, budding of the epithelial somites is not observed. This observation is consistent with the analysis of the *Mesp1* null mutant mice. These mice, though embryonic lethal, do not exhibit a strong segmentation defect in contrast to the *Mesp2* mutants.[89] Expression of the genes of the Notch pathway is not affected in the *Mesp1(-/-)* mice. Interestingly when the *Mesp1* gene is introduced into the *Mesp2* locus by a knock-in strategy, it rescues the *Mesp2(-/-)* segmentation defects in a dose dependent fashion. Thus, two copies of the ectopic *Mesp1* gene are able to restore the skeletal defects and the expression of *Notch1, Notch2* and *FGFR-1* in the PSM to almost normal levels. Even if the two proteins perform some different roles during development, these results suggest that Mesp1 and Mesp2 may also have some redundant functions that play an important role during somitogenesis via interaction with and regulation of Notch and FGF signalling. Such a dose dependent rescue event in the Mesp2 deficiency is also observed when zebrafish *mespb* is introduced into the mouse *Mesp2* locus by homologous recombination.[90] An exquisitely coordinated regulation of *Mesp1* and *Mesp2* maintains a specific level of Mesp activity that is required for normal somitogenesis and axial skeletogenesis along the body axis,[91] This was demonstrated by a comparative study of the A/P level of vertebrae affected along the body axis and the expression level of both *Mesp* genes in various combinations of the double-null allele of *Mesp* genes and the transgenic allele in which a minimum functional *Mesp2* is knocked-in. Chimera analysis using *Mesp2(-/-) cells* and the *Mesp1(-/-)Mesp2(-/-)* double-null cells confirms a minor contribution of *Mesp1* to the initial segmentation process and the AP patterning of the somites, however the data highlighted a rather significant role for *Mesp1* in epithelial somite and dermomyotome formation.[92] The relevance of cellular interactions in the AP patterning of the somites was also observed since the epithelialization and AP patterning of wild-type cell-derived somites was disrupted by the presence of mutant cells.

It is very important to understand the mechanisms underlying the transcriptional regulation of *Mesp2* since this gene governs many critical processes of somite formation as discussed above. Transgenic and biochemical approaches to decipher the upstream sequence of the mouse *Mesp2* gene have uncovered an essential interaction of periodic Notch signalling and direct binding of Tbx6 to the *Mesp2* gene, providing the restricted expression of Mesp2 in the anterior PSM.[93] Transgenic fish analyses characterized an enhancer of medaka fish *mesp-b*, an orthologue of mouse *Mesp2* and revealed a conserved regulatory mechanism.[94]

The chick *cMeso1* and *cMeso2* genes encode bHLH transcription factors closely related to the mouse Mesp proteins.[95,96] *cMeso1* and *cMeso2* show an overlapping striped expression pattern in the rostral PSM, which is dynamic and changes during the somite formation cycle. Initially their expression expands from the posterior region of prospective somite S_{-I} and the whole of S_{-II}. Later on this expression becomes restricted to the posterior half of the somite primordia. Low levels of *cMeso* transcripts remain in the posterior half somite. These results seem quite surprising, as *Mesp*-related transcripts in other organisms are restricted to the anterior region of the prospective somite. Loss of function of *cMeso1* brought about by antisense RNA or antisense oligonucleotide treatment produces attenuation of somitogenesis. cMeso2 exerts transcriptional inhibition when coexpressed with cMeso1, suggesting that cMeso1 and cMeso2 may function in an antagonistic manner during somitogenesis.[95,96]

A third factor related to the Mesp family called Thylacine1 has been described in *Xenopus*. *Thylacine1* is expressed only in the anterior half of somite primordia in the rostral PSM.[97] In vitro assays have shown that Thylacine1 is a potent transcriptional activator and its ectopic expression by RNA injection causes segmentation defects and perturbs the normal segmental expression of x-*Delta2* and *ESR-5*. Interestingly, alteration of the Notch signalling pathway by injection of x-*Su(H)* also affected *Thylacine1* expression, suggesting the existence of a feedback loop, similar to the situation described for Mesp2 and the Notch pathway in mouse.[97] It has been shown that in the Xenopus embryo, retinoic acid activates *Thylacine1* expression in two different ways in the rostral and caudal PSM. Treatment with cyclohexamide showed that in the rostral PSM retinoic acid activates *Thylacine* expression directly, whereas in the caudal PSM the activation is indirect via induction of MKP3, an inhibitor of the FGF signalling pathway.[98]

Paraxis

A number of experiments have demonstrated that border formation is not an essential requisite for metamerism and, therefore, segmentation and determination of A/P compartment identity are processes that can be uncoupled from somitic border formation.[78,99]

Paraxis was identified both by interaction cloning with bHLH proteins and by homologous cloning with the bHLH sequence of scleraxis.[100-102] Both proteins, paraxis and scleraxis, are nearly identical in the bHLH region but diverge in their amino and carboxyl termini. *Paraxis* is expressed in the rostral PSM and newly formed somites and as the somite matures its expression becomes restricted to the dermomyotome.[101] *Paraxis* homologues have also been identified in chicken, zebrafish and xenopus.[103-106] The importance of paraxis during the process of epithelialization was revealed through the analysis of the *paraxis(-/-)* homozygous null mouse.[101] The mutant mice embryos fail to form epithelial somites and exhibit an improperly segmented pattern of paraxial mesoderm derivatives, such as axial skeleton, skeletal muscle, or dermis. The critical implication of paraxis in the somite epithelialization process is corroborated by the fact that chicken embryos treated with antisense oligonucleotides to *paraxis* also fail to produce somites.[107] Palmeirim et al (1998) showed that isolated PSM from the chicken embryo is able to maintain *paraxis* expression after 4 h of incubation, although the explants do not form somite borders, suggesting that paraxis may be a necessary but not a sufficient component in the generation of somitic boundaries.[81] *Paraxis* function in somitic epithelialization has been further investigated in chick embryo.[108] Cells misexpressing *Paraxis* by means of electroporation preferentially colonize the epithelial somite, suggesting a role for this gene in directing cells to adopt an epithelial character and/or location. However, coelectroporation with dominant-negative *Rac1,* a Rho GTPase, abolishes this function of *paraxis.* Neither dominant-negative *Rac1* nor constitutively-active *Rac1* changed the expression of *Paraxis,* suggesting a posttranscriptional interaction between these two molecules.[108] Interestingly, *paraxis* expression seems to be affected when Eph or Notch signalling pathways are altered and thus, disruption of Eph signalling in zebrafish affects the normal downregulation of zebrafish *paraxis* (*par1*) in the anterior domain of the somites.[109] In principle, it is possible that this downregulation *of par1* is a direct effect of Eph signalling. Moreover, the expression *of paraxis* is reduced in *Dll1 (-/-)* mice,[110,111] raising the possibility that normal *paraxis* expression is also a target of the Notch signalling pathway. Neighbouring signals from the ectoderm and the neural tube maintain the *paraxis* expression in the paraxial mesoderm.[103] Wnt signalling constitutes one of the pathways regulating *paraxis* expression.[112] Surface ectoderm overlaying the somites maintains the epithelial structure of the dermomyotome. Wnt6 is the most likely candidate molecule responsible for this process by activating *paraxis* via β-Catenin-dependent signalling. The precise means by which Notch, Wnt and Rho GTPases interact to govern *paraxis* regulation remains to be studied.

Paraxis is also implicated in the maintaining of A/P somitic polarity.[111] As described above, specification of somitic A/P polarity is established at the rostral end of the PSM in response to both Mesp2 and Notch signalling. The *paraxis (-/-)* mice display defects in the axial skeleton and peripheral nerves that are consistent with a failure of somitic A/P patterning. Expression of *Mesp2* and genes in the Notch pathway including dynamic expression of *Lunatic fringe* are not altered in the PSM of these mutant mice. However, *Dll1* and *ephrin-B2,* which are normally restricted to the posterior half of somites, are present in a diffuse pattern in the homozygous null embryos, indicating a role for paraxis in maintaining somite polarity that is independent of cyclic Notch activity in the PSM.[111] Taken together with the model proposed by Takahashi et al (2003)[85] these results raise the possibility that paraxis is involved in the mechanism of restricting the *Dll1*-inducing, presenilin-dependent Notch pathway to the posterior somite half.

Myogenic Regulatory Factors (MRFs)

The skeletal muscle of the vertebrate body is derived from the somites. During the process of somitogenesis the newly formed somites rapidly mature and differentiate dorsally into the columnar epithelial cells of the dermomyotome. Myogenesis is initiated by delamination of cells from the lips of this structure.[113,114] Thus, cells from the medial part of the dermomyotome involutes beneath the dorsomedial dermomyotome to form the epaxial myotome. These cells rapidly become postmitotic and constitute the first differentiated skeletal muscle in the vertebrate embryo. The dorsal myotome later contributes to the back muscles. Cells derived from the lateral dermomyotome migrate to the limb bud to make the limb muscles, or involutes beneath the ventrolateral dermomyotome lip to form the hypaxial myotome, the precursors of the abdominal and intercostal muscles.[3,115,116]

The myogenic regulatory factors (MRFs) are a family of bHLH transcription factors that regulate muscle development. There is extensive data concerning the expression of *Myf5, MyoD, Myogenin* and *MRF4* and the transcriptional and signalling mechanisms that control their expression during the specification and differentiation of muscle progenitors. Most of these data were obtained using in situ hybridization techniques and, unfortunately, much less is known about MRF protein expression. Anatomical differences between vertebrate species may explain the diverse temporal and spatial pattern of expression of the myogenic factors early in development. In zebrafish, for example, both myogenic genes, *Myf-5* and *MyoD*, can be detected very early in development in adaxial cells, which are slow muscle progenitors, near the end of gastrulation.[116-118] Only *Myf-5* is expressed in more lateral cells of the PSM. In *Xenopus*, prior to gastrulation it is possible to detect maternal *MyoD* transcripts in the egg.[119] During early gastrulation *MyoD* is expressed in mesodermal progenitors before the initiation of somite formation. Single-cell transplantation studies have shown that cells from the myogenic regions of late gastrula embryos, when placed at ectopic ventral sites, can differentiate cell autonomously into muscle.[120] Analysis of gene expression by in situ hybridization techniques has shown that in both mouse and chick *MyoD* expression is delayed with respect to that of *Myf-5*. In the mouse embryo *Myf-5* can be detected in the PSM by RT-PCR[121] and from the onset of somitogenesis at day 8 in the dorsomedial aspect of every somite.[122] *MyoD*, on the other hand, follows *Myf-5* expression and is first detected at day 10.5.[123] In the chick embryo *Myf-5* is also expressed before somitogenesis. By in situ hybridization and RT-PCR *Myf-5* has been detected in the primitive streak at the gastrula stage.[124,125] From stage 7, *Myf-5* can be observed in the rostral PSM and the somites, starting from the first formed somite. The domain of expression shifts in somites I-IV from a posterior to a dorsomedial domain. Based on their results, Kiefer and Hauschka (2001) proposed a model whereby *Myf-5* expression in the PSM occurs in a periodic pattern that is coordinated with somitogenesis (Fig. 2).[124] In fact *Myf-5* expression in the rostral PSM resembles that of other dynamic genes, such as *mHes-5, mMesp1, mMesp2, cMeso1* and *cMeso2*.[45,83,84,95,96] The initially broad band of expression in the rostral PSM becomes restricted to the region in which the border is forming, suggesting that *Myf-5* expression at this time might be also implicated in the process of somite formation itself. Unlike *Myf-5, MyoD* is never expressed in the chick or mouse PSM and the expression in the somites is always restricted to the medial domain, suggesting that its only function is related to myogenesis.

All four of the MRFs are capable of activating transcription of muscle-specific genes and myogenic differentiation in some nonmuscle cell lines.[126,127] Genetic studies of MRFs in the mouse embryo have provided significant insights into their regulatory functions. Initial data suggested that MyoD and Myf-5 were redundant, since single-knockout mice did not show any muscle phenotype.[128,129] However the lack of myoblasts in the double homozygous null mice *MyoD(-/-)/Myf-5(-/-)* clearly indicated that MyoD and Myf-5 specifically regulate early myogenic determination.[130] A close examination of these mutants showed that the functions of MyoD and Myf-5 in myogenic determination are only partially redundant; each controls the early specification of distinct muscle cell subpopulations.[131,132] Consistent with this is the fact that the onset of their expression is seen in different myogenic compartments at different times in vertebrate development.[124,133] At least in the mouse the complex spatio-temporal expression pattern of *Myf-5* depends on many discrete regulatory elements that are dispersed over long distances throughout the *MRF4-Myf-5* locus.[134-138] Different enhancers distributed over a 90-kb region around the locus control lineage-specific transcription of *Myf-5* in the epaxial, hypaxial and limb and head muscle progenitors.

The precise mechanisms by which myogenesis is initiated and is then maintained in the somite has been described in length, but it is still highly controversial. It is widely accepted that the neural tube, surface ectoderm, lateral plate, notochord and floor plate are involved to some extent in these processes.[3,115] In particular, there are data which suggest that the neural tube, mediated by Wnt

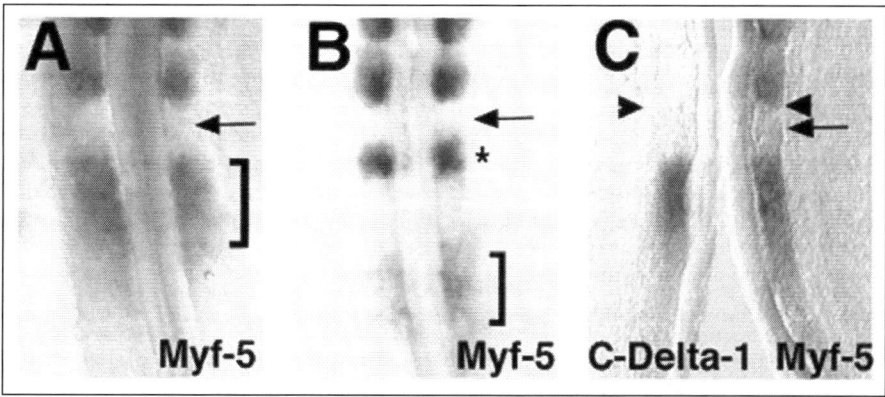

Figure 2. Dynamic *Myf-5* expression in the chick rostral PSM. A) At stage 13 *Myf-5* is expressed in the first somite but is absent in the most rostral end of the PSM, S$_O$, in an approximately somite-length segment (arrow). Caudal to the *Myf-5* negative domain, *Myf-5* is expressed broadly in a decreasing gradient toward the caudal PSM (bracket). B) In a second embryo the expression is slightly different. In this embryo there are two bands of *Myf-5* expression in the PSM. The first and most intense is a nearly somite-length domain located at the level of S$_{-1}$ (asterisk). The second segment of expression is less intense, more caudal and in a decreasing gradient (bracket). C) The expression of *Myf-5* in the chick PSM is similar to the strong domain of *c-Delta-1*. Embryo was bisected longitudinally and the halves hybridized with either *c-Delta-1* or *Myf-5* probes. Arrowheads point to the border between the rostral PSM and somite S$_I$. Arrow points to the negative *Myf-5*-negative segment of the most rostral PSM. Rostral is up. Reprinted from Kiefer and Hauschka (2001)[124] with permission from Elsevier.

proteins, and the notochord, mediated by Shh, act synergistically to promote somitic myogenesis. It appears that Wnt proteins and Shh, when individually applied to naïve mesoderm, are not sufficient to fully promote muscle development. However the situation is completely different when they are activated simultaneously and thus, when PSM is cocultured in the presence of both Shh and Wnt proteins, *Myf-5* and *MyoD* expression are robustly up-regulated.[139-142]

bHLH Factors Involved in the Control of Myogenesis

Much is known about mechanisms of control and inhibition of myogenesis in tissue culture. Growth factors and oncogenes inhibit myogenesis by preventing the cells from leaving the cell cycle and embarking on a differentiation program.[143] In other cases MRF proteins are direct targets for functional inhibition, e.g., by phosphorylation of the bHLH domain,[144] or by protein-protein interaction with factors not belonging to the bHLH family, such as jun.[145] In addition, proteins required for proper function of myogenic bHLH factors like the E-proteins can be targets for inhibition. This is the case with Id, an HLH protein lacking a basic region, but which is able to interact with the ubiquitous E-proteins, thereby preventing their interaction with myogenic factors.[146,147] Unfortunately our knowledge is not so complete concerning the mechanisms implicated in the spatial and temporal regulation of myogenesis in vivo during vertebrate development.

Wnt and Shh signals are permanently present along the whole length of the antero-posterior body axis, however, myogenesis is activated only at a specific moment of development starting in the most mature somites. This delay in the induction of the myogenic program can be explained by the action of different mechanisms of control. One of these mechanisms of control might involve the action of Twist, a bHLH transcription factor able to inhibit myogenic differentiation by interfering with the activity of the myogenic factors. At least in mouse *Twist* has been described to be expressed in the PSM before somite formation.[148,149] Several independent mechanisms have been proposed to explain the inhibitory effects of Twist on the muscle-specific transcriptional program. Twist, like Id, has the ability to repress MyoD function by titrating away E proteins, thus preventing the formation of functional MyoD-E protein heterodimers.[150] Twist is also able to inhibit muscle-specific gene activation by interfering with MEF2-mediated transactivation. A third mechanism of inhibition proposes direct interaction between Twist and the different MRFs, thereby making a transcriptionally inactive heterodimer.[151] However, while disruption of the *Twist* gene results in substantial neural tube defects as well as somitic abnormalities, there is no clear evidence of augmented or ectopic myogenesis, which one would predict if a suppressor function had been abrogated.[152] As in the case of MyoD and Myf-5, this lack of myogenic phenotype could be just due to functional redundancy.

Conclusions

The basic helix-loop-helix (bHLH) transcriptional regulatory proteins are key players in a wide array of developmental processes. As we have outlined in this review, several bHLH proteins play crucial roles in the context of vertebrate somitogenesis, at all stages of the process, thereby orchestrating a correctly spaced and patterned segmented body plan. The description of the cyclic expression of a number of hairy/E(Spl) bHLH factors in the PSM and their role as key components of the machinery of the segmentation clock is among the most important discoveries made in the last few years in the field of developmental biology. Subsequent studies described the implication of a number of other bHLH factors in later steps of somitogenesis. Thus, pMesogenin factors seem to be essential upstream regulators of trunk paraxial mesoderm development and segmentation, Mesp factors are implicated in the determination of somitic A/P compartmentalization, paraxis is important for both the maintaining of A/P polarity and the process of somitic border formation and finally, MRFs are critical regulators of myogenesis, a process that is controlled by, among others, a set of bHLH factors that includes Twist. The majority of studies performed to date have focused on loss of function phenotypes and in situ analyses of these bHLH factors to ascertain the knowledge we have of their function and regulation. Obviously additional knowledge at the protein level will provide an even deeper insight into the finer details of their roles in somitogenesis. It seems quite evident that the machinery that drives the segmentation clock and the different processes implicated in somitogenesis, such as metamerism, A/P determination and somitic border formation, are closely coordinated. This is an exciting open field and future studies can be expected to reveal additional insights into how this coordination works at the molecular level.

References

1. Gossler A, Hrabe de Angelis M. Somitogenesis. Curr Top Dev Biol 1998; 38:225-287.
2. Rawls A, Wilson-Rawls J, Olson EN. Genetic regulation of somite formation. Curr Top Dev Biol 2000; 47:131-154.
3. Hirsinger E, Jouve C, Dubrulle J et al. Somite formation and patterning. Int Rev Cytol 2000; 198:1-65.
4. Maroto M, Pourquie O. A molecular clock involved in somite segmentation. Curr Top Dev Biol 2001; 51:221-248.
5. Catala M, Teillet MA, Le Douarin NM. Organization and development of the tail bud analyzed with the quail-chick chimaera system. Mech Dev 1995; 51:51-65.
6. Psychoyos D, Stern CD. Fates and migratory routes of primitive streak cells in the chick embryo. Development 1996; 122:1523-1534.
7. Richardson MK, Allen SP, Wright GM et al. Somite number and vertebrate evolution. Development 1998; 125:151-160.
8. Cooke J, Zeeman EC. A clock and wavefront model for control of the number of repeated structures during animal morphogenesis. J Theor Biol 1976; 58:455-476.
9. Meinhardt H. Models of Biological Pattern Formation. London: Academic Press, 1982.
10. Meinhardt H. Models of segmentation. In: Bellairs R, Ede DA, Lash JW, eds. Somites in Developing Embryos. Plenum, New York: NATO ASI Series 1986; 118:179-189.
11. Primmett DR, Norris WE, Carlson GJ et al. Periodic segmental anomalies induced by heat shock in the chick embryo are associated with the cell cycle. Development 1989; 105:119-130.
12. Palmeirim I, Henrique D, Ish-Horowicz D et al. Avian hairy gene expression identifies a molecular clock linked to vertebrate segmentation and somitogenesis. Cell 1997; 91:639-648.
13. Lassar AB, Paterson BM, Weintraub H. Transfection of a DNA locus that mediates the conversion of 10T1/2 fibroblasts to myoblasts. Cell 1986; 47:649-656.

14. Davis RL, Weintraub H, Lassar AB. Expression of a single transfected cDNA converts fibroblasts to myoblasts. Cell 1987; 51:987-1000.

15. Massari ME, Murre C. Helix-loop-helix proteins: regulators of transcription in eucaryotic organisms. Mol Cell Biol 2000; 20:429-440.

16. Ellenberger T, Fass D, Arnaud M et al. Crystal structure of transcription factor E47: E-box recognition by a basic region helix-loop-helix dimer. Genes Dev 1994; 8:970-980.

17. Ephrussi A, Church GM, Tonegawa S et al. B lineage—specific interactions of an immunoglobulin enhancer with cellular factors in vivo. Science 1985; 227:134-140.

18. Lassar AB, Buskin JN, Lockshon D et al. MyoD is a sequence-specific DNA binding protein requiring a region of myc homology to bind to the muscle creatine kinase enhancer. Cell 1989; 58:823-831.

19. Gossett LA, Kelvin DJ, Sternberg EA et al. A new myocyte-specific enhancer-binding factor that recognizes a conserved element associated with multiple muscle-specific genes. Mol Cell Biol 1989; 9:5022-5033.

20. Buskin JN, Hauschka SD. Identification of a myocyte nuclear factor that binds to the muscle-specific enhancer of the mouse muscle creatine kinase gene. Mol Cell Biol 1989; 9:2627-2640.

21. Murre C, McCaw PS, Baltimore D. A new DNA binding and dimerization motif in immunoglobulin enhancer binding, daughterless, MyoD and myc proteins. Cell 1989; 56:777-783.

22. Davis RL, Turner DL. Vertebrate hairy and Enhancer of split related proteins: transcriptional repressors regulating cellular differentiation and embryonic patterning. Oncogene 2001; 20:8342-8357.

23. Sasai Y, Kageyama R, Tagawa Y et al. Two mammalian helix-loop-helix factors structurally related to Drosophila hairy and Enhancer of split. Genes Dev 1992; 6:2620-2634.

24. Fisher AL, Ohsako S, Caudy M. The WRPW motif of the hairy-related basic helix-loop-helix repressor proteins acts as a 4-amino-acid transcription repression and protein-protein interaction domain. Mol Cell Biol 1996; 16:2670-2677.

25. Paroush Z, Finley RL, Kidd T et al. Groucho is required for Drosophila neurogenesis, segmentation and sex determination and interacts directly with hairy-related bHLH proteins. Cell 1994; 79:805-815.

26. Castella P, Sawai S, Nakao K et al. HES-1 repression of differentiation and proliferation in PC12 cells: role for the helix 3-helix 4 domain in transcription repression. Mol Cell Biol 2000; 20:6170-6183.

27. Dawson SR, Turner DL, Weintraub H et al. Specificity for the hairy/enhancer of split basic helix-loop-helix (bHLH) proteins maps outside the bHLH domain and suggests two separable modes of transcriptional repression. Mol Cell Biol 1995; 15:6923-6931.

28. Kageyama R, Ishibashi M, Takebayashi K et al. bHLH transcription factors and mammalian neuronal differentiation. Int J Biochem Cell Biol 1997; 29:1389-1399.

29. Cooke J. A gene that resuscitates a theory—somitogenesis and a molecular oscillator. Trends Genet 1998; 14:85-88.

30. Dale KJ, Pourquie O. A clock-work somite. Bioessays 2000; 22:72-83.

31. Jouve C, Palmeirim I, Henrique D et al. Notch signaling is required for cyclic expression of the hairy-like gene HES1 in the presomitic mesoderm. Development 2000; 127:1421-1429.

32. Leimeister C, Dale JK, Fischer A et al. Oscillating expression of c-Hey2 in the presomitic mesoderm suggests that the segmentation clock may use combinatorial signaling through multiple interacting bHLH factors. Dev Biol 2000; 227:91-103.

33. Bessho Y, Sakata R, Komatsu S et al. Dynamic expression and essential functions of Hes7 in somite segmentation. Genes Dev 2001a; 15:2642-2647.

34. Dunwoodie SL, Clements M, Sparrow DB et al. Axial skeletal defects caused by mutation in the spondylocostal dysplasia/pudgy gene Dll3 are associated with disruption of the segmentation clock within the presomitic mesoderm. Development 2002; 129:1795-1806.

35. Gajewski M, Sieger D, Alt B et al. Anterior and posterior waves of cyclic her1 gene expression are differentially regulated in the presomitic mesoderm of zebrafish. Development 2003; 130:4269-4278.

36. Holley SA, Geisler R, Nüsslein-Volhard C. Control of her1 expression during zebrafish somitogenesis by a Delta-dependent oscillator and an independent wave-front activity. Genes Dev 2000; 14:1678-1690.

37. Li Y, Fenger U, Niehrs C et al. Cyclic expression of esr9 gene in Xenopus presomitic mesoderm. Differentiation 2003; 71:83-89.

38. Oates AC, Ho RK. Hairy/E(spl)-related (Her) genes are central components of the segmentation oscillator and display redundancy with the Delta/Notch signaling pathway in the formation of anterior segmental boundaries in the zebrafish. Development 2002; 129:2929-2946.

39. Sawada A, Fritz A, Jiang Y et al. Zebrafish Mesp family genes, mesp-a and mesp-b are segmentally expressed in the presomitic mesoderm and Mesp-b confers the anterior identity to the developing somites. Development 2000; 127:1691-1702.

40. Pasini A, Jiang YJ, Wilkinson DG. Two zebrafish Notch-dependent hairy/Enhancer-of-split-related genes, her6 and her4, are required to maintain the coordination of cyclic gene expression in the presomitic mesoderm. Development 2004; 131:1529-1541.

41. Jen WC, Gawantka V, Pollet N et al. Periodic repression of Notch pathway genes governs the segmentation of Xenopus embryos. Genes Dev 1999; 13:1486-1499.
42. Bae S, Bessho Y, Hojo M et al. The bHLH gene Hes6, an inhibitor of Hes1, promotes neuronal differentiation. Development 2000; 127:2933-2943.
43. Koyano-Nakagawa N, Kim J anderson D, Kintner C. Hes6 acts in a positive feedback loop with the neurogenins to promote neuronal differentiation. Development 2000; 127:4203-4216.
44. Kawamura A, Koshida S, Hijikata H et al. Zebrafish hairy/enhancer of split protein links FGF signaling to cyclic gene expression in the periodic segmentation of somites. Genes Dev 2005; 19:1156-1161.
45. Barrantes IB, Elia AJ, Wunsch K et al. Interaction between Notch signaling and Lunatic fringe during somite boundary formation in the mouse. Curr Biol. 1999; 9:470-480.
46. Sieger D, Tautz D, Gajewski M. The role of Suppressor of Hairless in Notch mediated signaling during zebrafish somitogenesis. Mech Dev 2003; 120:1083-1094.
47. Takke C, Campos-Ortega JA. her1, a zebrafish pair-rule like gene, acts downstream of notch signaling to control somite development. Development 1999; 126:3005-3014.
48. Aulehla A, Johnson RL. Dynamic expression of lunatic fringe suggests a link between notch signaling and an autonomous cellular oscillator driving somite segmentation. Dev Biol 1999; 207:49-61.
49. Forsberg H, Crozet F, Brown NA. Waves of mouse Lunatic fringe expression, in four-hour cycles at two-hour intervals, precede somite boundary formation. Curr Biol 1998; 8:1027-1030.
50. McGrew MJ, Dale JK, Fraboulet S et al. The lunatic fringe gene is a target of the molecular clock linked to somite segmentation in avian embryos. Curr Biol 1998; 8:979-982.
51. Bettenhausen B, Hrabe de Angelis M, Simon D et al. Transient and restricted expression during mouse embryogenesis of Dll1, a murine gene closely related to Drosophila Delta. Development 1995; 121:2407-2418.
52. Elmasri H, Liedtke D, Lucking G et al. her7 and hey1, but not lunatic fringe show dynamic expression during somitogenesis in medaka (Oryzias latipes). Gene Expr Patterns 2004; 4:553-559.
53. Morimoto M, Takahashi Y, Endo M et al. The Mesp2 transcription factor establishes segmental borders by suppressing Notch activity. Nature 2005; 435:354-359.
54. Jiang YJ, Aerne BL, Smithers L et al. Notch signaling and the synchronization of the somite segmentation clock. Nature 2000; 408:475-479.
55. Horikawa K, Ishimatsu K, Yoshimoto E et al. Noise-resistant and synchronized oscillation of the segmentation clock. Nature 2006; 441:719-723.
56. Aulehla A, Wehrle C, Brand-Saberi B et al. Wnt3a plays a major role in the segmentation clock controlling somitogenesis. Dev Cell 2003; 4:395-406.
57. Dale JK, Malapert P, Chal J et al. Oscillations of the snail genes in the presomitic mesoderm coordinate segmental patterning and morphogenesis in vertebrate somitogenesis. Dev Cell 2006; 10:355-366.
58. Ishikawa A, Kitajima S, Takahashi Y et al. Mouse Nkd1, a Wnt antagonist, exhibits oscillatory gene expression in the PSM under the control of Notch signaling. Mech Dev 2004; 121:1443-1453.
59. Henry CA, Urban MK, Dill KK et al. Two linked hairy/Enhancer of split-related zebrafish genes, her1 and her7, function together to refine alternating somite boundaries. Development 2002; 129:3693-3704.
60. Holley SA, Julich D, Rauch GJ et al. her1 and the notch pathway function within the oscillator mechanism that regulates zebrafish somitogenesis. Development 2002; 129:1175-1183.
61. Bessho Y, Miyoshi G, Sakata R et al. Hes7 a bHLH-type repressor gene regulated by Notch and expressed in the presomitic mesoderm. Genes Cells 2001b; 6:175-185.
62. Bessho Y, Hirata H, Masamizu Y et al. Periodic repression by the bHLH factor Hes7 is an essential mechanism for the somite segmentation clock. Genes Dev 2003; 17:1451-1456.
63. Morales AV, Yasuda Y, Ish-Horowicz D. Periodic Lunatic fringe expression is controlled during segmentation by a cyclic transcriptional enhancer responsive to notch signaling. Dev Cell 2002; 3:63-74.
64. Takebayashi K, Sasai Y, Sakai Y et al. Structure, chromosomal locus and promoter analysis of the gene encoding the mouse helix-loop-helix factor HES-1. Negative autoregulation through the multiple N box elements. J Biol Chem 1994; 269:5150-5156.
65. Strom A, Castella P, Rockwood J et al. Mediation of NGF signaling by posttranslational inhibition of HES-1, a basic helix-loop-helix repressor of neuronal differentiation. Genes Dev 1997; 11:3168-3181.
66. Cole SE, Levorse JM, Tilghman SM et al. Clock regulatory elements control cyclic expression of Lunatic fringe during somitogenesis. Dev Cell 2002; 3:75-84.
67. Hirata H, Yoshiura S, Ohtsuka T et al. Oscillatory expression of the bHLH factor Hes1 regulated by a negative feedback loop. Science 2002; 298:840-843.
68. Hirata H, Bessho Y, Kokubu H et al. Instability of Hes7 protein is crucial for the somite segmentation clock. Nat Genet 2004; 36:750-754.
69. Dale JK, Maroto M. A Hes1-based oscillator in cultured cells and its potential implications for the segmentation clock. Bioessays 2003; 25:200-203.

70. Lewis J. Autoinhibition with transcriptional delay: a simple mechanism for the zebrafish somitogenesis oscillator. Curr Biol 2003; 13:1398-1408.
71. Monk NA. Oscillatory expression of Hes1, p53 and NF-kappaB driven by transcriptional time delays. Curr Biol 2003; 13:1409-1413.
72. Dale JK, Maroto M, Dequeant ML et al. Periodic Notch inhibition by Lunatic Fringe underlies the chick segmentation clock. Nature 2003; 421:275-278.
73. Joseph EM, Cassetta LA. Mespo: a novel basic helix-loop-helix gene expressed in the presomitic mesoderm and posterior tailbud of Xenopus embryos. Mech Dev 1999; 82:191-194.
74. Yoon JK, Moon RT, Wold B. The bHLH class protein pMesogenin1 can specify paraxial mesoderm phenotypes. Dev Biol 2000; 15:376-391.
75. Buchberger A, Bonneick S, Arnold HH. Expression of the novel basic-helix-loop-helix transcription factor cMespo in presomitic mesoderm of chicken embryos. Mech Dev 2000; 97:223-226.
76. Yoo KW, Kim CH, Park HC et al. Characterization and expression of a presomitic mesoderm- specific mespo gene in zebrafish. Dev Genes Evol 2003; 213:203-206.
77. Yoon JK, Wold B. The bHLH regulator pMesogenin1 is required for maturation and segmentation of paraxial mesoderm. Genes Dev 2000; 14:3204-3214.
78. Aoyama H, Asamoto K. Determination of somite cells: independence of cell differentiation and morphogenesis. Development 1988; 104:15-28.
79. Bronner-Fraser M, Stern C. Effects of mesodermal tissues on avian neural crest cell migration. Dev Biol 1991; 143:213-217.
80. Dubrulle J, McGrew MJ, Pourquie O. FGF signaling controls somite boundary position and regulates segmentation clock control of spatiotemporal Hox gene activation. Cell 2001; 106:219-232.
81. Palmeirim I, Dubrulle J, Henrique D et al. Uncoupling segmentation and somitogenesis in the chick presomitic mesoderm. Dev Genet 1998; 23:77-85.
82. Keynes RJ, Stern CD. Mechanisms of vertebrate segmentation. Development 1988; 103:413-429.
83. Saga Y, Hata N, Kobayashi S et al. MesP1: a novel basic helix-loop-helix protein expressed in the nascent mesodermal cells during mouse gastrulation. Development 1996; 122:2769-2778.
84. Saga Y, Hata N, Koseki H et al. Mesp:2 a novel mouse gene expressed in the presegmented mesoderm and essential for segmentation initiation. Genes Dev 1997; 11:1827-1839.
85. Takahashi Y, Inoue T, Gossler A et al. Feedback loops comprising Dll1, Dll3 and Mesp2 and differential involvement of Psen1 are essential for rostrocaudal patterning of somites. Development 2003; 130:4259-4268.
86. Takahashi Y, Koizumi K, Takagi A et al. Mesp2 initiates somite segmentation through the Notch signaling pathway. Nat Genet 2000; 25:390-396.
87. Morimoto M, Takahashi Y, Endo M et al. The Mesp2 transcription factor establishes segmental borders by suppressing Notch activity. Nature 2005; 435:354-359.
88. Nakajima Y, Morimoto M, Takahashi Y et al. Identification of Epha4 enhancer required for segmental expression and the regulation by Mesp2. Development 2006; 133:2517-2525.
89. Saga Y. Genetic rescue of segmentation defect in MesP2-deficient mice by MesP1 gene replacement. Mech Dev 1998; 75:53-66.
90. Nomura-Kitabayashi A, Takahashi Y, Kitajima S et al. Hypomorphic Mesp allele distinguishes establishment of rostrocaudal polarity and segment border formation in somitogenesis. Development 2002; 129:2473-2481.
91. Morimoto M, Kiso M, Sasaki N et al. Cooperative Mesp activity is required for normal somitogenesis along the anterior-posterior axis. Dev Biol 2006; 300:687-698.
92. Takahashi Y, Kitajima S, Inoue T et al. Differential contributions of Mesp1 and Mesp2 to the epithelialization and rostro-caudal patterning of somites. Development 2005; 132:787-796.
93. Yasuhiko Y, Haraguchi S, Kitajima S et al. Tbx6-mediated Notch signaling controls somite-specific Mesp2 expression. Proc Natl Acad Sci USA 2006; 103:3651-3656.
94. Terasaki H, Murakami R, Yasuhiko Y et al. Transgenic analysis of the medaka mesp-b enhancer in somitogenesis. Dev Growth Differ 2006; 48:153-68.
95. Buchberger A, Seidl K, Klein C et al. cMeso1, a novel bHLH transcription factor, is involved in somite formation in chicken embryos. Dev Biol 1998; 199:201-215.
96. Buchberger A, Bonneick S, Klein C et al. Dynamic expression of chicken cMeso2 in segmental plate and somites. Dev Dyn 2002; 223:108-118.
97. Sparrow DB, Jen WC, Kotecha S et al. Thylacine 1 is expressed segmentally within the paraxial mesoderm of the Xenopus embryo and interacts with the Notch pathway. Development 1998; 125:2041-2051.
98. Moreno TA, Kintner C. Regulation of segmental patterning by retinoic acid signaling during Xenopus somitogenesis. Dev Cell 2004; 6:205-218.
99. Burgess R, Rawls A, Brown D et al. Requirement of the paraxis gene for somite formation and musculoskeletal patterning. Nature 1996; 384:570-573.

100. Blanar MA, Crossley PH, Peters KG et al. Meso1, a basic-helix-loop-helix protein involved in mammalian presomitic mesoderm development. Proc Natl Acad Sci USA 1995; 20:5870-5874.

101. Burgess R, Cserjesi P, Ligon KL et al. Paraxis: a basic helix-loop-helix protein expressed in paraxial mesoderm and developing somites. Dev Biol 1995; 168:296-306.

102. Quertermous EE, Hidai H, Blanar MA et al. Cloning and characterization of a basic helix-loop-helix protein expressed in early mesoderm and the developing somites. Proc Natl Acad Sci USA 1994; 19:7066-7070.

103. Sosic D, Brand-Saberi B, Schmidt C et al. Regulation of paraxis expression and somite formation by ectoderm- and neural tube-derived signals. Dev Biol 1997; 185:229-243.

104. Shanmugalingam S, Wilson SW. Isolation, expression and regulation of a zebrafish paraxis homologue. Mech Dev 1998; 78:85-89.

105. Carpio R, Honoré SM, Araya C et al. Xenopus paraxis homologue shows novel domains of expression. Dev Dyn 2004; 231:609-613.

106. Tseng HT, Jamrich M. Identification and developmental expression of Xenopus paraxis. Int J Dev Biol 2004; 48:1155-1158.

107. Barnes GL, Alexander PG, Hsu CW et al. Cloning and characterization of chicken Paraxis: a regulator of paraxial mesoderm development and somite formation. Dev Biol 1997; 189:95-111.

108. Nakaya Y, Kuroda S, Katagiri YT et al. Mesenchymal-epithelial transition during somitic segmentation is regulated by differential roles of Cdc42 and Rac.1 Dev Cell 2004; 7:425-438.

109. Durbin L, Brennan C, Shiomi K et al. Eph signaling is required for segmentation and differentiation of the somites. Genes Dev 1998; 12:3096-3109.

110. Hrabe de Angelis M, McIntyre J, Gossler A. Maintenance of somite borders in mice requires the Delta homologue DII1. Nature 1997; 386:717-721.

111. Johnson J, Rhee J, Parsons SM et al. The anterior/posterior polarity of somites is disrupted in paraxis-deficient mice. Dev Biol 2001; 229:176-187.

112. Linker C, Lesbros C, Gros J et al. beta-Catenin-dependent Wnt signalling controls the epithelial organisation of somites through the activation of paraxis. Development 2005; 132:3895-905.

113. Gros J, Scaal M, Marcelle C. A two-step mechanism for myotome formation in chick, Dev Cell 2004; 6:875-882.

114. Kalcheim C, Ben-Yair R. Cell rearrangements during development of the somite and its derivatives, Curr Opin Genet Dev 2005; 15:371-380.

115. Ordahl CP, Williams BA, Denetclaw W. Determination and morphogenesis in myogenic progenitor cells: an experimental embryological approach. Curr Top Dev Biol 2000; 48:319-367.

116. Hollway G, Currie P. Vertebrate myotome development. Birth Defects Res C Embryo Today 2005; 75:172-179.

117. Weinberg ES, Allende ML, Kelly CS et al. Developmental regulation of zebrafish MyoD in wild-type, no tail and spadetail embryos. Development 1996; 122:271-280.

118. Coutelle O, Blagden CS, Hampson R et al. Hedgehog signaling is required for maintenance of myf5 and myoD expression and timely terminal differentiation in zebrafish adaxial myogenesis. Dev Biol 2001; 236:136-510.

119. Harvey RP. The Xenopus MyoD gene: an unlocalised maternal mRNA predates lineage-restricted expression in the early embryo. Development 1990; 108:669-680.

120. Kato K, Gurdon JB. Single-cell transplantation determines the time when Xenopus muscle precursor cells acquire a capacity for autonomous differentiation. Proc Natl Acad Sci USA 1993; 90:1310-1314.

121. Kopan RR, Nye JS, Weintraub H. The intracellular domain of mouse Notch: A constitutively activated repressor of myogenesis directed at the basic helix-loop-helix region of MyoD. Development 1994; 120:2385-2396.

122. Ott MO, Bober E, Lyons G et al. Early expression of the myogenic regulatory gene, myf-5, in precursor cells of skeletal muscle in the mouse embryo. Development 1991; 111:1097-1107.

123. Sassoon D, Lyons G, Wright WE et al. Expression of two myogenic regulatory factors myogenin and MyoD1 during mouse embryogenesis. Nature 1989; 341:303-307.

124. Kiefer JC, Hauschka SD. Myf-5 Is Transiently Expressed in Nonmuscle Mesoderm and Exhibits Dynamic Regional Changes within the Presegmented Mesoderm and Somites I-IV. Dev Biol 2001; 232:77-90.

125. Lin-Jones J, Hauschka SD. Myogenic determination factor expression in the developing avian limb bud: An RT-PCR analysis. Dev Biol 1996; 174:407-422.

126. Tapscott SJ, Davis RL, Thayer MJ et al. MyoD1: a nuclear phosphoprotein requiring a Myc homology region to convert fibroblasts to myoblasts. Science 1988; 242:405-141.

127. Weintraub H, Tapscott SJ, Davis RL et al. Activation of muscle-specific genes in pigment, nerve, fat, liver and fibroblast cell lines by forced expression of MyoD. Proc Natl Acad Sci USA 1989; 86:5434-5438.

128. Rudnicki MA, Braun T, Hinuma S et al. Inactivation of MyoD in mice leads to up-regulation of the myogenic HLH gene Myf-5 and results in apparently normal muscle development. Cell 1992; 71:383-390.

129. Braun T, Rudnicki MA, Arnold HH et al. Targeted inactivation of the muscle regulatory gene Myf-5 results in abnormal rib development and perinatal death. Cell 1992; 71:369-382.
130. Rudnicki MA, Schnegelsberg PN, Stead RH et al. MyoD or Myf-5 is required for the formation of skeletal muscle. Cell 1993; 75:1351-1359.
131. Kablar B, Krastel K, Ying C et al. MyoD and Myf-5 differentially regulate the development of limb versus trunk skeletal muscle. Development 1997; 124:4729-4738.
132. Kablar B, Asakura A, Krastel K et al. MyoD and Myf-5 define the specification of musculature of distinct embryonic origin. Biochem Cell Biol 1998; 76:1079-1091.
133. Kablar B, Krastel K, Tajbakhsh S et al. Myf5 and MyoD activation define independent myogenic compartments during embryonic development. Dev Biol 2003; 258:307-318.
134. Carvajal JJ, Cox D, Summerbell D et al. A BAC transgenic analysis of the Mrf4/Myf5 locus reveals interdigitated elements that control activation and maintenance of gene expression during muscle development. Development 2001; 128:1857-1868.
135. Gustafsson MK, Pan H, Pinney DF et al. Myf5 is a direct target of long-range Shh signaling and Gli regulation for muscle specification. Genes Dev 2002; 16:114-126.
136. Hadchouel J, Tajbakhsh S, Primig M et al. Modular long-range regulation of Myf5 reveals unexpected heterogeneity between skeletal muscles in the mouse embryo. Development 2000; 127:4455-4467.
137. Summerbell D, Ashby PR, Coutelle O et al. The expression of Myf5 in the developing mouse embryo is controlled by discrete and dispersed enhancers specific for particular populations of skeletal muscle precursors. Development 2000; 127:3745-3757.
138. Teboul L, Hadchouel J, Daubas P et al. The early epaxial enhancer is essential for the initial expression of the skeletal muscle determination gene Myf5 but not for subsequent, multiple phases of somitic myogenesis. Development 2002; 129:4571-4580.
139. Munsterberg AE, Kitajewski J, Bumcrot DA et al. Combinatorial signaling by Sonic hedgehog and Wnt family members induces myogenic bHLH gene expression in the somite. Genes Dev 1995; 9:2911-2922.
140. Maroto M, Reshef R, Munsterberg AE et al. Ectopic Pax-3 activates MyoD and Myf-5 expression in embryonic mesoderm and neural tissue. Cell 1997; 89:139-148.
141. Reshef R, Maroto M, Lassar AB. Regulation of dorsal somitic cell fates: BMPs and Noggin control the timing and pattern of myogenic regulator expression. Genes Dev 1998; 12:290-303.
142. Tajbakhsh S, Borello U, Vivarelli E et al. Differential activation of Myf5 and MyoD by different Wnts in explants of mouse paraxial mesoderm and the later activation of myogenesis in the absence of Myf5. Development 1998; 125:4155-4162.
143. Olson EN. Proto-oncogenes in the regulatory circuit for myogenesis. Semin Cell Biol 1992; 3:127-136.
144. Li L, Zhou J, James G et al. FGF inactivates myogenic helix-loop-helix proteins through phosphorylation of a conserved protein kinase C site in their DNA-binding domains. Cell 1992; 71:1181-1194.
145. Bengal E, Ransone L, Scharfmann R et al. Functional antagonism between c-Jun and MyoD proteins: a direct physical association. Cell 1992; 68:507-519.
146. Benezra R, Davis RL, Lockshon D et al. The protein Id: a negative regulator of helix-loop-helix DNA binding proteins. Cell 1990; 61:49-59.
147. Ruzinova MB, Benezra R. Id proteins in development, cell cycle and cancer. Trends Cell Biol 2003; 13:410-418.
148. Fuchbauer EM. Expression of M-twist during postimplantation development of the mouse. Dev Dyn 1995; 204:316-322.
149. Spicer DB, Rhee J, Cheung WL et al. Inhibition of myogenic bHLH and MEF2 transcription factors by the bHLH protein Twist. Science 1996; 272:1476-1480.
150. Stoetzel C, Weber B, Bourgeois P et al. Dorso-ventral and rostro-caudal sequential expression of M-twist in the postimplantation murine embryo. Mech Dev 1995; 51:251-263.
151. Hamamori Y, Wu HY, Sartorelli V et al. The basic domain of myogenic basic helix-loop-helix (bHLH) proteins is the novel target for direct inhibition by another bHLH protein, Twist. Mol Cell Biol 1997; 17:6563-6573.
152. Chen ZF, Behringer RR. Twist is required in head mesenchyme for cranial neural tube morphogenesis. Genes Dev 1995; 9:686-699.

CHAPTER 8

Mouse Mutations Disrupting Somitogenesis and Vertebral Patterning

Kenro Kusumi,* William Sewell and Megan L. O'Brien

Introduction

The mouse was one of the first model organisms used in genetic analysis, beginning in 1902 with the studies of inheritance carried out by William E. Castle, Director of the Bussey Institute at Harvard.[1] The first mutations identified derived from mouse fanciers, who primarily selected coat color variants or neurobehavioral traits. However, disruptions affecting the axial skeleton were also reported early in the century. For example, the classic brachyury (*T*) short tail mutant was identified in a laboratory stock by Dobrovolskaïa-Zavadskaïa in 1927 and was subsequently cloned and found to be a member of the T-box family of transcription factors, required for the formation and differentiation of paraxial mesoderm.[2] Spontaneous mutations causing vertebral defects, including undulated (*Pax1^un*) and pudgy (*Dll3^pu*), have also been cloned and found to encode genes involved in somite patterning.[3,4] More recently, advances in genetic technologies have greatly expanded the number of mouse mutations with somite defects. These approaches include use of homologous recombination in embryonic stem (ES) cell lines to generate "knock-out" and "knock-in" mice, transgenic insertion of dominant-negative alleles and chemical mutagenesis by agents such as N-ethyl-N-nitrosourea (ENU). Mouse mutant phenotypes and signaling pathways have been studied and characterized through analysis of double mutants. These genetic studies in the mouse have yielded a tremendous amount of information about the process of mammalian somitogenesis.

The process of somitogenesis is iterative and begins at approximately 7.75 days post coitum (dpc) when the most cranial somites begin to coalesce and ends with the segmentation of the final 63rd-65th somite around 13.25-13.5 dpc.[5] Somitogenesis is not a singular developmental event (Fig. 1). Somite formation begins at the end of gastrulation, a process that produces the paraxial mesodermal substrate for somites. Prepatterning events within the presomitic mesoderm appear to control the regular production of somites and the resulting highly organized spine. Recent findings point to a mechanism, which has been called the "clock and wave-front" model of somitogenesis. This mechanism requires an internal oscillator or clock underlying the regular periodicity of patterning events.[6-8]

The physical process of segmentation of the somite, which is specified in the presomitic mesoderm (PSM), is associated with a drastic cellular cytoskeletal rearrangement from mesenchymal to columnar epithelioid cells. When first formed, a somite consists of a single-layered epithelial sphere, bound together through tight junctions at the basal surface and a lumen filled with mesenchymal cells.[9] The development of enhanced videomicroscopic technology has allowed for real-time analysis of somitogenesis and has demonstrated significant mobility of cells prior to segmentation, which is characterized by a "ball-and-socket" morphology.[10] By the time a somite forms, rostral and caudal domains appear to be clearly delineated. Early microdissection experiments in the chick

*Corresponding Author: Kenro Kusumi—School of Life Sciences, Arizona State University
PO Box 874501, Tempe, AZ 85287-4501, USA. Email: kenro@asu.edu

Somitogenesis, edited by Miguel Maroto and Neil V. Whittock. ©2008 Landes Bioscience and Springer Science+Business Media.

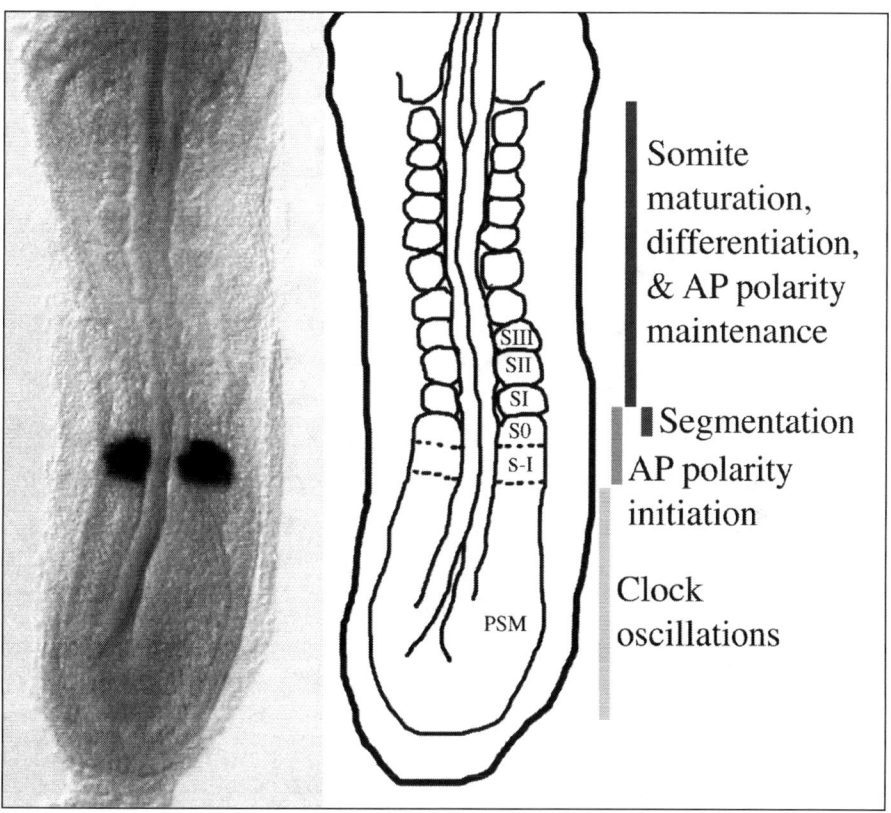

Figure 1. Developmental processes in mouse somitogenesis. At left is an image of a mouse embryo at 8.5 dpc (10 somite-pair stage) with expression of *Mesp2* detected by whole mount *in situ* hybridization. At right, a diagram highlights the morphological borders in this embryo and developmental components of somitogenesis. SI, newly formed somite; S0, PSM region to form next somite; S-I, PSM region immediately caudal to S0; PSM, presomitic mesoderm.

have indicated that rostral-caudal polarity can be reoriented if tissue is rotated 180° in the caudal presomitic mesoderm, but this polarity is fixed by the time of somite formation.[11,9]

As they mature, somites subdivide into three sections, the sclerotome, myotome and dermatome.[12] The sclerotome forms the ossified bones of the vertebrae and the ribs, the myotome forms the axial skeletal musculature and the dermatome forms the dermis of the back. The vertebral column is finally formed from the spinal cord and nerves (derived from neuroectoderm), the ossified vertebral and costal bones (derived from sclerotomal mesoderm), the spinal muscles and ligaments (derived from myotomal mesoderm) and the intervertebral discs (derived from notochordal mesoderm). The myotome-derived musculature connecting the vertebrae corresponds to each somite, but the vertebral bones derive from resegmentation of the rostral compartment of one somite and the caudal compartment of the adjacent somite (Fig. 2).

There are a large number of spontaneous and targeted mutations that have been generated which exhibit defects in one or more of the various stages of somitogenesis. These defects include gastrulation and early paraxial mesodermal formation, somite clock regulation, anterior-posterior patterning within somites, segmental border formation and postsegmentation. Postsegmentation includes the production of differentiated cell types and lineages. The diversity of mutations makes

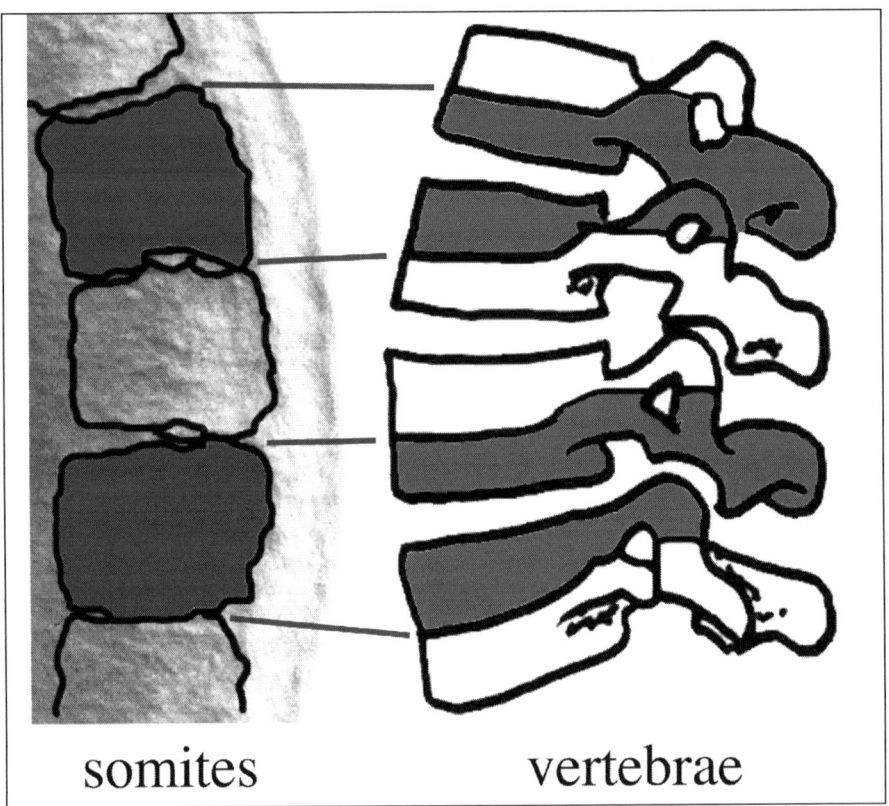

somites   vertebrae

Figure 2. Diagram illustrating the correspondence between embryonic somite segments and adult vertebral bones. At left is an image of a 9.5 dpc embryo, with somites highlighted. At right is a diagram illustrating lumbar vertebrae. Somites contribute the caudal part of one vertebra and the rostral part of the immediately adjacent caudal vertebra.

it difficult to organize these genetic resources. In this chapter, mutations are organized by gene pathways and families as discussed below or by individual genes, as shown in Table 1.

Notch Signaling Pathway

Notch genes are large transmembrane receptors, which recognize delta and serrate/jagged classes of ligands and regulate cell-fate determination and embryonic patterning in animals. Genes in the notch-signaling pathway have been identified which are expressed in an oscillatory pattern, in synchrony with each somite cycle.[13-22] The cycling of notch pathway genes, including *Lfng, Hes1, Hes5, Hes7, Hey2, Nrarp, Nkd1* and *Bcl9l* in the mouse and *hairy1* and *hairy2* in the chick, is thought to regulate the periodic activation of the notch signaling pathway, which is required for the segmentation process.

The disruption of genes in the notch pathway results in somite segmentation defects and vertebral anomalies in mice. In addition to the previously mentioned cycling notch pathway genes, the receptors *Notch1* and *Notch2*, the ligands *Dll1, Dll3* and *Jag1*, the ligand regulator mind bomb 1 (*Mib1*), the gamma secretase *Psen,1* the notch modifier Protein O-fucosyltransferase 1 (*Pofut1*) and the effectors *Hey1* and *Heyl/Hey3*, are all expressed in unique regions during mouse paraxial mesodermal development.[13,14,16,18,4,20,23-33] Homozygous mutations in mouse *Notch1* result in irregular segmentation and early embryonic lethality at 10.0-11.0 dpc.[34,35] Although the receptor

Table 1. Selected mouse mutations disrupting somitogenesis and vertebral patterning

Gene	Name/Synonym	Mutation Allele(s)	Molecular Description	Phenotype Description	Reference(s)
Btg2	B-cell translocation gene 2	Btg2^{tm1Spo} targeted	Antiproliferative (APRO) homology boxes. Coactivator-corepressor of BMP signaling and/or adaptor molecule that modulates activity of interacting proteins.	Posterior homeotic transformations of the axial vertebrae.	Park et al (2004)
Cdx2	Caudal type homeo box 2	Cdx2^{tm1Mmt} targeted, Cdx2^{tm1Fbe} targeted	Homeobox transcription factor	Semidominant vertebral phenotype.	Chawengsaksophak et al (2004)
Dll1	Delta-like 1	Dll1^{tm1Cos} targeted, Dll1^{tm1Mjo} targeted	Notch ligand	Somite AP patterning defect.	Hrabe de Angelis et al (1997) Koizumi et al (2001)
Dll3	Delta-like 3	Dll3^{pu} X-ray induced, Dll3^{tm1Rbe} targeted, Dll3^{oma} spontaneous	Notch ligand	Somite AP patterning defect. *Lfng, Hes1* cycling defect.	Kusumi et al (1998) Dunwoodie et al (2002) Shinkai et al (2004)
Dkk1	Dickkopf homolog 1	Dkk1^{dblr} transgenic	Wnt antagonist	vertebral fusions and segmental defects.	MacDonald et al (2004)
Dmrt2	Doublesex and mab-3 related transcription factor 2	Dmrt2^{tm1Rjo} targeted	DM domain transcription factor	Dermomyotome and myotome fail to adopt a normal epithelial morphology.	Seo et al (2006)
Foxc1	Forkhead box C1	Foxc1^{tm1Blh} targeted, Foxc1^{ch} spontaneous	Forkhead domain transcription factor	Defective chondrogenic differentiation.	Kume et al (1998, 2001)
Foxc2	Forkhead box C2	Foxc2^{tm1Blh} targeted, Foxc2^{tm1Miu} targeted	Forkhead domain transcription factor	Reduced sclerotome proliferation.	Kume et al (2001)

continued on next page

Table 1. *Continued*

Gene	Name/Synonym	Mutation Allele(s)	Molecular Description	Phenotype Description	Reference(s)
Fgfr1	Fibroblast growth factor receptor 1	*Fgfr1*^{Δfrs}	Fgf receptor	Eliminates binding site for Fgf receptor-specific substrates 2 & 3. Defects in neural tube closure and in the development of the tail bud.	Hoch and Soriano (2006)
Gbx2	Gastrulation brain homeobox 2	*Gbx2*^{tm1Mrt} targeted	Homeobox transcription factor	Produced Hox-like phenotypes in the axial skeleton without affecting Hox gene expression.	Carapuco et al (2005)
Gdf11	Growth differentiation factor 11	*Gdf11*^{tm1Sjl} targeted, *Gdf11*^{tm1Clf} targeted	Growth factor	Anterograde homeotic transformations throughout the axial skeleton and resulting displacement of the hindlimbs caudally.	McPherron et al (1999)
Hes7	Hairy/enhancer of split homolog 7	*Hes7*^{tm1Rka} targeted, Hes7 (GOF)	bHLH transcription factor	Somite segmentation and oscillatory expression becomes severely disorganized.	Bessho et al (2001b); Hirata et al (2004)
Hoxd3	Homeobox D3	*Hoxd3*^{tm1Mrc} targeted	Homeobox transcription factor	Lack the anterior arch of the atlas and the dens of the axis.	Condie and Capecchi (1993)
Hox10	Homeobox group 10	*Hox10* triple mutant, *Dll1-Hoxa10* transgenic	Homeobox transcription factor	No lumbar vertebrae are formed. Instead, ribs project from all posterior vertebrae, extending caudally from the last thoracic vertebrae to beyond the sacral region.	Wellik and Capecchi (2003); Carapuco et al (2005)
Hoxa10 (GOF)	Homeobox A10 (gain of function)			Expression in presomitic mesoderm is sufficient to confer a Hox group 10 patterning program to the somite, producing vertebrae without ribs.	

continued on next page

Table 1. *Continued*

Gene	Name/Synonym	Mutation Allele(s)	Molecular Description	Phenotype Description	Reference(s)
Hox11	Homeobox group 11	Hox11 triple mutant, Dll1-Hoxa11 transgenics	Homeobox transcription factor	Sacral vertebrae are not formed and instead these vertebrae assume a lumbar identity.	Wellik and Capecchi (2003); Carapuco et al (2005)
Hoxa11 (GOF)	Homeobox A11 (gain of function)			Vertebral sacralization requires Hoxa11 expression in the presomitic mesoderm, while their caudal differentiation requires that Hoxa11 is expressed in the somites.	
Lfng	Lunatic fringe	Lfng^{tm1Rjo} targeted, Lfng^{tm1Grid} targeted	Notch signaling modifier	Somite AP patterning defect.	Evrard et al (1998) Zhang and Gridley (1998)
Lrp6	LDL receptor related protein 6	Lrp6^{rs} spontaneous, Lrp6$^{Gt(pGT1.8TM)187Wcs}$/ Lrp6^{Cd} targeted	Wnt coreceptor	Cranial neural tube defect. Mutant LRP6 binds to Wnt and Dickkopf1 (Dkk1). Dkk1 cannot antagonize Wnt.	Kokubu et al (2004); Carter et al (2005)
Meox1	Mesenchyme homeobox 1, Mox1	Meox1^{tm1Bmk} targeted, Meox1$^{Tg(Mx-1TAX)2627Amh}$ transgenic	Homeobox transcription factor	Hemi-vertebrae and rib, vertebral and cranial-vertebral fusions.	Mankoo et al (2003)
Meox2	Mesenchyme homeobox 2, Mox2	Meox2^{tm1Pach} targeted	Homeobox transcription factor	Mild defects of rib and vertebrae development.	Mankoo et al (2003)
Mesp2	Mesoderm posterior 2	Mesp2^{tm1Ysa} targeted and others	bHLH transcription factor	Somite AP patterning defect.	Saga et al (1997)

continued on next page

Table 1. *Continued*

Gene	Name/Synonym	Mutation Allele(s)	Molecular Description	Phenotype Description	Reference(s)
Mib1	Mind bomb 1	Mib1^{tm1Art}/Mib1^{tm1Art} targeted	Ubiquitin ligase that promotes Delta endocytosis	Die prior to embryonic day 11.5, with pan-Notch defects in somitogenesis, neurogenesis, vasculogenesis and cardiogenesis.	Koo et al (2005)
Msgn1	Mesogenin 1	*Msgn1*^{tm1Wb} targeted	bHLH transcription factor	Failure to form somites and animals completely lack axial structures.	Yoon and Wold (2000)
Ncstn	Nicastrin	*Ncstn*^{tm1Jli} targeted, *Ncstn*^{tm1Pcw} targeted	Component of the multimeric gamma-secretase complex	Display abnormal somite segmentation.	Li et al (2003)
Nog	Noggin	*Nog*^{tm1Amc} targeted	BMP antagonist	Reduction of Bmp4 dosage results in an extensive rescue of the axial skeleton of Noggin mutant embryos. Formation of small mispatterned somites. Human subjects with fibrodysplasia ossificans progressiva (FOP) show malformations similar to mice with homozygous Noggin deletions.	Wijgerde et al (2005); Schaffer et al (2005)
Notch1	Notch gene homolog 1	*Notch1*^{tm1Con} targeted, *Notch1*^{tm1Rko} targeted and others	Receptor	Irregular segmentation and early embryonic lethality at 10.0-11.0 dpc.	Conlon et al (1995)
Pax1	Paired box gene 1	*Pax1*^{un} spontaneous and targeted	Paired box transcription factor	Severe malformations of the vertebral bodies, intervertebral discs and ribs.	Balling et al (1988)
Pax3	Paired box gene 3, Sp, splotch	*Pax3*^{Sp} spontaneous and targeted	Paired box transcription factor	Somite AP patterning defect.	Schubert et al (2001)

continued on next page

Table 1. *Continued*

Gene	Name/Synonym	Mutation Allele(s)	Molecular Description	Phenotype Description	Reference(s)
Pcdh8	Protocad-herin 8, *papc*	Explants treated with soluble papc fusion protein	Calcium-dependent cell adhesion molecule	Mice deficient for *papc* are viable and develop a normal skeletal system, however the PSM fails to epithelialize at the segmental border and distinct segmental boundaries fail to form in explants treated with soluble papc.	Rhee et al (2003)
Pofut1	Protein O-fucosyltransferase 1	*Pofut1*^tm1Pst targeted	Modifies Notch receptor	Die at midgestation with severe defects in somitogenesis, vasculogenesis, cardiogenesis and neurogenesis.	Shi and Stanely (2003)
Ptpn12	protein tyrosine phosphatase, non-receptor type 12, PTP-PEST	*Ptpn12* targeted	Regulator of integrin signaling	Embryo turning, truncated caudal region.	Sirois et al (2006)
Psen1 *Raldh2*	Presenilin 1, PS1 Retinaldehyde dehydrogenase-2	*Psen1*^tm1Shn targeted *Aldh1a2*^tm1Dll / *Aldh1a2*^tm1Ipc targeted	Notch pathway Enzyme essential for retinoic acid (RA) biosynthesis for RA signaling	Somite AP patterning defect. Left-right asymmetry.	Koizumi et al (2001) Niederreither et al (1999); Sirbu and Duester (2006); Vermot et al (2005)
Rbpsuh	Recombining binding protein sup. of hairless	*Rbpsuh*^tm1Kyo targeted, *Rbpsuh*^tm2Kyo targeted, *Rbpsuh*^tm3Kyo targeted	Transcription regulator, Notch signaling mediator	Irregular somite formation and early lethality at 9.0-10.0 dpc.	Oka et al (1995)
Ror2	Receptor tyrosine kinase	*Ror2*^tm1Ymi targeted	Receptor tyrosine kinase	Vertebral malformations in Ror2^-/- mice are due to reductions in the presomitic mesoderm and defects in somitogenesis.	Takeuchi et al (2000); Schwabe et al (2004).

continued on next page

Table 1. *Continued*

Gene	Name/Synonym	Mutation Allele(s)	Molecular Description	Phenotype Description	Reference(s)
Sfrp1 & Sfrp2	Secreted Fz-related protein 1 & 2	Sfrp1^tm1Aksh; Sfrp2^tm1Aksh targeted	Wnt antagonist	Homozygous exerts no effect on embryogenesis. Sfrp1/2 homozygous produce shortened AP axis.	Satoh et al (2006)
Shp2	SH2 domain-containing protein tyrosine phosphatase-2, Ptpn11	Ptpn11^tm1Paw	Tyrosine specific phosphatase	Homozygous null mutants exhibit abnormal mesoderm patterning resulting in gastrulation failure and death by embryonic day 10.5.	Saxton et al (1997)
Sip1	Smad-interacting protein 1	SIP1^Δex7	Two-handed zinc finger transcription factor	Somites with roughly half rostro-caudal length.	Higashi et al (2002); Maruhashi et al (2005)
Snai1	Snail homolog 1	Snai1^tm1Grid targeted	Zinc finger transcription factor	Embryos form a mesodermal layer, but the cells are abnormal in morphology with epithelial characteristics. Lethal prior to 8.5 dpc.	Carver et al (2001)
Sox9	SRY (sex determining region Y)-box 9	Ck19-Cre; Sox9^flox/flox conditional	HMG-box transcription factor	Severe malformations of the vertebral column.	Barrionuevo et al (2006)
Tbx6	T-box homolog 6, rib-vertebrae	Tbx6^rv spontaneous, Tbx6^tm1Pa targeted	T-box transcription factor	Somite AP patterning defect.	Watabe-Rudolph et al (2002)
Tbx18	T-box 18	Tbx18^tm1Akis targeted	T-box transcription factor	Somite caudal compartment decrease.	Bussen et al (2004)
Tcf15	Transcription factor 15, paraxis	Tcf15^tm1Eno targeted	bHLH transcription factor	Defects in the axial skeleton, including poor formation of vertebral bodies and elements and rib fusions. Defects in peripheral nerves that are consistent with a disruption of AP patterning.	Johnson et al (2001)

continued on next page

Table 1. *Continued*

Gene	Name/Synonym	Mutation Allele(s)	Molecular Description	Phenotype Description	Reference(s)
Twist1	Twist gene homolog 1	Twist1[tm1Bhr] targeted and others	bHLH transcription factor	Neural tube defects.	Chen and Behringer (1995)
Uncx4.1	Unc4.1 homeobox	Uncx4.1[tm1Pgr] targeted	Homeobox transcription factor	Mice die at birth and have defects in the axial skeleton and ribs, with no formation of pedicles of the neural arches and proximal ribs. Dorsal root ganglia and spinal nerve fibers are disorganized.	Mansouri et al (2000)
Wnt3a	Wingless-related MMTV integration site 3A	Wnt3a[tm1Amc] targeted, Wnt3a[vt] spontaneous	Soluble ligand	Segmental defects include reduced numbers of vertebrae, abnormal vertebral body ossification and a shortened tail due to vertebral loss.	Greco et al (1996); Takada et al (1994)
Uncloned Mutations					
am	amputated	am radiation induced	Lack tail and vertebrae in the lumbosacral region.		Flint and Ede (1978); Flint et al (1978)
Cd	crooked tail	Cd spontaneous	Mutation in the Lrp6 gene is linked via complementation. Mutants exhibit growth retardation, crooked tail, abnormal vertebrae and a small skull.		Morgan (1954); Carter et al (2005)
Mv	malformed vertebrae	Mv spontaneous	N/A		Theiler et al (1975)
rh	rachiterata	rh spontaneous	N/A		Theiler et al (1974)
tk	tail-kinks	tk spontaneous	N/A		Grüneberg (1955)
Rf	rib fusions	Rf spontaneous	Extinct.		Theiler and Stevens (1960)

Notch2 is expressed in rostral compartments of newly formed somites and within the presomitic mesoderm, no somite or vertebral phenotypes have been observed in homozygous *Notch2* hypomorphic mutations.[36]

Disruption of notch ligands or regulators can also severely disturb somite patterning. Mutations in the notch ligands *Dll1*[37] and *Dll3*[38,39] produce severe phenotypes, including poorly segmented epithelial somites and defects in somite antero-posterior (AP) polarity, which are discussed in greater detail in the chapter by S. Dunwoodie. Another notch ligand, Jag1, is expressed in the PSM, but neither heterozygous nor homozygous targeted mutations have been reported to display any defects in somitogenesis.[40] This contrasts with findings in humans, where heterozygous mutations in *JAG1* cause Alagille syndrome, a multi-organ disorder with 67% of cases displaying vertebral anomalies.[41] Additionally, mutation of the gene mind bomb 1 (*Mib1*), a regulator that promotes Notch ligand endocytosis, exhibits pan-Notch defects in somitogenesis.[27]

Presenilin has been identified as the gamma-secretase enzyme that carries out the third and final cleavage of notch receptors, leading to the release of the intracellular notch-signaling domain.[42] In mammals, two homologous presenilin genes (Psen1, Psen2) have been reported. *Psen1* homozygous mice show defects in the development of caudal somite compartments and altered expression of several genes in the notch pathway (*Dll3, Hes5* and *Lfng*). In contrast, *Psen2* homozygous mice have no obvious defects.[43] In analysis of double mutants, mice homozygous for mutation in *Psen1* and heterozygous for *Psen2* die between 9.5 dpc and 13.5 dpc, similar to *Psen1* homozygous mutation alone. Conversely, mice lacking *Psen2* and heterozygous for *Psen1* have no obvious defects, which is similar to *Psen2* homozygous mutants. However, double homozygous *Psen1-Psen2* mice display a phenotype more severe than *Psen1* mutants alone. The double homozygous *Psen1-Psen2* embryos die before 9.5 dpc and exhibit a severe, complex phenotype including a lack of somite segmentation, a lack of mesenchyme cells in the midbrain, disorganization of the trunk ventral neural tube, anterior neuropore closure delays and abnormal heart and second brachial arch development. Also in these double null embryos, *Hes5* expression was undetectable and *Dll1* expression was ectopic in the neural tube and brain.[44] Presenilins have recently been found to form complexes with the transmembrane glycoprotein, nicastrin (NCSTN), which binds to the membrane-tethered form of NOTCH1. *Ncstn* is critical for *Psen1* and *Psen2* mediated notch signaling.[45] *Ncstn* homozygous mutant embryos display abnormal somite segmentation, similar to defects observed in *Psen1* and *Notch1* homozygous mutant embryos. Together, these results suggest that *Psen1* and *Psen2* have partially overlapping but non-identical functions and are part of a gamma-secretase complex that includes NCSTN.

The notch pathway modulator *Lfng* is a cycling gene that displays oscillatory gene expression within the PSM.[13] Initially, targeted mutations in *Lfng* were shown to display defects in epithelial somite segmentation and vertebral malformations similar to those seen in the *Dll3* mutants.[46,33] Subsequently, transgenic mice have been generated that constitutively express *Lfng* in the presomitic mesoderm along with normal cyclic *Lfng* activity.[47] These transgenic mutants exhibit defects in somite patterning and vertebral organization that are comparable to defects seen in *Lfng* homozygous null mutants. This finding suggests that cyclic presence and absence of *Lfng* activity is essential and that ectopic *Lfng* expression disrupts the normal pattern of notch signaling in the PSM. Regulatory studies have identified several conserved regions surrounding *Lfng* that drive or regulate periodic expression in the PSM.[48] Cole et al identified a 2.3kb region that is needed for activation and repression of cycling *Lfng* transcription,[49] and they have identified two other conserved *cis*-acting regions; the first, Fringe Clock Element 1 (FCE1), appears to direct cyclic *Lfng* RNA expression in the posterior PSM and the second element directs *Lfng* expression to the prospective rostral area of the forming somite. Furthermore, deficiency in the gene Protein O-fucosyltransferase 1 (*Pofut1*) produces defects similar to *Lnfg* null embryos.[30] POFUT1 attaches a fucose moiety to Notch receptors, thereby generating a substrate for LNFG.

A number of downstream effector genes in the notch pathway, notably members of the hairy/ enhancer of split family (*Hes* and *Hey/Hrt/Hesr*), have been targeted for mutation. Reports of somitic defects have not been reported for *Hes1* mutants[50,51] nor for *Hes5* nulls.[52,53] However, *Hes7* displays oscillatory expression within the PSM and mutations in the gene result in a severe failure of epithelial somite formation, similar to that observed in *Lfng* and *Dll3* mutants.[23,14] Interestingly, increasing the half-life of HES7 leads to disruption of oscillatory expression of genes essential for somitogenesis and the formation of segmental boundaries.[54] This finding suggests that even apparently minor temporal perturbations to the segmentation clock can have dire consequences.

Recently, *Hey* genes, originally referred to as *Hrt* or *Hesr* genes, have been shown to share high homology with the *Hes* family but have distinct characteristics.[55,19,28,56] *Hey1* and *Heyl/Hey3* are expressed in the PSM of mouse embryos, with *Hey1* in the dermomytome and *Heyl* in the paraxial mesoderm and sclerotome. *Hey1* does not display any somite phenotype[57] and *Hey2* mutant mice display vascular defects but no segmental anomalies.[58] The intracellular notch signaling domain acts to convert the suppressor of Hairless gene (*Su(H)*) from repressor to activator of *Hes* expression. The mammalian homologue of *Su(H)*, *Rbpjsuh*, is required for somitogenesis and homozygous disruption leads to irregular somite formation and early lethality at 9.0-10.0 dpc, similar to *Notch1* mutants.[59]

Canonical Wnt Signaling Pathway

Genes in other developmental pathways have also been found to display cycling gene expression within the PSM. *Axin2* and *Dact1* (dapper homolog 1/Frodo), both negative regulators of the wnt pathway, display oscillatory expression out of phase with notch pathway genes.[60,61] Furthermore, while expression of *Wnt3a* is unaffected by disruption of notch signaling in the *Dll1* homozygous mutant, the vestigial tail (*Wnt3a^vt*) mutation causes failure of *Lfng* cycling expression in the PSM. Both the targeted mutation and *Wnt3a^vt/Wnt3a^vt* display numerous developmental defects.[62,63] Segmental defects include reduced numbers of vertebrae, abnormal vertebral body ossification and a shortened tail due to vertebral loss. These defects may be due to interactions between the wnt and notch signaling pathways. In *Drosophila*, the PSD-95 disk-large ZO-128 (PDZ) domain of disheveled (*Dvl*) has been found to bind to the intracellular signaling domain of notch (*N-ic*), potentially inhibiting notch signaling.[64] Mammalian *Axin2* also binds to the PDZ domain of *Dvl* and may serve to trigger notch signaling.[65] The interactions between the notch and wnt pathways are an area of active investigation.

Lrp6 is a coreceptor for the wnt ligand and recently, the spontaneous mutation ringelschwanz has been found to be allelic to this gene (*Lrp6^rs*).[66] An additional spontaneous mutation, crooked tail (*Cd*), appears to be an allele of the *Lrp6* gene as well. This discovery is based on mutant phenotypes observed in *Lrp6^Gt(pGT1.8TM)187Wcs/Lrp6^Cd* compound allele animals and genetic mapping of the *Cd* locus.[67] *Lrp6^rs* mutant mice display vertebral segmental defects and have disrupted expression of notch pathway markers, with failure to compartmentalize *Dll1* expression in newly formed somites and failure to maintain *Lfng* cycling. Interestingly, the homologue *dickkopf 1* (*Dkk1*) inhibits wnt signaling by binding to the *Lrp5/6* receptor and the mouse doubleridge mutation (*Dkk1^dblr*), a transgenic insertion into *Dkk1*, causes vertebral fusions and segmental defects.[68]

SFRP1 and 2 are secreted wnt antagonists, which contain a domain similar to the wnt receptor, Frizzled.[69] This domain interacts with the wnt ligand and, as a result, wnt signaling is inhibited. Homozygous mutations of the *Sfrp1* or *Sfrp2* genes alone yield no discernable phenotypes; however, double homozygous mutants have severe shortening of the thoracic region.[70]

Noncanonical Wnt Signaling Pathway

Ror2 encodes a tyrosine kinase receptor that possesses a cysteine-rich domain similar to Wnt-binding domain on the Frizzled receptor.[71] Both *Ror2* and *Wnt5a* homozygous mutant mice display similar phenotypes, characterized by dwarfism, facial abnormalities, vertebral malformations and short tails.[71-73] The vertebral defects in *Ror2^-/-* mice are due to reduced PSM and abnormalities in somitogenesis.[72]

FGF (Fibroblast Growth Factor) Signaling Pathway

Genes with cycling expression profiles in the PSM are also found in the fibroblast growth factor signaling pathway.[15] Pequeant et al discovered that the genes *Spry2, Efna1, Hspg2, Egr1, Dusp6, Bcl2l11* and *Shp2* demonstrate oscillatory expression during a microarray expression screen of the PSM of 9.0 dpc embryos. Targeted mutants have only been reported for *Shp2* and null mutant embryos exhibit abnormal mesoderm patterning, culminating in gastrulation failure.[73]

The fibroblast growth factor gene, *Fgf8*, is expressed in the caudal PSM and has been shown to play a key role in determining the region where somite compartmental boundaries are determined.[74,75] However, targeted mutations in *Fgf8* display numerous developmental defects, but no reported segmental disruptions.[76] On the other hand, embryos with mutations in the FGF receptor 1 (*Fgfr1*), which abolishes binding of FGF receptor-specific substrates 2 and 3 (FRS2 and FRS3), display defects in neural tube closure and tail bud development.[77]

RA (Retinoic Acid) Signaling Pathway

RA is synthesized in the PSM by RALDH2 and is essential for antagonizing *Fgf8* expression as well as maintaining bilateral symmetry of left and right somite columns.[78-81] *Raldh2*[-/-] embryos display aberrant asymmetric expression in the PSM of the oscillatory Notch pathway genes *Hes7* and *Lnfg*.[80] However, normal asymmetric expression of *Nodal* and *Pitx2*, which are required for left-right asymmetry, is preserved in the lateral plate mesoderm of *Raldh2*[-/-] embryos.[82]

BMP/TGF-Beta Signaling Pathway

Gdf11 (Bmp11) plays a critical role in establishing skeletal pattern.[83] *Gdf11* homozygous mutant mice show anterograde homeotic transformations throughout the axial skeleton and resulting displacement of the hindlimbs caudally. Analysis of *Gdf11* mutants found posterior displacement or expansion of expression of *Hoxc6, Hoxc8, Hoxc10* and *Hoxc11*. These findings show that *Gdf11* regulates *Hox* genes and thus plays a role in specification of positional identity along the AP border.

The BMP antagonist, Noggin (*Nog*), is expressed in the notochord, node, dorsal spinal cord and dorsal lip of the rostral somites.[84] Neural tube and somite mesoderm growth and patterning are severely affected in Noggin null embryos and as a consequence, have an extremely malformed axial skeleton.[84] *Bmp4* haploinsufficiency suppresses the Noggin axial skeleton phenotype.[85] Thus, Noggin regulation of Bmp4 signaling is crucial for patterning and development of the somites and the axial skeleton.

B-cell translocation gene 2 (*Btg2*) is a coactivator-corepressor of the BMP signaling pathway.[86] *Btg2* enhances the activity of BMP signaling by interacting with BMP-activated Smads.[86] During somitogenesis, *Btg2* is expressed in the PSM, tail bud and somites. *Btg2* null mice exhibit posteriograde homeotic transformation of axial vertebrae, suggesting that BMP signaling may be interacting with the *Hox* complexes during somitogenesis.[86](see section on Homeobox transcription factors).

Hedgehog Signaling Pathway

In *Drosophila*, hedgehog was identified in a saturation mutagenesis screen as a gene involved in body segmentation. Hedgehog is essential in creating differences between anterior and posterior parts of individual body segments.[87] Within the sonic hedgehog (*Shh*) pathway, the vertebrate hedgehog homologue, the *Gli* genes play a role in sclerotomal development. Double homozygous *Gli2-/-;Gli3-/-* mutants have severely reduced sclerotomal gene expression, abnormal somite morphology and abnormal gene expression in dermomyotome and myotome.[88] These sclerotomal and myotomal phenotypes are more severe than those observed in *Shh-/-Gli3-/-* mutants. Thus, *Gli2* and *Gli3* both appear to be required for mediating Shh-induced sclerotomal gene expression.

Snail Transcriptional Repressors

The *Drosophila* zinc finger protein *snail* is required for mesodermal development and the mouse homologues snail (*Snai1*) and slug (*Snai2*) are both expressed in the primitive streak and mesoderm at gastrulation.[89] In addition, *Snai1* exhibits cyclic expression in the PSM, which requires Wnt and FGF signaling but is independent of Notch signaling.[90] Targeted disruption of *Snai1* leads to lethality prior to 8.5 dpc.[91] *Snai1* mutant embryos form a mesodermal layer, but the cells are abnormal in morphology with epithelial characteristics. In contrast, disruption of *Snai2* does not lead to any defects in mesoderm formation or patterning, suggesting that there may be overlap with *Snai1* function.[92]

T-Box Transcription Factors

The T-box gene *Tbx18* is expressed in two stripes in anterior compartments at the rostral end the PSM. Targeted disruption of *Tbx18* leads to a decrease in the size of the caudal compartment of the sclerotome and results in reduction of pedicles and proximal ribs.[93] Transgenic overexpression of *Tbx18* also leads to disruption of rostral-caudal compartments within the formed somites. In *Tbx18* mutants, newly formed somites appear to have normal rostral-caudal compartmentalization, as assayed by somite gene markers, suggesting that *Tbx18* is downstream of the regulatory mechanism determining AP boundaries. However, the caudal compartmental marker *Uncx4.1* expands during somite maturation in *Tbx18* mutants, suggesting that *Tbx18* may be required for maintenance of the integrity of these compartmental regions.

Rib-vertebrae (rv) is a spontaneous, autosomal recessive mutation in the T-box gene *Tbx6* that affects somite formation and patterning.[94-96] This mutation leads to malformation of the axial skeleton. In mice, hemivertebrae, butterfly vertebrae, vertebral fusions (both vertebral bodies and lateral masses) and rib fusions are present and in embryos, somites are irregular in shape and size, epithelial morphology is disrupted and AP patterning is abnormal. Genetic mapping and positional cloning have identified that *rv* is a regulatory mutation of *Tbx6* causing the hypomorphic phenotype seen in these mice.[95,96] Homozyous targeted mutations in *Tbx6* cause a more severe phenotype (more severe rib fusion and extensive fusion of lateral masses), with posterior paraxial mesoderm being replaced by neural tubes and severely reduced *Dll1* expression in the PSM, suggesting that *Dll1* may be a target of *Tbx6*. Double heterozygous *rv/+; Dll1*^{lacZ}*/+* mutants show some vertebral defects, but *rv/rv;Dll1*^{lacZ}*/+* show significantly more severe vertebral malformations, including severe vertebral and rib fusions. Disruption of *Dll1* results in loss of posterior somite compartments, as observed in *Dll1*^{lacZ}*/Dll1*^{lacZ} mutant embryos. While *Dll1* expression was absent in *rv/rv* mutants, posterior compartments were expanded rather than reduced.[97] These findings suggest that *Dll1* expression within the segmented mesoderm is not required to confer posterior identities to somite halves, but is required for establishing and maintaining posterior half-segment identity in either the PSM or nascent somites (or both) but not in mature somites.

Homeobox Transcription Factors

During somitogenesis, the segments generated appear to be morphologically identical, whereas the vertebrae have distinct shapes, depending on their location along the rostral-caudal axis. Even at formation, somite segments are molecularly encoded to direct the development of unique morphology of each vertebra. The *Hox* complexes play a key role in determination of rostral-caudal identity, due to the unique patterns of expression of component genes.[98] Genes in the *Hox* complexes are expressed in the paraxial mesoderm, with the anterior border of expression of the *Hox* gene colinear with its genomic localization within the complex. Thus, mutations of mouse *Hoxd3* cause occipitalization, or fusion of the atlas (first cervical vertebra) to the immediately rostral occipitum[99] and *Hoxd4* disruption causes replacement of the occipital region by supernumerary cervical vertebrae.[100] Further caudally, disruption of all paralogous group 10 *Hox* genes results in the transformation of lumbar vertebrae into vertebrae having a thoracic appearance with attached ribs and loss of all group 11 *Hox* genes leads to sacral vertebrae having a lumbar phenotype.[101] On the other hand, ectopic expression of *Hoxa10* in the PSM conferred a group 10 *Hox* patterning program to the somites, thereby producing vertebrae without ribs, similar to lumbar vertebrae.[102] Furthermore, ectopic *Hoxa11* expression in the PSM led to fusions of thoracic vertebrae, similar to fusions observed in sacral vertebrae and an anteriorized position of the sacrum.[102] Ectopic expression of *Hoxa11* in the somitic mesoderm resulted in anterior projections in lumbar vertebrae similar to those observed in caudal vertebrae. In summary, *Hoxa11* ectopic expression produces a posteriograde transformation of vertebral morphology.

Hox genes have been shown to have stage-specific expression that changes with each somite cycle and mutation in the notch pathway effector *Rbpjsuh* has been shown to lead to the disruption of *Hoxd1* and *Hoxd3* dynamic expression.[103] Cordes et al have generated transgenic mice with dominant negative alleles of *Dll1*, which have reduced notch signaling in the PSM.[104] This disruption leads to the absence of *Hes5* expression and defects in somites and vertebrae similar to those seen in other notch pathway mutants. However, these *Dll1* dominant negative mutants display an additional phenotype of altered vertebral identities resembling homeotic transformations and subtle changes of *Hox* gene expression in day 12.5 embryos. In depth analysis of *Dll1* heterozygous animals showed previously undetected ribs on C7 cervical vertebrae, indicating a posterograde shift of identity. In homozygous *Lfng* null mutants and in transgenic animals overexpressing *Lfng*, vertebral identities were changed and numbers of segments in the cervical and thoracic regions were reduced. Combined with observations that the expression of *Hoxb6* was shifted rostrally in *Lfng* mutants, these findings suggest that anterior shifts of axial identity are occurring and that precisely regulated levels of *Notch1* and *Lfng* activity are critical for positional specification of the anteroposterior body axis in the paraxial mesoderm.

Cdx2 is a caudal-type homeobox gene needed for trophoblastic development, vasculogenesis in the yolk sac mesoderm, allantoic growth, chorioallantoic fusion, completion of gastrulation and tail bud elongation.[105] *Cdx2* heterozygous animals have tail abnormalities and posterograde shifts in identify of cervical and upper thoracic vertebrae and ribs. Loss of *Cdx2* also causes a posterior shift in the expression of *Hox* genes. Homozygous *Cdx2* mutant embryos die between 3.5 and 5.5 dpc, the onset of gastrulation. In *Cdx2* homozygous mutants "rescued" by tetraploid aggregation, gastrulation takes place normally but axial segmentation halts at approximately somite 17 and somite morphology is abnormal after somite 5. Thus, *Cdx2* likely plays an important role in regulating both the production of paraxial mesoderm required for embryonic axial elongation and the AP patterning through *Hox* complexes.

The mesenchyme homeobox gene 1 (*Meox1/Mox1*) null mice have mild defects in vertebrae and ribs and *Meox2* homozygous mutant mice have limb musculature abnormalities.[106] *Meox1-/-;Meox2-/-* double mutant mice show severely disrupted somite morphogenesis, patterning and severely deficient skeletal muscles, which can be traced back to early defects in ventral somite specification and patterning. Overall, these double mutants have severely disrupted vertebrae formation. These malformations grow progressively worse in the caudal direction; multiple tail vertebrae are absent. In addition, no ribs form in these mutants. The transcription factors *Pax1*, *Pax2*, *Meox1*, *Meox2* and *Bapx1/Nkx3.1* are required for differentiation of the sclerotome.[107] In *Meox1-Meox2* double mutants, *Bapx1/Nkx3.1* expression in sclerotome is lost, indicating that *Bapx1*, which regulates chondrogenic differentiation, is downstream of *Meox* genes. Further analysis shows that *Meox1* activates the *Bapx1* promoter in a dose dependent manner and this activation is enhanced by *Pax1* and/or *Pax9*.

Unxc4.1 is a homeobox gene whose expression is restricted to the caudal half of the newly formed somite and sclerotome. Homozygous *Uncx4.1* mutant mice die at birth and have defects in the axial skeleton and ribs, with no formation of pedicles of the neural arches and proximal ribs, disorganization of the dorsal root ganglia and disorganization of the spinal nerve fibers.[108,109] Given the fact that the missing skeletal elements are derived from the caudal half-sclerotome (Fig. 2), the authors conclude that *Uncx4.1* is needed for its condensation, but not for somite segmentation. *Uncx4.1* is likely to be upstream of *Pax9* in the caudolateral sclerotome and may play a role in the differential cell adhesion properties of the somite.

Gastrulation brain homeobox 2 (*Gbx2*) is expressed in the PSM but not the somites.[102] Skeletal preps from *Gbx2*[-/-] embryos possessed homeotic transformations of the axial skeleton.[110] However, these null mutations in *Gbx2* did not alter the spatiotemporal expression of other *Hox* genes in the somites, suggesting that its activity in the PSM is sufficient to provide patterning instructions to the ensuing somites.[102]

Paired-Box Transcription Factors

The paired-box genes have also been shown to regulate somite segmentation in the mouse. Mouse *Pax1* mutations, *undulated* and *Danforth's short-tail*, have severe malformations of the vertebral bodies, intervertebral discs and ribs.[111,112] *Splotch* mutants (*Pax3^Sp*) display defects in spinal muscle and vertebrae and *Pax3* is expressed in the anterior PSM and newly formed somites, becoming localized to the ventral myotomal region during somite maturation.[113] During somitogenesis, the *Pax3^Sp* mutant embryos display disruption of rostral-caudal compartmentalization of *Epha4* and irregular expression of *Uncx4.1* and *Lfng* in newly formed somites. These finding suggests that *Pax3* plays a role in the rostral-caudal compartmentalization within somites before it starts functioning as a regulator of dermomyotome, resulting in segmental vertebral defects.

Forkhead Transcription Factors

Foxc1 and *Foxc2* are forkhead/winged helix genes, which encode closely related transcription factors required in paraxial mesoderm. *Foxc1* and *Foxc2* homozygous mutants display defects in chondrogenesis and sclerotome formation. Double homozygous *Foxc1-Foxc2* mutants display even more severe defects, including the complete absence of segmented paraxial mesoderm. These mutants die *in utero*.[114,115] *Foxc1-Foxc2* double mutants also display diffused boundaries of expression of notch pathway genes, including *Lfng* and *Dll1* and *Mesp2*, suggesting that prepatterning within the presomitic mesoderm is disrupted.

Basic Helix-Loop-Helix (bHLH) Transcription Factors

At the time of formation of somite segments at the anterior end of the PSM, tissue boundaries delineating anterior and posterior compartments of each segment are already specified. This process of anterior/posterior (AP) patterning is regulated by the *Mesp* genes and is discussed further in the chapter by Y. Saga.

In brief, during the process of segmentation, rostral caudal differences within each somite or precursor region become determined. In interactions with notch pathway genes, *Mesp2* appears to play a key role in this AP determination. *Mesp2* knockout mice have abnormal segmentation of somites, abnormal rostral identity of sclerotomes and abnormal rostral-caudal polarity of somites.[116] Disruptions of the related gene *Mesp1* produce mutants with morphogenetic abnormalities of the heart, however, these mutants have normal somites.[117] *Mesp1* and *Mesp2* double knockout mice show a complete lack of mesodermal migration.

Takahashi et al[118] have proposed that feedback loops of *Dll1* and *Mesp2* are essential for the establishment of rostrocaudal polarity. The authors propose that *Dll3* is essential for localization and integration of expression of *Dll1* and that *Mesp2* may be needed for the coordination of the *Dll1-Mesp2* loop. The *Psen1* independent *Dll3-Notch1* pathway can counteract the *Psen1* dependent *Dll1-Notch* signaling pathway and the authors question whether *Dll1* and *Dll3* have counteracting functions. They also suggest that *Mesp2* plays a central role in AP patterning through regulation of *Dll1* or *Dll3*.

Interestingly, some mouse mutants appear to have normal presomitic mesodermal gene expression boundaries but fail to properly produce segments. *Tcf15* (Paraxis) is a gene required for somite epithelialization. Despite normal expression of *Mesp2* and notch pathway genes such as *Notch1* and *Dll1*, *Tcf15* homozygous mutant mice still have defects in the axial skeleton, including poor formation of vertebral bodies and elements and rib fusions and have defects in peripheral nerves that are consistent with a disruption of AP patterning.[119,120] *Tcf15* appears to play a role as a factor promoting epithelialization of somites that is independent of notch signaling.

Homozygous mutations in pMesogenin1 (*Msgn1*) leads to a failure of formation of somites and animals completely lack axial structures, including skeletal muscle, vertebrae and ribs. The nonaxial skeleton, such as hindlimbs, develops normally in these animals.[121] Loss of *Msgn1* disrupts expression of notch pathway genes important in somitogenesis, including *Notch1, Dll1, Dll3, Hes1* and *Lfng*, affecting multiple components of the notch-signaling pathway. Given the severe defects seen in these mutants, *Msgn1* may play a key role in early paraxial mesoderm specification, which is an essential step preceding somitogenesis. More information on bHLH genes and their roles in somitogenesis can be found in other chapters.

Protocadherins

The protocadherin *(Pcdh8/Papc)* gene is required for cell-cell adhesion and was observed to be expressed in an anterior-specific manner in the presumptive somites in the PSM of mouse, *Xenopus* and zebrafish. In *Pcdh8* homozygous mutant mice, the PSM fails to epithelialize at the segmental border and distinct segmental boundaries fail to form.[122] The authors suggest that transcription of *Pcdh8* depends on *Mesp2*, which specifies the anterior domain of the somite and *Lfng*, which regulates the segmentation clock. They also suggest that *Pcdh8* is a morphogenic gene required for directing epithelialization at the site of border formation during somitogenesis.

Other Genes

Dmrt2, a member of the DM (Doublesex/MAB-3) domain family, was first associated with sex determination in flies and worms and later found expressed in developing somites.[123,124] While a targeted mutation of *Dmrt2* displays normal sex determination, null embryos fail to form normal epithelial morphology in the dermotome and myotome leading to sternal and rib defects.[125] Interestingly, normal axial muscle mass forms suggesting a *Dmrt2* independent mechanism of myogenesis.

Sip1, a member of the ZFHX1 family of two-handed zinc finger transcription factors, is expressed in the neuroectoderm, paraxial mesoderm and neural crest cells.[126] Homozygous *Sip1* knockout mice produce somites with approximately 50% of the rostral-caudal length of wild type embryo somites.[126] Furthermore, the altered somitogenic periodicity is accompanied by rostral expansion of the expression of *fgf8, Wnt3a, Dll3* and *Tbx6*, which are involved in maintenance of unsegmented paraxial mesoderm.[116] Additionally, the oscillatory genes *Lnfg, Dll1* and *Hes7*, as well as the termination of *Raldh2* expression, is shifted rostrally.[126] Human mutations in *SIP1* are associated with a form of Hirschsprung disease.[127]

The notochord plays an important role during somitogenesis and Barrinuevo et al further demonstrate this by investigating the role of *Sox9* in axial skeletogenesis.[128] Expression of *Sox9* in the notochord and sclerotome suggests a role in axial skeletal development. Absence of *Sox9* results in severe malformations of the vertebral column. Before manifestation of the vertebral phenotype, 9.5 dpc embryos with deficiencies in *Sox9* exhibit a cranial to caudal disintegration of the notochord. This uncovers a function of notochord-derived signals in promoting segmentation of the ventral sclerotome and a requirement of *Sox9* in axial skeletogenesis by regulating notochord survival.

Protein-tyrosine phosphatase nonreceptor 12 (*Ptpn12/PTP-PEST*) is a ubiquitously expressed regulator of integrin signaling and its product has been demonstrated to bind several proteins participating downstream of the focal adhesion kinase pathway.[129] Therefore, *Ptpn12* plays a vital role in cell-matrix interactions, which are essential for proper embryogenesis. Embryos null for *Ptpn12* are able to reach gastrulation but display abnormalities in turning and somitogenesis. In addition, null embryos surviving to 10.5 dpc exhibit caudal truncations and mesenchyme deficiencies.[129]

Conclusion

Not all genes expressed in the presomitic mesoderm prior to segmentation have phenotypes in targeted mutations. The cerberus 1 (*Cer1*) gene is a cytokine and inhibitor of the Wnt and BMP pathways[130] and is expressed in the rostral compartments of newly formed somites and in the anterior border of the PSM. However, homozygous targeted mutations do not display any segmental defects.[131] The *Twist1* gene is expressed in developing sclerotome, but homozygous disruption of the gene leads to lethality by 11.5 dpc with neural tube defects and heterozygous animals display craniofacial but no segmental defects.[132] *Epha4,* a member of the Eph family of tyrosine kinase receptors that is important in maintaining developmental compartments, is expressed in the rostral region of the newly formed somite and in the PSM. However, disruption of *Epha4* does not produce any reported segmental defects,[133] suggesting that the function of *Epha4* may overlap with that of other genes.

Significant advances in our understanding of somitogenesis have derived from studies in the mouse genetic system. However, despite the resources available from the Mouse Genome Project, many genes involved in the regulation of somitogenesis remain to be studied. First, many spontaneous mutations have yet to be cloned. These include the classic mouse mutations amputated (*am*), jumbled spine and ribs (*Jsr*), malformed vertebrae (*Mv*), rachiterata (*rh*) and tail kinks (*tk*), as described further in Table 1.[134-140] Some mutations, such as rib fusions (*Rf*) have become extinct.[141] However, new mutants have been generated that have phenotypes similar to some of these spontaneous mutations and may provide further insight into the genetics underlying the phenotypes. For example, Okano et al showed that *Jsr* and *Lnfg* are in the same chromosomal location as well as demonstrating similar mutant phenotypes.[46,143,47,33,142] Therefore, *Lnfg* is a likely candidate for *Jsr*. Furthermore, several international efforts are underway to generate additional ENU-induced mutation alleles[144] and mice with tail kinks and vertebral phenotypes have been reported. Genes that are expressed within somites continue to be described and many genes still do not have mutational alleles for further analysis.

Identification and characterization of mouse somitogenesis mutants are an essential model for understanding the causes of human spinal birth defects. Already, disruptions in three segmentation genes have been identified in the etiology of the severe vertebral disorder, spondylocostal dysostosis (SCD). Mutations in *DLL3* have been reported to cause SCD type 1, disruption of *MESP2* has been identified in a family with SCD type 2 and recently a missense mutation in *LNFG* has been reported in a family with autosomal recessive SCD.[145-150] *ROR2* has been implicated in the pathology of Robinow syndrome, a form of dwarfism characterized by vertebral and craniofacial malformations.[72]

Klippel-Feil (KF) anomaly is a human segmental disorder characterized by cervical vertebral fusions that limit flexion of the neck. The mouse mutant undulated (*Pax1^{un}*) has cervical vertebral defects similar to patients with KF anomaly.[151] Sequencing of these patients identified *PAX1* variant alleles that may play a role in the development of KF anomaly.[152]

Patients with fibrodysplasia ossificans progressiva (FOP) have been described to display KF anomaly. However, radiographic analysis of patients with FOP revealed a progressive cervical vertebral fusion, distinct from KF anomaly, but similar to the cervical fusions observed in mice with mutations in the BMP/TGF-beta antagonist Noggin.[153] FOP has been cloned and found to be a defect in the BMP/TGF-beta receptor *ACVR1*.[154] Since Noggin and *ACVR1* are components of the same signaling pathway, similarities in phenotypes are not unexpected.

Given that human congenital vertebral defects appear to be genetically heterogeneous, mouse models are a powerful tool to use in conjunction with human genetic approaches. Microarray based screens have recently been utilized to characterize the transcriptome of the PSM. These assays reveal what genes are expressed in the PSM and whether those genes are up or down regulated, as well as highlight patterns to the changes in expression. Many genes important in early development and patterning of the somites have been identified by microarray analysis of PSM and somites[155] and *Dll1* null mutants.[156] Microarray analysis of mouse embryonic PSM has identified novel wnt pathway cycling genes, including *Myc, Has2, Dkk1, Sp5, Tnfrsf19* and *Phlda1.*[15] Genetic analysis of mouse mutations disrupting somitogenesis and vertebral patterning will continue to be a valuable source of candidate genes for identifying the causes of human spinal birth defects.

Acknowledgements

We thank Alyssa Schaffer for helpful comments. KK is a Hitching-Elion Fellow of the Burroughs Wellcome Fund.

Bibliography

1. Beck JA, Lloyd S, Hafezparast M et al. Genealogies of mouse inbred strains. Nat Genet 2000; 24:23-5.
2. Kispert A, Hermann BG. The Brachyury gene encodes a novel DNA binding protein. EMBO J 1993; 12:4898-9.
3. Balling R, Deutsch U, Gruss P. undulated, a mutation affecting the development of the mouse skeleton, has a point mutation in the paired box of Pax 1. Cell 1988; 55:531-5.
4. Kusumi K, Sun ES, Kerrebrock AW et al. The mouse pudgy mutation disrupts Delta homologue Dll3 and initiation of early somite boundaries. Nat Genet 1998; 19:274-8.
5. Tam PP. The control of somitogenesis in mouse embryos. J Embryol Exp Morphol 1981; 65 Suppl:103-28.
6. Cooke J. The problem of periodic patterns in embryos. Philos Trans R Soc Lond B Biol Sci 1981; 295:509-24.
7. Cooke J. A gene that resuscitates a theory--somitogenesis and a molecular oscillator. Trends Genet 1998; 14:85-8.
8. Dale KJ, Pourquie O. A clock-work somite. Bioessays 2000; 22:72-83.
9. Keynes RJ, Stern CD. Mechanisms of vertebrate segmentation. Development 1988; 103:413-29.
10. Kulesa PM, Fraser SE. Cell dynamics during somite boundary formation revealed by time-lapse analysis. Science 2002; 298:991-5.
11. Keynes RJ, Stern CD. Segmentation in the vertebrate nervous system. Nature 1984; 310:786-9.
12. Brand-Saberi B, Christ B. Evolution and development of distinct cell lineages derived from somites. Curr Top Dev Biol 2000; 48:1-42.
13. Aulehla A, Johnson RL. Dynamic expression of lunatic fringe suggests a link between notch signaling and an autonomous cellular oscillator driving somite segmentation. Dev Biol 1999; 207:49-61.
14. Bessho Y, Sakata R, Komatsu S, et al. Dynamic expression and essential functions of Hes7 in somite segmentation. Genes Dev 2001b; 15:2642-7.
15. Dequeant ML, Glynn E, Gaudenz K et al. A complex oscillating network of signaling genes underlies the mouse segmentation clock. Science 2006; 314:1595-8.
16. Forsberg H, Crozet F, Brown NA. Waves of mouse Lunatic fringe expression, in four-hour cycles at two-hour intervals, precede somite boundary formation. Curr Biol 1998; 8:1027-30.
17. Jiang YJ, Aerne BL, Smithers L et al. Notch signalling and the synchronization of the somite segmentation clock. Nature 2000; 408:475-9.
18. Jouve C, Palmeirim I, Henrique D et al. Notch signalling is required for cyclic expression of the hairy-like gene HES1 in the presomitic mesoderm. Development 2000; 127:1421-9.
19. Leimeister C, Dale K, Fischer A et al. Oscillating expression of c-Hey2 in the presomitic mesoderm suggests that the segmentation clock may use combinatorial signaling through multiple interacting bHLH factors. Dev Biol 2000; 227:91-103.
20. Leimeister C, Externbrink A, Klamt B et al. Hey genes: a novel subfamily of hairy- and Enhancer of split related genes specifically expressed during mouse embryogenesis. Mech Dev 1999; 85:173-7.
21. McGrew MJ, Dale JK, Fraboulet S et al. The lunatic fringe gene is a target of the molecular clock linked to somite segmentation in avian embryos. Curr Biol 1998; 8:979-82.
22. Palmeirim I, Henrique D, Ish-Horowicz D et al. Avian hairy gene expression identifies a molecular clock linked to vertebrate segmentation and somitogenesis. Cell 1997; 91:639-48.
24. Bettenhausen B, Hrabe de Angelis M, Simon D et al. Transient and restricted expression during mouse embryogenesis of Dll1, a murine gene closely related to Drosophila Delta. Development 1995; 121:2407-18.
25. de la Pompa JL, Wakeham A, Correia KM et al. Conservation of the Notch signalling pathway in mammalian neurogenesis. Development 1997; 124:1139-48.
27. Koo BK, Lim HS, Song R et al. Mind bomb 1 is essential for generating functional Notch ligands to activate Notch. Development 2005; 132:3459-70.
28. Nakagawa O, McFadden DG, Nakagawa M et al. Members of the HRT family of basic helix-loop-helix proteins act as transcriptional repressors downstream of Notch signaling. Proc Natl Acad Sci USA 2000; 97:13655-60.
26. Dunwoodie SL, Henrique D, Harrison SM et al. Mouse Dll:3 a novel divergent Delta gene which may complement the function of other Delta homologues during early pattern formation in the mouse embryo. Development 1997; 124:3065-76.

29. Reaume AG, Conlon RA, Zirngibl R et al. Expression analysis of a Notch homologue in the mouse embryo. Dev Biol 1992; 154:377-87.
23. Bessho Y, Miyoshi G, Sakata R et al. Hes: 7a bHLH-type repressor gene regulated by Notch and expressed in the presomitic mesoderm. Genes Cells 2001a; 6:175-85.
30. Shi S, Stanley P. Protein O-fucosyltransferase 1 is an essential component of Notch signaling pathways. Proc Natl Acad Sci USA 2003; 100:5234-9.
31. Williams R, Lendahl U, Lardelli M et al. Complementary and combinatorial patterns of Notch gene family expression during early mouse development. Mech Dev 1995; 53:357-68.
32. Wong PC, Zheng H, Chen H et al. Presenilin 1 is required for Notch1 and DII1 expression in the paraxial mesoderm. Nature 1997; 387:288-92.
33. Zhang N, Gridley T. Defects in somite formation in lunatic fringe-deficient mice. Nature 1998; 394:374-7.
34. Conlon RA, Reaume AG, Rossant J. Notch1 is required for the coordinate segmentation of somites. Development 1995; 121:1533-45.
35. Huppert SS, Le A, Schroeter EH et al. Embryonic lethality in mice homozygous for a processing-deficient allele of Notch1. Nature 2000; 405:966-70.
36. McCright B, Gao X, Shen L et al. Defects in development of the kidney, heart and eye vasculature in mice homozygous for a hypomorphic Notch2 mutation. Development 2001; 128:491-502.
38. Dunwoodie SL, Clements M, Sparrow DB et al. Axial skeletal defects caused by mutation in the spondylocostal dysplasia/pudgy gene Dll3 are associated with disruption of the segmentation clock within the presomitic mesoderm. Development 2002; 129:1795-806.
37. Hrabe de Angelis M, McIntyre J, 2nd and Gossler A. Maintenance of somite borders in mice requires the Delta homologue DII1. Nature 1997; 386:717-21.
39. Shinkai Y, Tsuji T, Kawamoto Y et al. New mutant mouse with skeletal deformities caused by mutation in delta like 3 (Dll3) gene. Exp Anim 2004; 53:129-36.
40. Xue Y, Gao X, Lindsell CE et al. Embryonic lethality and vascular defects in mice lacking the Notch ligand Jagged.1 Hum Mol Genet 1999; 8:723-30.
41. Emerick KM, Rand EB, Goldmuntz E et al. Features of Alagille syndrome in 92 patients: frequency and relation to prognosis. Hepatology 1999; 29:822-9.
42. Fortini ME. Gamma-secretase-mediated proteolysis in cell-surface-receptor signalling. Nat Rev Mol Cell Biol 2002; 3:673-84.
44. Donoviel DB, Hadjantonakis AK, Ikeda M et al. Mice lacking both presenilin genes exhibit early embryonic patterning defects. Genes Dev 1999; 13:2801-10.
43. Koizumi K, Nakajima M, Yuasa S. The role of presenilin 1 during somite segmentation. Development 2001; 128:1391-402.
45. Li T, Ma G, Cai H et al. Nicastrin is required for assembly of presenilin/gamma-secretase complexes to mediate Notch signaling and for processing and trafficking of beta-amyloid precursor protein in mammals. J Neurosci 2003; 23:3272-7.
46. Evrard YA, Lun Y, Aulehla A et al. lunatic fringe is an essential mediator of somite segmentation and patterning. Nature 1998; 394:377-81.
47. Serth K, Schuster-Gossler K, Cordes R et al. Transcriptional oscillation of lunatic fringe is essential for somitogenesis. Genes Dev 2003; 17:912-25.
48. Morales AV, Yasuda Y, Ish-Horowicz D. Periodic Lunatic fringe expression is controlled during segmentation by a cyclic transcriptional enhancer responsive to notch signaling. Dev Cell 2002; 3:63-74.
49. Cole SE, Levorse JM, Tilghman SM et al. Clock regulatory elements control cyclic expression of Lunatic fringe during somitogenesis. Dev Cell 2002; 3:75-84.
50. Ishibashi M, Moriyoshi K, Sasai Y et al. Persistent expression of helix-loop-helix factor HES-1 prevents mammalian neural differentiation in the central nervous system. EMBO J 1994; 13:1799-805.
51. Tomita K, Ishibashi M, Nakahara K et al. Mammalian hairy and Enhancer of split homolog 1 regulates differentiation of retinal neurons and is essential for eye morphogenesis. Neuron 1996; 16:723-34.
52. Cau E, Gradwohl G, Casarosa S et al. Hes genes regulate sequential stages of neurogenesis in the olfactory epithelium. Development 2000; 127:2323-32.
53. Ohtsuka T, Ishibashi M, Gradwohl G et al. Hes1 and Hes5 as notch effectors in mammalian neuronal differentiation. EMBO J 1999; 18:2196-207.
54. Hirata H, Bessho Y, Kokubu H et al. Instability of Hes7 protein is crucial for the somite segmentation clock. Nat Genet 2004; 36:750-4.
55. Kokubo H, Lun Y, Johnson RL. Identification and expression of a novel family of bHLH cDNAs related to Drosophila hairy and enhancer of split. Biochem Biophys Res Commun 1999; 260:459-65.
56. Steidl C, Leimeister C, Klamt B et al. Characterization of the human and mouse HEY1, HEY2 and HEYL genes: cloning, mapping and mutation screening of a new bHLH gene family. Genomics 2000; 66:195-203.

57. Fischer A, Schumacher N, Maier M et al. The Notch target genes Hey1 and Hey2 are required for embryonic vascular development. Genes Dev 2004; 18:901-11.
58. Gessler M, Knobeloch KP, Helisch A et al. Mouse gridlock: no aortic coarctation or deficiency, but fatal cardiac defects in Hey2 -/- mice. Curr Biol 2002; 12:1601-4.
59. Oka C, Nakano T, Wakeham A et al. Disruption of the mouse RBP-J kappa gene results in early embryonic death. Development 1995; 121:3291-301.
60. Aulehla A, Wehrle C, Brand-Saberi B et al. Wnt3a plays a major role in the segmentation clock controlling somitogenesis. Dev Cell 2003; 4:395-406.
61. Suriben R, Fisher DA, Cheyette BN et al. Dact1 presomitic mesoderm expression oscillates in phase with Axin2 in the somitogenesis clock of mice. Dev Dyn 2006; 235:3177-83.
62. Greco TL, Sussman DJ, Camper SA. Dishevelled-2 maps to human chromosome 17 and distal to Wnt3a and vestigial tail (vt) on mouse chromosome 11. Mamm Genome 1996; 7:475-6.
63. Takada S, Stark KL, Shea MJ et al. Wnt-3a regulates somite and tailbud formation in the mouse embryo. Genes Dev 1994; 8:174-89.
64. Axelrod JD, Matsuno K, Artavanis-Tsakonas S et al. Interaction between Wingless and Notch signaling pathways mediated by dishevelled. Science 1996; 271:1826-32.
65. Seidensticker MJ, Behrens J. Biochemical interactions in the wnt pathway. Biochim Biophys Acta 2000; 1495:168-82.
66. Kokubu C, Heinzmann U, Kokubu T et al. Skeletal defects in ringelschwanz mutant mice reveal that Lrp6 is required for proper somitogenesis and osteogenesis. Development 2004; 131: 5469-80.
67. Carter M, Chen X, Slowinska B et al. Crooked tail (Cd) model of human folate-responsive neural tube defects is mutated in Wnt coreceptor lipoprotein receptor-related protein 6. Proc Natl Acad Sci USA 2005; 102:12843-8.
68. MacDonald BT, Adamska M, Meisler MH. Hypomorphic expression of Dkk1 in the doubleridge mouse: dose dependence and compensatory interactions with Lrp6. Development 2004; 131:2543-52.
69. Kawano Y, Kypta R. Secreted antagonists of the Wnt signalling pathway. J Cell Sci 2003; 116:2627-34.
70. Satoh W, Gotoh T, Tsunematsu Y et al. Sfrp1 and Sfrp2 regulate anteroposterior axis elongation and somite segmentation during mouse embryogenesis. Development 2006; 133:989-99.
71. Oishi I, Suzuki H, Onishi N et al. The receptor tyrosine kinase Ror2 is involved in noncanonical Wnt5a/JNK signalling pathway. Genes Cells 2003; 8:645-54.
72. Schwabe GC, Trepczik B, Suring K et al. Ror2 knockout mouse as a model for the developmental pathology of autosomal recessive Robinow syndrome. Dev Dyn 2004; 229:400-10.
73. Saxton TM, Henkemeyer M, Gasca S et al. Abnormal mesoderm patterning in mouse embryos mutant for the SH2 tyrosine phosphatase Shp-2. EMBO J 1997; 16:2352-64.
73. Takeuchi S, Takeda K, Oishi I et al. Mouse Ror2 receptor tyrosine kinase is required for the heart development and limb formation. Genes Cells 2000; 5:71-8.
74. Dubrulle J, McGrew MJ, Pourquie O. FGF signaling controls somite boundary position and regulates segmentation clock control of spatiotemporal Hox gene activation. Cell 2001; 106:219-32.
75. Dubrulle J, Pourquie O. fgf8 mRNA decay establishes a gradient that couples axial elongation to patterning in the vertebrate embryo. Nature 2004; 427:419-22.
76. Meyers EN, Lewandoski M, Martin GR. An Fgf8 mutant allelic series generated by Cre- and Flp-mediated recombination. Nat Genet 1998; 18:136-41.
77. Hoch RV, Soriano P. Context-specific requirements for Fgfr1 signaling through Frs2 and Frs3 during mouse development. Development 2006; 133:663-73.
78. Diez del Corral R, Olivera-Martinez I, Goriely A. Opposing FGF and retinoid pathways control ventral neural pattern, neuronal differentiation and segmentation during body axis extension. Neuron 2003; 40:65-79.
79. Sirbu IO, Duester G. Retinoic-acid signalling in node ectoderm and posterior neural plate directs left-right patterning of somitic mesoderm. Nat Cell Biol 2006; 8:271-7.
80. Vermot J, Gallego Llamas J, Fraulob V et al. Retinoic acid controls the bilateral symmetry of somite formation in the mouse embryo. Science 2005; 308:563-6.
81. Vermot J, Pourquie O. Retinoic acid coordinates somitogenesis and left-right patterning in vertebrate embryos. Nature 2005; 435:215-20.
82. Niederreither K, Subbarayan V, Dolle P et al. Embryonic retinoic acid synthesis is essential for early mouse post-implantation development. Nat Genet 1999; 21:444-8.
83. McPherron AC, Lawler AM, Lee SJ. Regulation of anterior/posterior patterning of the axial skeleton by growth/differentiation factor 11. Nat Genet 1999; 22:260-4.
84. McMahon JA, Takada S, Zimmerman LB et al. Noggin-mediated antagonism of BMP signaling is required for growth and patterning of the neural tube and somite. Genes Dev 1998; 12: 1438-52.

85. Wijgerde M, Karp S, McMahon J et al. Noggin antagonism of BMP4 signaling controls development of the axial skeleton in the mouse. Dev Biol 2005; 286:149-57.
87. Nusslein-Volhard C, Wieschaus E. Mutations affecting segment number and polarity in Drosophila. Nature 1980; 287:795-801.
88. Buttitta L, Mo R, Hui CC et al. Interplays of Gli2 and Gli3 and their requirement in mediating Shh-dependent sclerotome induction. Development 2003a; 130:6233-43.
86. Park S, Lee YJ, Lee HJ et al. B-cell translocation gene 2 (Btg2) regulates vertebral patterning by modulating bone morphogenetic protein/smad signaling. Mol Cell Biol 2004; 24:10256-62.
89. Sefton M, Sanchez S, Nieto MA. Conserved and divergent roles for members of the Snail family of transcription factors in the chick and mouse embryo. Development 1998; 125:3111-21.
90. Dale JK, Malapert P, Chal J et al. Oscillations of the snail genes in the presomitic mesoderm coordinate segmental patterning and morphogenesis in vertebrate somitogenesis. Dev Cell 2006; 10:355-66.
91. Carver EA, Jiang R, Lan Y et al. The mouse snail gene encodes a key regulator of the epithelial-mesenchymal transition. Mol Cell Biol 2001; 21:8184-8.
92. Jiang R, Lan Y, Norton CR et al. The Slug gene is not essential for mesoderm or neural crest development in mice. Dev Biol 1998; 198, 277-85.
93. Bussen M, Petry M, Schuster-Gossler K et al. The T-box transcription factor Tbx18 maintains the separation of anterior and posterior somite compartments. Genes Dev 2004; 18:1209-21.
94. Nacke S, Schafer R, Habre de Angelis M et al. Mouse mutant rib-vertebrae (rv): a defect in somite polarity. Dev Dyn 2000; 219:192-200.
95. Watabe-Rudolph M, Schlautmann N, Papaioannou VE et al. The mouse rib-vertebrae mutation is a hypomorphic Tbx6 allele. Mech Dev 2002; 119:251-6.
96. White PH, Farkas DR, McFadden EE et al. Defective somite patterning in mouse embryos with reduced levels of Tbx6. Development 2003; 130:1681-90.
97. Beckers J, Schlautmann N, Gossler A. The mouse rib-vertebrae mutation disrupts anterior-posterior somite patterning and genetically interacts with a Delta1 null allele. Mech Dev 2000; 95:35-46.
98. Krumlauf, R. Hox genes in vertebrate development. Cell 1994; 78:191-201.
99. Condie BG, Capecchi MR. Mice homozygous for a targeted disruption of Hoxd-3 (Hox-4.1) exhibit anterior transformations of the first and second cervical vertebrae, the atlas and the axis. Development 1993; 119:579-95.
100. Lufkin T, Mark M, Hart CP et al. Homeotic transformation of the occipital bones of the skull by ectopic expression of a homeobox gene. Nature 1992; 359:835-41.
101. Wellik DM, Capecchi MR. Hox10 and Hox11 genes are required to globally pattern the mammalian skeleton. Science 2003; 301:363-7.
102. Carapuco M, Novoa A, Bobola N et al. Hox genes specify vertebral types in the presomitic mesoderm. Genes Dev 2005; 19:2116-21.
103. Zakany J, Kmita M, Alarcon P et al. Localized and transient transcription of Hox genes suggests a link between patterning and the segmentation clock. Cell 2001; 106:207-17.
104. Cordes R, Schuster-Gossler K, Serth K et al. Specification of vertebral identity is coupled to Notch signalling and the segmentation clock. Development 2004; 131:1221-33.
105. Chawengsaksophak K, de Graaff W, Rossant J et al. Cdx2 is essential for axial elongation in mouse development. Proc Natl Acad Sci USA 2004; 101:7641-5.
106. Mankoo BS, Skuntz S, Harrigan I et al. The concerted action of Meox homeobox genes is required upstream of genetic pathways essential for the formation, patterning and differentiation of somites. Development 2003; 130:4655-64.
107. Rodrigo I, Bovolenta P, Mankoo BS et al. Meox homeodomain proteins are required for Bapx1 expression in the sclerotome and activate its transcription by direct binding to its promoter. Mol Cell Biol 2004; 24:2757-66.
108. Leitges M, Neidhardt L, Haenig B et al. The paired homeobox gene Uncx4.1 specifies pedicles, transverse processes and proximal ribs of the vertebral column. Development 2000; 127:2259-67.
109. Mansouri A, Voss AK, Thomas T et al. Uncx4.1 is required for the formation of the pedicles and proximal ribs and acts upstream of Pax9. Development 2000; 127:2251-8.
110. Wassarman KM, Lewandoski M, Campbell K et al. Specification of the anterior hindbrain and establishment of a normal mid/hindbrain organizer is dependent on Gbx2 gene function. Development 1997; 124:2923-34.
111. Balling R. The undulated mouse and the development of the vertebral column. Is there a human PAX-1 homologue? Clin Dysmorphol 1994; 3:185-91.
112. Chi N, Epstein JA. Getting your Pax straight: Pax proteins in development and disease. Trends Genet 2002; 18:41-7.
113. Schubert FR, Tremblay P, Mansouri A et al. Early mesodermal phenotypes in splotch suggest a role for Pax3 in the formation of epithelial somites. Dev Dyn 2001; 222:506-21.

114. Kume T, Deng KY, Winfrey V et al. The forkhead/winged helix gene Mf1 is disrupted in the pleiotropic mouse mutation congenital hydrocephalus. Cell 1998; 93:985-96.

115. Kume T, Jiang H, Topczewska JM et al. The murine winged helix transcription factors, Foxc1 and Foxc,2 are both required for cardiovascular development and somitogenesis. Genes Dev 2001; 15:2470-82.

117. Saga Y, Miyagawa-Tomita S, Takagi A et al. MesP1 is expressed in the heart precursor cells and required for the formation of a single heart tube. Development 1999; 126:3437-47.

116. Saga Y, Hata N, Koseki H et al. Mesp:2 a novel mouse gene expressed in the presegmented mesoderm and essential for segmentation initiation. Genes Dev 1997; 11:1827-39.

118. Takahashi Y, Inoue T, Gossler A et al. Feedback loops comprising Dll1, Dll3 and Mesp2 and differential involvement of Psen1 are essential for rostrocaudal patterning of somites. Development 2003; 130:4259-68.

119. Burgess R, Rawls A, Brown D et al. Requirement of the paraxis gene for somite formation and musculoskeletal patterning. Nature 1996; 384:570-3.

120. Johnson J, Rhee J, Parsons SM et al. The anterior/posterior polarity of somites is disrupted in paraxis-deficient mice. Dev Biol 2001; 229:176-87.

121. Yoon JK, Wold B. The bHLH regulator pMesogenin1 is required for maturation and segmentation of paraxial mesoderm. Genes Dev 2000; 14:3204-14.

122. Rhee J, Takahashi Y, Saga Y et al. The protocadherin papc is involved in the organization of the epithelium along the segmental border during mouse somitogenesis. Dev Biol 2003; 254:248-61.

123. Kim S, Kettlewell JR, Anderson RC et al. Sexually dimorphic expression of multiple doublesex-related genes in the embryonic mouse gonad. Gene Expr Patterns 2003; 3:77-82.

124. Meng A, Moore B, Tang H et al. A Drosophila doublesex-related gene, terra, is involved in somitogenesis in vertebrates. Development 1999; 126:1259-68.

125. Seo KW, Wang Y, Kokubo H et al. Targeted disruption of the DM domain containing transcription factor Dmrt2 reveals an essential role in somite patterning. Dev Biol 2006; 290:200-10.

126. Maruhashi M, Van De Putte T, Huylebroeck D et al. Involvement of SIP1 in positioning of somite boundaries in the mouse embryo. Dev Dyn 2005; 234:332-8.

127. Cacheux V, Dastot-Le Moal F, Kaariainen H et al. Loss-of-function mutations in SIP1 Smad interacting protein 1 result in a syndromic Hirschsprung disease. Hum Mol Genet 2001; 10:1503-10.

128. Barrionuevo F, Taketo MM, Scherer G et al. Sox9 is required for notochord maintenance in mice. Dev Biol 2006; 295:128-40.

129. Sirois J, Cote JF, Charest A et al. Essential function of PTP-PEST during mouse embryonic vascularization, mesenchyme formation, neurogenesis and early liver development. Mech Dev doi:10.1016/j.mod.2006.08.011.

130. Biben C, Stanley E, Fabri L et al. Murine cerberus homologue mCer-1: a candidate anterior patterning molecule. Dev Biol 1998; 194:135-51.

131. Simpson EH, Johnson DK, Hunsicker P et al. The mouse Cer1 (Cerberus related or homologue) gene is not required for anterior pattern formation. Dev Biol 1999; 213:202-6.

132. Chen ZF, Behringer RR. twist is required in head mesenchyme for cranial neural tube morphogenesis. Genes Dev 1995; 9:686-99.

133. Helmbacher F, Schneider-Maunoury S, Topilko P et al. Targeting of the EphA4 tyrosine kinase receptor affects dorsal/ventral pathfinding of limb motor axons. Development 2000; 127: 3313-24.

134. Flint OP, Ede DA. Facial development in the mouse; a comparison between normal and mutant (amputated) mouse embryos. J Embryol Exp Morphol 1978; 48:249-67.

136. Flint OP, Ede DA, Wilby OK et al. Control of somite number in normal and amputated mutant mouse embryos: an experimental and a theoretical analysis. J Embryol Exp Morphol 1978; 45:189-202.

136. Grüneberg H. Genetical studies on the skeleton of the mouse. XVI. Tail-kinks. J Genet 1955; 53:536-550.

137. Miyoshi H, Kon Y, Seo KW et al. Jumbled spine and ribs (Jsr): a new mutation on mouse chromosome 5. Mamm Genome 1999; 10:213-7.

138. Morgan WC. A new crooked tail mutation involving distinctive pleiotropism. J Genet 1954; 52:354-373.

139. Theiler K, Varnum D, Stevens LC. Development of rachiterata, a mutation in the house mouse with 6 cervical vertebrae. Z Anat Entwicklungsgesch 1974; 145:75-80.

140. Theiler K, Varnum DS, Southard JL et al. Malformed vertebrae: a new mutant with the wirbel-rippen syndrom in the mouse. Anat Embryol (Berl) 1975; 147:161-6.

141. Theiler K, Stevens LC. The development of rib fusions, a mutation in the house mouse. Am J Anat 1960; 106:171-83.

142. Okano S, Asano A, Kon Y et al. Genetic analysis of jumbled spine and ribs (Jsr) mutation affecting the vertebral development in mice. Biochem Genet 2002; 40:311-22.

143. Okano S, Asano A, Sasaki N et al. Examination of the Lunatic fringe and Uncx4.1 expression by whole-mount in situ hybridization in the embryo of the CKH-Jsr (jumbled spine and ribs) mouse. Jpn J Vet Res 2005; 52:145-9.

144. Rossant J, McKerlie C. Mouse-based phenogenomics for modelling human disease. Trends Mol Med 2001; 7:502-7.

145. Bulman MP, Kusumi K, Frayling TM et al. Mutations in the human delta homologue, DLL3, cause axial skeletal defects in spondylocostal dysostosis. Nat Genet 2000; 24:438-41.

146. Sparrow DB, Chapman G, Wouters MA et al. Mutation of the LUNATIC FRINGE gene in humans causes spondylocostal dysostosis with a severe vertebral phenotype. Am J Hum Genet 2006; 78:28-37.

147. Sparrow DB, Clements M, Withington SL et al. Diverse requirements for Notch signalling in mammals. Int J Dev Biol 2002; 46:365-74.

149. Turnpenny PD, Whittock N, Duncan J et al. Novel mutations in DLL3, a somitogenesis gene encoding a ligand for the Notch signalling pathway, cause a consistent pattern of abnormal vertebral segmentation in spondylocostal dysostosis. J Med Genet 2003; 40:333-9.

150. Whittock NV, Sparrow DB, Wouters MA et al. Mutated MESP2 causes spondylocostal dysostosis in humans. Am J Hum Genet 2004; 74:1249-54.

151. McGaughran JM, Oates A, Donnai D et al. Mutations in PAX1 may be associated with Klippel-Feil syndrome. Eur J Hum Genet 2003; 11:468-74.

148. Turnpenny PD, Kusumi K. Delta-like 3 and spondylocostal dysostosis. New York: Oxford University Press 2004.

152. Schaffer AA, Kaplan FS, Tracy MR et al. Developmental anomalies of the cervical spine in patients with fibrodysplasia ossificans progressiva are distinctly different from those in patients with Klippel-Feil syndrome: clues from the BMP signaling pathway. Spine 2005; 30:1379-85.

153. Shore EM, Xu M, Feldman GJ et al. A recurrent mutation in the BMP type I receptor ACVR1 causes inherited and sporadic fibrodysplasia ossificans progressiva. Nat Genet 2006; 38:525-7.

154. Buttitta L, Tanaka TS, Chen A et al. Microarray analysis of somitogenesis reveals novel targets of different WNT signaling pathways in the somitic mesoderm. Dev Biol 2003b; 258:91-104.

155. Machka C, Kersten M, Zobawa M et al. Identification of Dll1 (Delta1) target genes during mouse embryogenesis using differential expression profiling. Gene Expr Patterns 2005; 6:94-101.

157. Higashi Y, Maruhashi M, Nelles L et al. Generation of the floxed allele of the SIP1 (Smad-interacting protein 1) gene for Cre-mediated conditional knockout in the mouse. Genesis 2002; 32:82-4.

CHAPTER 9

Defective Somitogenesis and Abnormal Vertebral Segmentation in Man

Peter D. Turnpenny*

Abstract

In recent years molecular genetics has revolutionized the study of somitogenesis in developmental biology and advances that have taken place in animal models have been applied successfully to human disease. Abnormal segmentation in man is a relatively common birth defect and advances in understanding have come through the study of cases clustered in families using DNA linkage analysis and candidate gene approaches, the latter stemming directly from knowledge gained through the study of animal models. Only a minority of abnormal segmentation phenotypes appear to follow Mendelian inheritance but three genes—*DLL3*, *MESP2* and *LNFG*—have now been identified for spondylocostal dysostosis (SCD), a spinal malformation characterized by extensive hemivertebrae, trunkal shortening and abnormally aligned ribs with points of fusion. In affected families autosomal recessive inheritance is followed. These genes are all important components of the Notch signaling pathway. Other genes within the pathway cause diverse phenotypes such as Alagille syndrome (AGS) and CADASIL, conditions that may have their origin in defective vasculogenesis. This review deals mainly with SCD, with some consideration of AGS. Significant future challenges lie in identifying causes of the many abnormal segmentation phenotypes in man but it is hoped that combined approaches in collaboration with developmental biologists will reap rewards.

Introduction

In medicine the problems associated with abnormal spinal segmentation are of great interest to a variety of disciplines. Radiologists seek to describe abnormal patterns on imaging, spinal surgeons have to make difficult decisions about surgery on affected children and adults, pediatricians have to care for the wider consequences such as respiratory insufficiency, and geneticists try to make specific diagnoses, consider genetic testing and offer recurrence risk figures when appropriate. Despite the fact that abnormal vertebral segmentation (AVS) in man is relatively common (though precise incidence and prevalence figures are not available), this is a field that has moved forward only slowly, though it has begun to accelerate recently. There is substantial confusion over nomenclature for the various radiological phenotypes, the cause in the majority of cases is not understood and genetic counseling is often, of necessity, somewhat vague. It is hoped that multidisciplinary approaches and large data collections can make inroads into these areas of difficulty and unknown.

In vertebrate species somites are symmetrically aligned paired blocks of mesoderm formed from the segmentation of paraxial presomitic mesoderm. The process begins shortly after gastrulation and continues until the preprogrammed number of somite blocks is formed; in man 31 blocks

*Peter D. Turnpenny—Clinical Genetics Department, Royal Devon & Exeter Hospital, Gladstone Road, Exeter EX1 2ED, United Kingdom. Email: peter.turnpenny@rdeft.nhs.uk

Somitogenesis, edited by Miguel Maroto and Neil V. Whittock. ©2008 Landes Bioscience and Springer Science+Business Media.

of paired tissue are formed but the number is specific for each species. In human embryonic development this process takes place between 20 and 35 days post conception. Somites are laid down in a rostro-caudal direction, first forming the most rostral somites and progressively laying down more caudal somites. Somites ultimately give rise to three substructures: sclerotome, which forms the axial skeleton and ribs; dermotome, which forms the dermis; and myotome, which forms the axial musculature. The potential number and diversity of human conditions due to defective somitogenesis is therefore large, but this chapter will concentrate on well-defined conditions of the axial skeleton and their genetic basis where this is known, as well as highlighting the large number of relevant phenotypes for which no cause has yet been identified. AVS, in all its various and diverse manifestations, is a common birth defect and an important 'handle' in clinical dysmorphology. However, the proportion of cases for which a cause can be confidently ascribed is small.

Nomenclature and Terminology

Before the modern era of molecular genetics and cell biology the classification of human disease relied upon visible differences in phenotype at the gross (naked eye) level, characteristic radiological patterns and/or histopathological features. Molecular genetics and cell biology have begun to complement and inform this basic framework and at the same time provide a new dimension in nomenclature. It is now accepted that a single clinical condition may not only follow more than one pattern of inheritance but also demonstrate genetic heterogeneity. Conversely, apparently diverse clinical phenotypes may be due to mutations in the same gene.

In segmentation disorders of the axial skeleton radiological features are crucial in syndrome delineation, which in turn is essential for offering accurate genetic counseling. However, in the medical literature and in clinical practice nomenclature is very confused, such that terms are used interchangeably between phenotypes. There is a need to introduce consistency and standardization. In describing segmentation abnormalities of the spine and ribs the terms *Jarcho-Levin syndrome*, *costovertebral/spondylocostal/spondylothoracic dysostosis/dysplasia* all feature. Strictly speaking these disorders are probably *dysostoses* rather than *dysplasias*. A *dysplasia* refers to a developmental and ongoing abnormality of chondro-osseous tissues during (pre- and) postnatal life, whilst a *dysostosis* is a stable condition resulting from a formation abnormality early in morphogenesis.

The eponymous *Jarcho-Levin syndrome* (JLS) is frequently used across the entire spectrum of radiological phenotypes that include abnormal vertebral segmentation (AVS) and rib alignment. In 1938 Jarcho and Levin reported two siblings of Puerto Rican origin with AVS of the entire vertebral column, though most severe in the thoracic region.[1] Fusion of several ribs was present and both subjects died in infancy of respiratory failure. Close scrutiny suggests the phenotype is closest to the form of *spondylocostal dysostosis* that we now call 'type 2', due to mutations in *MESP2* (see below). To prevent confusion we believe it is preferable to avoid the use of eponymous designations in this field (including the use of 'Klippel-Feil') in favour of a more rigorous descriptive system of radiological abnormalities.

In the early literature the term *costovertebral dysplasia* can be found[2-4] but is less often used today. There is a preference for *spondylocostal dysostosis* (SCD) for those phenotypes with extensive segmental vertebral involvement (≥10 contiguous segments) plus rib abnormalities with points of fusion and *multiple segmentation defects of the spine* (MSDS) when >1 <10 segments are involved. However, in practice a consensus on the use of terminology has yet to be achieved. As with JLS, the term SCD is used for a wide variety of abnormal axial radiological phenotypes,[5] including those with gross asymmetry that are probably sporadic (Fig. 1).

The term *spondylothoracic dysostosis/dysplasia* (STD) was first proposed by Moseley and Bonforte[6] and is best reserved for the distinctive condition, most commonly reported in Puerto Ricans, characterized by severe trunkal shortening and a radiological appearance of the ribs fanning out from a crowded vertebro-costal origin in a 'crab-like' fashion (Fig. 2). The ribs are fused posteriorly, though otherwise they are neatly aligned and packed tightly together and in contrast to the appearance in SCD, they show no points of fusion other than at their posterior origins. The thoracic vertebrae appear most severely affected and 'telescoped' together and scoliotic curves in

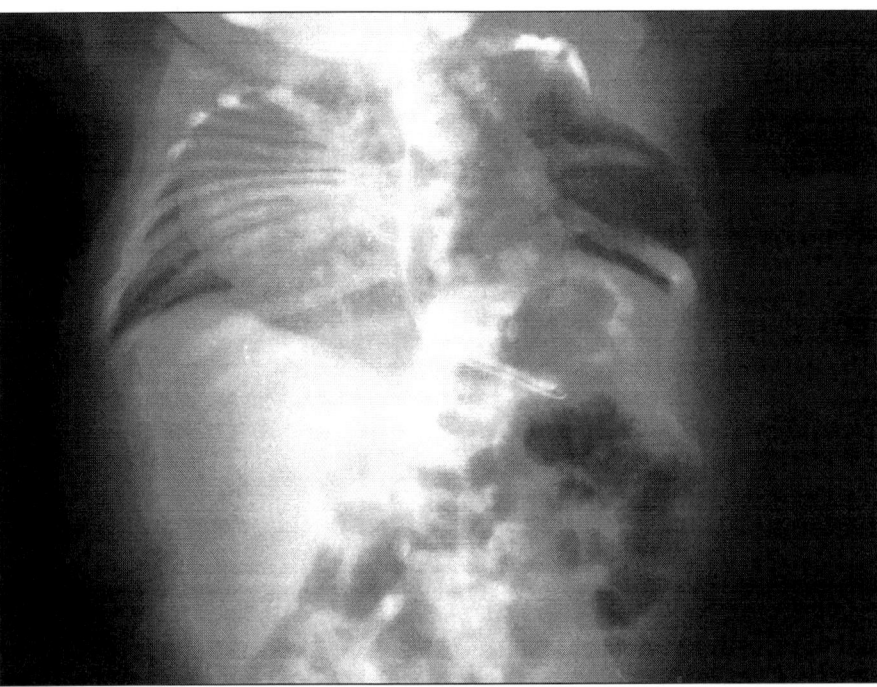

Figure 1. An example of a child with a severe segmentation malformation of the spine causing gross asymmetry and almost complete absence of ribs on one side. These cases are probably best not referred to as 'spondylocostal dysostosis' but there is no concensus on alternative nomenclature. Mutations in Notch signaling pathway genes have not so far been found in these cases.

early life are uncommon. This distinctive phenotype has been well characterised in recent studies.[7,8] Infant mortality can be as high as 50% due to restrictive respiratory insufficiency, though prognosis depends on the quality of medical care available. As illustration of the confusion in nomenclature, a number of case reports in the literature that demonstrate the typical STD phenotype, predominantly in Puerto Ricans, have been reported as examples of JLS.[6,9-18] Other case reports[19-24,25] designated as JLS are neither very similar to those described by Jarcho and Levin[1] nor consistent with STD, whilst others would be better classified as forms of SCD.[26-28]

The Diversity of Phenotypes Manifesting Axial Skeletal Defects

AVS is a feature of a broad range of dysmorphic syndromes, most of them rare. Table 1 gives a list of conditions that include MSDS, for only a small proportion of which is the cause known. This multiplicity of syndromes, with their different genetic and nongenetic causes, highlights the sensitivity and susceptibility of axial skeletal development to perturbations of normal somitogenesis. Apart from the Mendelian forms of SCD due to mutated genes in the Notch signaling pathway, the functions of other genes implicated in syndromes that include AVS are not necessarily so well understood. *ROR2*, mutated in Robinow syndrome, is a receptor tyrosine kinase required for transcriptional regulation within the Wnt signaling pathway, which is integral to somitogenesis.[29] *CHD7*, mutated in CHARGE syndrome, has mainly been studied in the developing central nervous system. However, a polymorphism within the gene was recently been shown to be significantly over-transmitted in a large series of cases of familial idiopathic scoliosis (without obvious segmentation defects).[30]

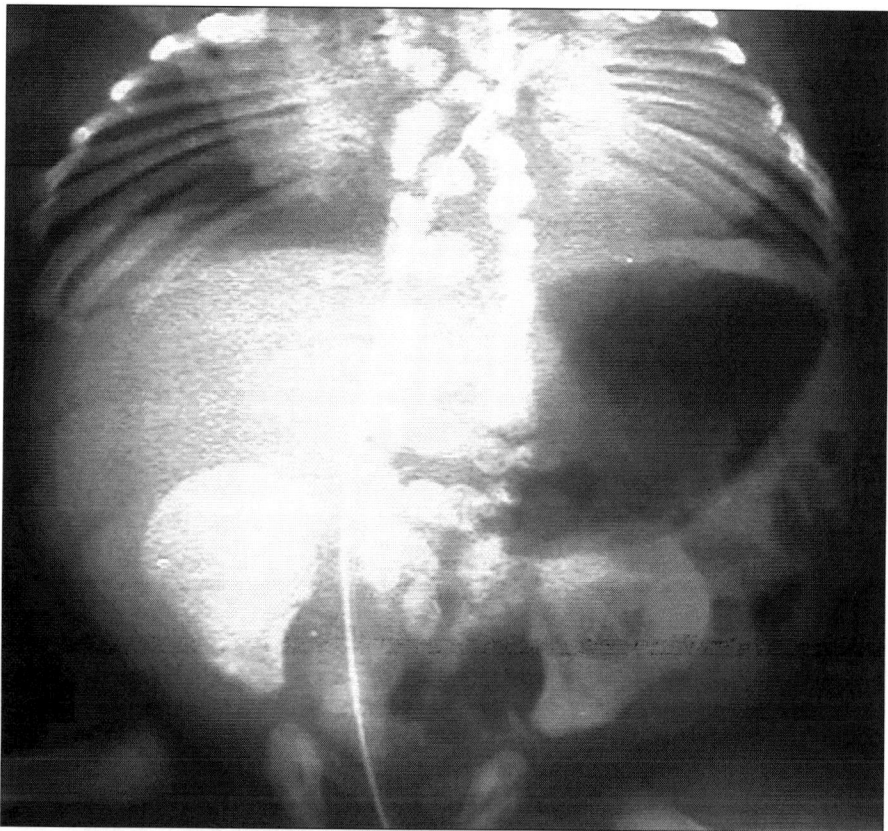

Figure 2. An example of a child with 'spondylothoracic dysplasia'. Severe shortening of the spine occurs with poorly formed vertebrae. The ribs have very 'crowded' origins but do not show points of fusion along their length. Reproduced with permission from Oxford University Press.

The literature includes familial autosomal recessive (AR) SCD whose genetic basis is not known,[2,3,31-36] as well as SCD families apparently following autosomal dominant (AD) inheritance.[37-43] Additional features have been reported in some families—urogenital anomalies,[44] congenital heart disease[21,22,45,46] and inguinal herniae in males.[47] Limb anomalies have been described in COVESDEM syndrome[48] but this condition is almost certainly identical to Robinow syndrome.

In clinical practice sporadically occurring cases of AVS are far more common than familial, and sporadic cases are more likely to be associated with additional anomalies. The isolated case reported by Young and Moore[49] may well be a case of AR SCD. However, the affected twin in a case of monozygotic twins discordant for SCD showed a similar phenotype,[50] raising intriguing speculation about the mechanism. There is wide clinical heterogeneity within the sporadic group[51] and the literature has been well reviewed by Mortier et al.[18] Anal and urogenital anomalies occur most frequently[12,15,18-20,47,52-58] followed by a variety of congenital heart disease.[18,22,45,46,54-59] Limb abnormalities occur but are generally of a minor nature, e.g., talipes, oligodactyly or polydactyly.[18,20] Infrequently, diaphragmatic hernia is a feature.[25,51] As a minor anomaly, inguinal and abdominal herniae are frequently reported in association with MSDS.

Many case reports could reasonably be assigned a diagnosis of the VATER (Vertebral defects, Anal atresia, Tracheo-Esophageal fistula, Radial defects and Renal anomalies) or VACTERL (Vertebral defects, Anal atresia, Cardiac defects, Tracheo-Esophageal fistula, Radial defects and

Table 1. Some syndromes and disorders with abnormal vertebral segmentation

Syndromes/Disorders	OMIM Reference	Gene(s)
Acrofacial dysostosis*	263750	
Alagille	118450	JAGGED1, NOTCH2
Anhalt*	601344	
Atelosteogenesis III	108721	FLNB
Campomelic Dysplasia	211970	SOX9
Casamassima-Morton-Nance*	271520	
Caudal Regression*	182940	
Cerebro-facio-thoracic Dysplasia*	213980	
CHARGE	214800	CHD7
'Chromosomal'		
Currarino	176450	HLXB9
De La Chapelle*	256050	
DiGeorge/Sedláčková	188400	Chromosomal
Dysspondylochondromatosis*		
Femoral hypoplasia-unusual facies*	134780	
Fibrodysplasia ossificans progressiva	135100	ACVR1
Fryns-Moerman*		
Goldenhar*	164210	
Holmes-Schimke*		
Incontinentia Pigmenti	308310	NEMO
Kabuki*	147920	
Kaufman-McKusick	236700	MKKS
KBG syndrome*	148050	
Klippel-Feil*	148900	? PAX1
Larsen	150250	FLNB
Lower Mesodermal Agenesis*		
Maternal Diabetes*		
Mathieu		
MURCS Association*	601076	
Multiple Pterygium Syndrome	265000	CHRNG
OEIS syndrome*	258040	
Phaver*	261575	
Rapadilino	266280	RECQL4
Robinow	180700	ROR2
Rolland-Desbuquois*	224400	
Rokitansky Sequence*	277000	? WNT4
Silverman	224410	HSPG2
Simpson-Golabi-Behmel	312870	GPC3
Sirenomelia*	182940	
Spondylocarpotarsal Synostosis	269550	FLNB
Spondylocostal Dysostosis	277300	DLL3, MESP2, LNFG
Spondylothoracic Dysostosis	277300	
Thakker-Donnai*	227255	
Toriello*		
Urioste*		
VATER/VACTERL*	192350	
Verloove-Vanhorick*	215850	
Wildervanck*	314600	
Zimmer*	301090	

*Underlying cause not known

Renal anomalies, non-radial Limb defects) Associations[54] and this would appear to be a very heterogeneous group with few clues regarding causation at the present time. Figure 3 shows the case of a child with abnormal vertebral segmentation affecting mainly the thoracic region, progressive scoliosis, anal stenosis, unilateral renal agenesis and tricuspid regurgitation; VATER or VACTERL is the likely diagnosis.

A frequent association with severely disorganised vertebrae, affecting rib number and alignment, is neural tube defect (NTD).[51-53,55,60-64] This NTD-associated group should be classified separately from the SCD group because the primary developmental pathology presumably lies in the processes determining neural tube closure as distinct from somitogenesis and it is valid to classify on the basis of underlying cause rather than the radiological features.[65] Similarly, an association between spina bifida occulta and/or diastematomyelia and AVS has been reported,[19,66-68] strongly suggesting a causal link or sequence, though the mechanisms remain to be elucidated.

AVS may be a consequence of maternal diabetes syndrome, which can give rise to multiple congenital anomalies. Classically, the malformation is caudal regression to a varying degree, i.e., absent sacrum or agenesis of the lower vertebral column,[69] features that overlap with the ill-defined disorder known as axial mesodermal dysplasia. However, there are patients with hemivertebrae[70] and various forms of axial skeletal defect (Fig. 4) following poorly controlled diabetes throughout pregnancy in an insulin dependent mother.

Some clues to genetic causes of AVS may come from patients with axial skeletal defects and chromosomal abnormalities. These cases are relatively rare and apart from trisomy 8 mosaicism[71] there is no clear consistency to the group with chromosome abnormalities. Deletions affecting both 18q[72] and 18p[73] have been reported, a supernumerary dicentric 15q marker[74] and an apparently balanced translocation between chromosomes 14 and 15.[75] It can be postulated that haploinsufficiency may be unmasking a new Mendelian locus for SCD but the paucity and diversity of these cases may indicate that the resulting MSDS in association with MCA represents a common pathway of complex pleiotropic developmental mechanisms that are sensitive to a range of unbalanced karyotypes. It is also possible that chromosome *mosaicism* accounts for some cases where there is marked asymmetry in the radiological phenotype, which would also explain sporadic occurrence, but skin or tissue biopsy is rarely undertaken.

Notch Signaling Pathway Genes and Spondylocostal Dysostosis

Notch signaling is a key cascade pathway in somitogenesis and the function of many genes and their products, together with their complex interactions, has been at least partially elucidated through research in animal models. The consequences of many different heterozygous and homozygous gene knock-outs in animal models have been well described elsewhere in this volume. In man the functions of orthologous genes and their proteins obviously cannot be studied in the same way. Nevertheless, the association of a number of genes in the pathway with certain phenotypes has become established through linkage studies carried out on affected families, followed by candidate gene sequencing to identify pathogenic mutations. The diseases that result from mutations in Notch signaling genes are diverse and would not readily be considered to link to a common pathway. The affected organ systems include the vascular and central nervous systems; the skeleton, face and limbs; haematopoiesis, the determination of laterality, and the liver, heart, kidney and eye. Notch pathway genes and their associated diseases appear in Table 2 and good review articles are found in references 76-80. As this chapter deals primarily with abnormalities of somitogenesis affecting the axial skeleton, consideration is now given to the *DLL3, MESP2, LNFG, JAGGED1* and *NOTCH2* genes.

Delta-Like 3 (DLL3) and Spondylocostal Dysostosis

The key breakthrough in the search for a gene linked to SCD came with autozygosity mapping studies in a large inbred Arab kindred with 7 affected individuals.[81,82] The locus identified, 19q13.1, is syntenic with mouse chromosome 7,[83] which harbors the *Dll3* gene. A mutation in this gene was shown to be the cause of MSDS in a radiation-damaged mouse known as *Pudgy*.[84-86] *DLL3*

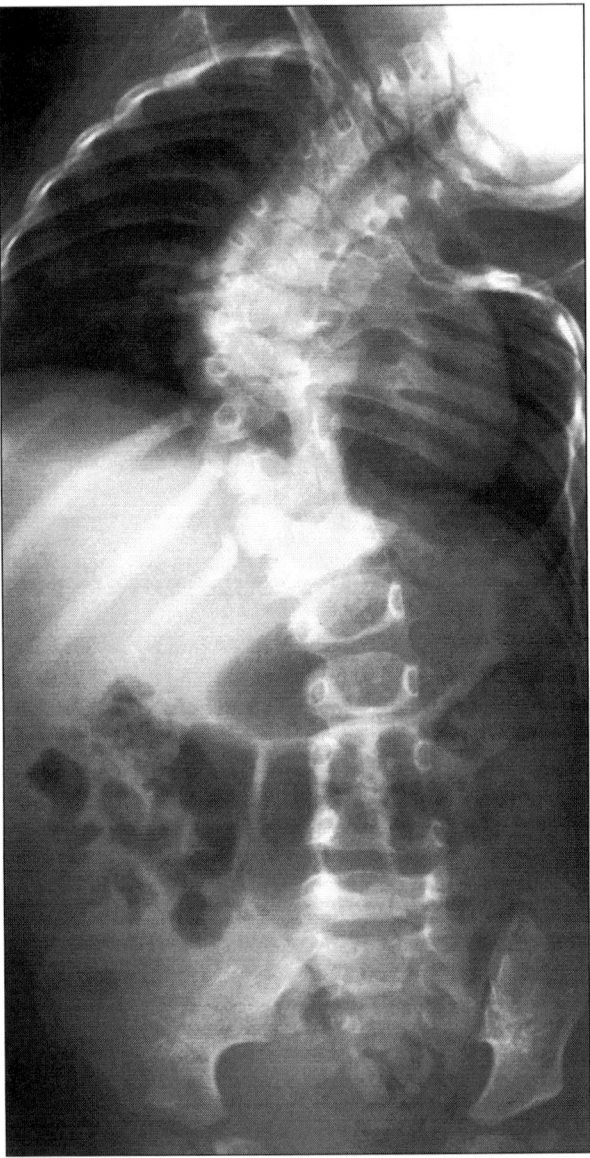

Figure 3. The X-ray of a child with multiple congenital abnormalities that best fits a form of the VACTERL association. She had abnormal vertebral segmentation affecting mainly the thoracic region, progressive scoliosis, anal stenosis, unilateral renal agenesis and tricuspid regurgitation.

was therefore the obvious candidate for the SCD families demonstrating linkage to 19q13.1 and sequencing initially identified mutations in three consanguineous affected families, including the large Arab kindred referred to.[87]

Human *DLL3* encodes a ligand for Notch signaling. The gene comprises 8 exons and spans approximately 9.2 kb of chromosome 19. A 1.9 kb transcript encodes a protein of 618 amino acids.

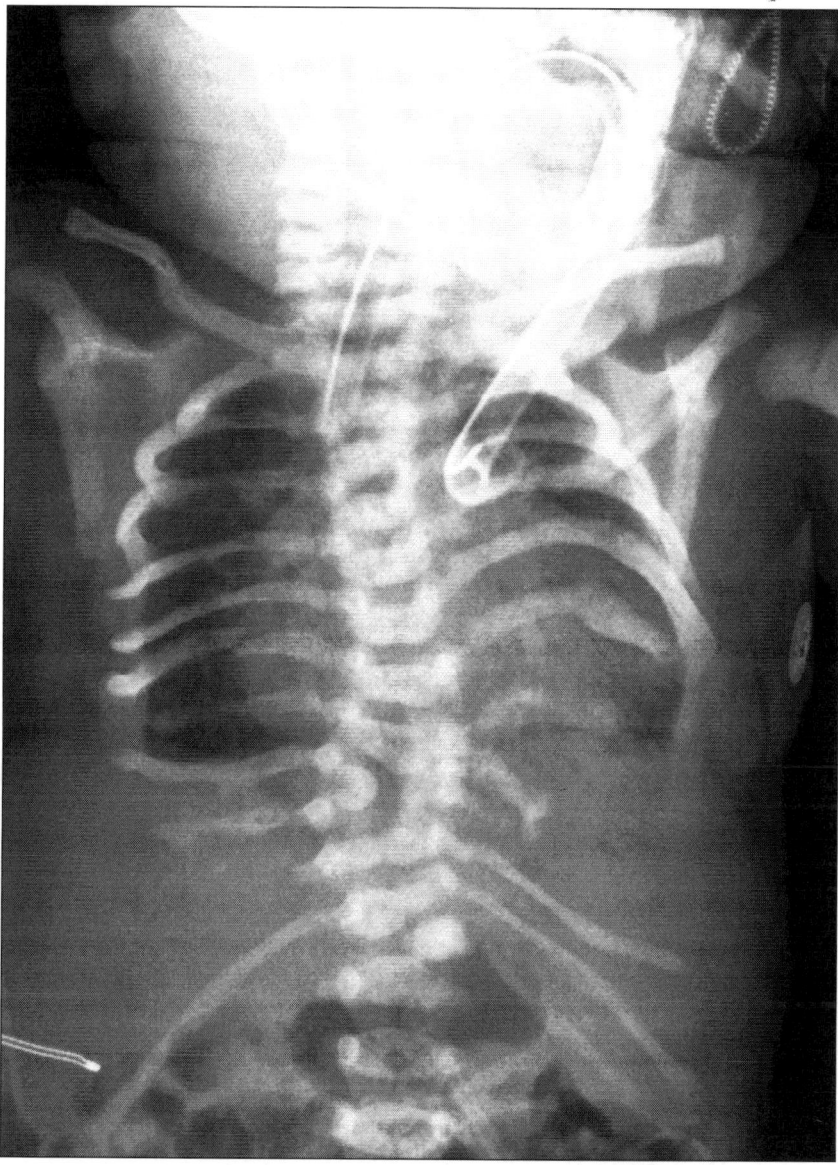

Figure 4. The X-ray of a child with hemivertebrae, axial skeletal defects and left renal agenesis. The mother was a poorly controlled insulin dependent diabetic and it is likely that this is causally related to the malformations.

The protein consists of a signal sequence, a Delta-Serrate-Lag2 (DSL, receptor interacting) domain, six epidermal growth factor (EGF) like domains and a transmembrane domain (TM). Studies in animal models have found that *DLL3* shows spatially restricted patterns of expression during somite formation and it is believed to have a key role in the cell signaling processes giving rise to somite boundary formation, which proceeds in a rostral-caudal direction with a precise temporal periodicity driven by an internal oscillator, or molecular 'segmentation clock.'[88,89]

Table 2. *Notch signaling pathway genes and human disease*

Gene	Chromosomal Locus	Condition/Disease	System
Notch 1	9q34	T-cell ALL/lymphoma	Lymphoid development/
		Aortic valve disease	neoplasia Angiogenesis
Notch 3	19p13	CADASIL*	Angiogenesis
Notch 4§	6p21	? Schizophrenia	? Neural maintenance
JAGGED 1	20p12	Alagille syndrome	Hepatic/angiogenesis/ocular
NOTCH2	1p12	Alagille syndrome	Hepatic/angiogenesis/ocular
DLL3	19q13	Spondylocostal dysostosis—type 1	Axial skeleton
MESP2	15q26	Spondylocostal dysostosis—type 2	Axial skeleton
LNFG	7p22	Spondylocostal dysostosis—type 3	Axial skeleton
Presenilin 1	14q24	Presenile dementia	Neural maintenance
Presenilin 2	1q31	Presenile dementia	Neural maintenance
NIPBL	5p13	Cornelia de Lange syndrome	Upper limb/central nervous system/growth

*Cerebral Autosomal Dominant Arteriopathy with Sub-cortical Infarcts and Leukoencephalopathy
§Association studies only, which remain controversial

Mutated *DLL3* results in abnormal vertebral segmentation throughout the entire spine with all vertebrae losing their normal form and regular three-dimensional shape. The most dramatic changes, radiologically, affect the thoracic vertebrae and the ribs are mal-aligned with a variable number of points of fusion along their length (Fig. 5a). There is an overall symmetry of the thoracic cage and minor scoliotic curves that are nonprogressive and therefore do not require surgery. These are the features that clearly define *spondylocostal dysostosis* (SCD). In early childhood, before ossification is complete, the vertebrae have smooth, rounded outlines—especially in the thoracic region—and for this we have suggested the term 'pebble beach' sign (Fig. 5b). We designate *DLL3*-associated SCD, SCD 'type 1', though 'SCD—*DLL3* type' may be preferred.[90] Additional anomalies appear to be rare. However, in one case abdominal situs inversus was present but the link with mutated *DLL3* is uncertain. In another family affected siblings homozygous for the exon 8 mutation 1369delCGCTCCCGGCTACATGG (C655M660del17) manifested a form of distal arthrogryposis in keeping with fetal akinesia sequence and both succumbed in early childhood (C. McKeown, personal communication). Multiple inbreeding occurred in this particular family and it is possible that a separate AR condition segregated coincidentally to SCD. Spinal cord compression and associated neurological features have not been observed and intelligence and cognitive performance are normal. This suggests that *DLL3* is not expressed in human brain, which contrasts to findings in the mouse, where central nervous system defects have been found,[91] including defects in the neuroventricles of the *Pudgy* mouse.[86,92]

To date, 24 mutations (Table 3) have been identified in SCD patients from 26 families (Table 4). Only five of these families are definitely nonconsanguineous and mutations have been identified in a wide range of ethnic groups. The majority of the mutations are protein truncating but six are missense. Some missense mutations may give rise to a slightly milder phenotype and protein modeling studies may help explain these differential effects in due course. With the one exception of a mutation in the transmembrane domain, all mutations affect the extracellular domain of the gene (Fig. 6) and are clustered in exons 4-8. Three mutations have been identified in more than one family. The 949delAT (T315C316del2) mutation has been found in two ethnic Pakistani kindreds originating from Kashmir and DNA marker analysis flanking the *DLL3* locus has identified a common haplotype, supporting a common ancestry between these two kindreds. Similarly, the 614insGTCCGGGACTGCG (R205ins13) mutation is found in two families, both consanguineous,

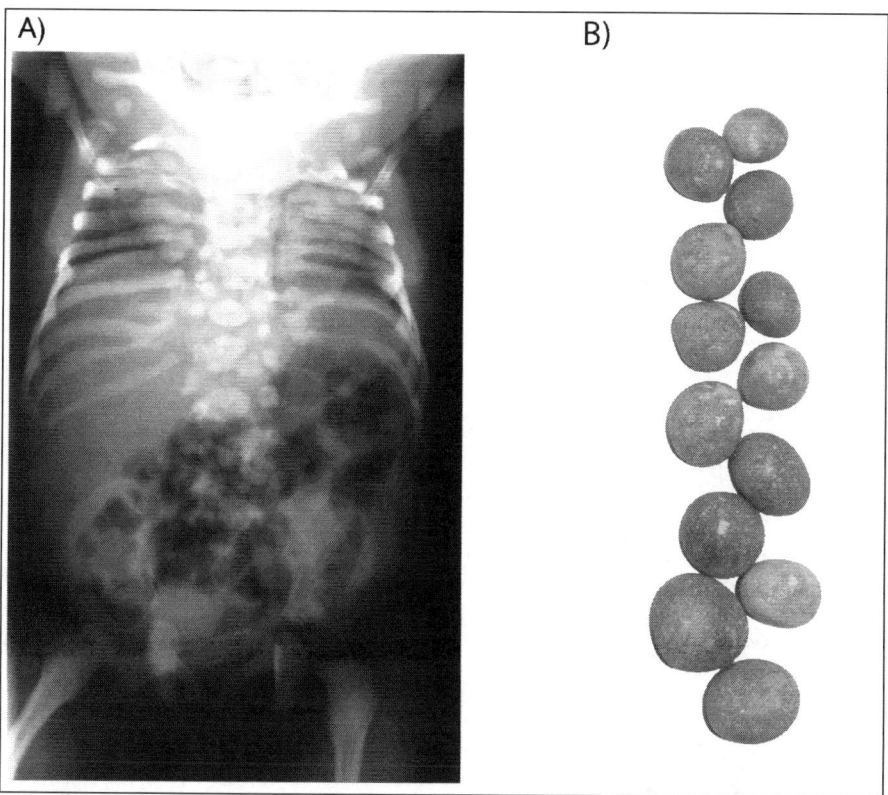

Figure 5. A) Spondylocostal dysostosis due to homozygous mutations in *DLL3*. All vertebrae show abnormal segmentation and the ribs show irregular points of fusion along their length. However, there is an overall symmetry to the thoracic cage. B) Because of the similarity to smooth, eroded pebbles on a beach, we have suggested calling the radiological appearance the 'pebble beach' sign.

one ethnic Lebanese Arab and the other ethnic Turkish. Haplotype analysis again supports a common ancestry for these two Levant kindreds. The 593insGCGGT (S198ins5) mutation is present in the original Arab kindred,[81,87] homozygous in those affected, but in a Spanish family the affected child is heterozygous for the same mutation. Haplotype analysis on these two pedigrees does not support a common ancestry and the 593insGCGGT mutation is therefore believed to be recurrent, occurring as it does within a region of the gene with multiple repeat GCGGT sequences. Slipped mispairing during DNA replication is the likely explanation of this insertion mutation.

The missense mutation G504D was been found in a Northern European family originally reported as demonstrating autosomal dominant SCD[93] but recently shown to be an example of pseudo-dominant inheritance.[94] The same mutation was identified in one other Northern European patient who was heterozygous for this allele. No other mutation was found in *DLL3* in this subject but the radiological phenotype was typical for SCD and another pathogenic allele is presumed to be present, though unidentified. The missense mutation G404C, which is one of the missense mutations associated with a milder phenotype, has been found in only one patient, who is homozygous for this allele. At present there is no confirmed case of AD SCD due to mutated *DLL3*. However, in the large Arab family reported by Turnpenny et al[81] one female heterozygous for the 593insGCGGT (S198ins5) mutation had a mild thoracic scoliosis, but no associated segmentation abnormality in

Table 3. DLL3 *mutations, reported and unreported*

Gene Domain	Protein Truncating Mutations	Missense Mutations
N-terminus	215del28, 395delG	
DSL	602delG**, 603ins5*, 614ins13**, 615delC***, C207X**	
EGF 1-6	712C > T***, 868del11**, 945delAT*, 948delTG**, Q360X, C362X**, 1256ins18, 1285-1301dup17***, 1365del17**	C309Y §, C309R, G325S, G385D*, G404C
C-terminus, pre-TM	1418delC**	
TM	1440delG¥	G504D¥
Total	**18**	**6**

*Published (Bulman et al 2000; ref. 87); ¥Published (Whittock et al 2004; ref. 94); **Published (Turnpenny et al 2003; ref. 96); §Published (Sparrow et al 2002; ref. 141); ***Published (Bonafé et al 2003; ref. 142)

the thoracic region, and a very localised segmentation anomaly in the lower lumbar vertebrae. It is possible her scoliosis and lumbar segmentation anomaly were coincidental to her *DLL3* carrier status, as no other obligate carrier in the kindred is known to have had similar features. It is the possible she carried a mutation in a separate somitogenesis gene and was therefore a manifesting 'double heterozygote', an example of which is known in an animal model.[95]

In general, we have observed a remarkable consistency in the radiological phenotype in mutation positive cases[96] and with experience scrutiny of the radiograph is usually possible to identify those patients who will prove to have *DLL3* mutations. Importantly, *DLL3* mutations have not been found in the wide variety of more common, though diverse, phenotypes that include MSDS and abnormal ribs.[97,98] Therefore, it appears that there is remarkably little clinical heterogeneity for the axial skeletal malformation due to mutated *DLL3*, which has significant implications for the application of genetic testing in the clinical setting.

In relation to defects of the axial skeleton in man, identification of the *DLL3* gene in SCD has represented a breakthrough in understanding the causative basis of this group of malformations, as well as highlighting another example of cross-species biological homology. It has become the paradigm for searching for the genetic basis of other SCD phenotypes, which led directly to the identification of *MESP2* and later *LNFG*, in cases of SCD.

Table 4. *Origins of* DLL3 *mutation positive cases*

Ethnic Origin—Families	N	Consanguineous	Nonconsanguineous
Pakistan	6	6	-
Middle-East—Arab	4	4	-
Turkey	5	5	-
Northern Europe	9	6	3
Northern Europe—Turkey	1	-	1
Southern Europe	1	-	1
Total	**26**	**21**	**5**

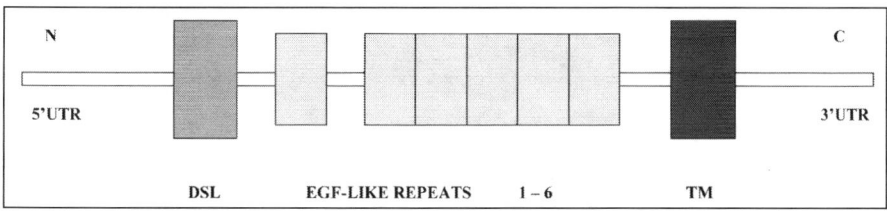

Figure 6. Schematic of the *DLL3* gene in man, showing the mutations identified to date. DSL—delta-serrate-lag; EGF—epidermal growth factor; TM—transmembrane.

Mesoderm Posterior 2 (MESP2) and Spondylocostal Dysostosis

The identification of mutated *MESP2* in association with SCD arose from the study of two small families in whom no *DLL3* mutations were found and linkage to 19q13.1 was excluded. In these families the radiological phenotype was subtly different to SCD type 1. Thoracic vertebrae were severely affected and similar to the appearance in SCD due to mutated *DLL3*. However, the lumbar vertebrae were only mildly affected, as illustrated by MR imaging (Fig. 7).

Genome wide homozygosity mapping in one family of Lebanese Arab origin revealed 84 homozygous markers scattered throughout the genome, of which 6 were concentrated in a block on 15q (D15S153 to D15S120) and 5 were concentrated in a block on 20q (D20S117 to D20S186). Subsequent mapping excluded linkage to the 20q region but demonstrated linkage to the 15q markers D15S153, D15S131, D15S205 and D15S127. Fine mapping using additional markers demonstrated a 36.6 Mb region (according to The Human Genome Working Draft) on 15q21.3-15q26.1, between markers D15S117 and D15S1004, with a maximum two-point lod score of 1.588 at θ = 0 for markers D15S131, D15S205, D15S1046 and D15S127. The region between markers D15S117 and D15S1004 contains in excess of 50 genes and is syntenic to mouse chromosome 7 that contains the *Mesp2* gene. The *Mesp2* knockout mouse manifests altered rostro-caudal polarity, resulting in axial skeletal defects.[99] The predicted human gene, *MESP2*, comprises two exons spanning approximately 2 kb of genomic DNA at 15q26.1. Direct sequencing of *MESP2* gene in the two affected siblings demonstrated a homozygous 4-bp (ACCG) duplication mutation in exon 1, termed 500-503dup.[100] The parents were shown to be heterozygous and the unaffected sibling homozygous normal, consistent with the duplication segregating with SCD in the family. Fluorescent PCR excluded this mutation from 68 ethnically matched control chromosomes. Analysis of the genomic structure of the *MESP2* gene highlighted a discrepancy between the Sanger Centre and NCBI human genomic assembly databases. In the latter, there is an additional short intron located after base 502 of the *MESP2* coding region. This does not appear in the Ensembl gene prediction and due to a lack of consensus splice sites within this proposed intronic sequence, we concluded that the intron does not exists. However, the presence or absence of such an intron does not effect our conclusions concerning the effects of the 4 bp insertion on MESP2 protein production. The insertion is predicted to interrupt splicing, leading to a frameshift at the same point in the MESP2 protein.

In the second, nonconsanguineous family under study, haplotype data was consistent with linkage to the *MESP2* locus but sequencing of the gene failed to identify mutations. Furthermore, sequencing of the promotor region has not identified a mutation (S. Dunwoodie, personal communication).

The *MESP2* gene is predicted to produce a transcript of 1,191 bp encoding a protein of 397 amino acids with a predicted molecular weight of 41,744 Da and isoelectric point (pI) of 7.06. The human MESP2 protein has 58.1 % identity with mouse MesP2 and 47.4 % identity with human *MESP1*. Human *MESP2* amino terminus contains a basic helix-loop-helix (bHLH) region encompassing 51 amino acids divided into an 11 residue basic domain, a 13 residue helix I domain, an 11 residue loop domain and a 16 residue helix II domain. The loop region is slightly longer than that found in homologues such as *paraxis*. The length of the loop region is conserved between

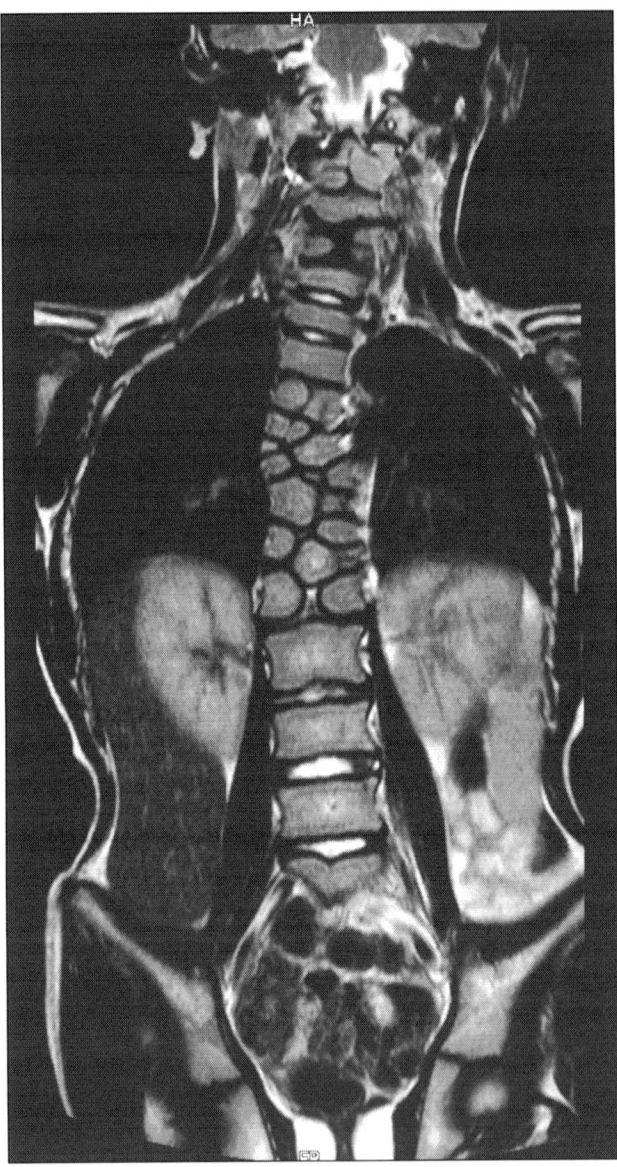

Figure 7. MR image of the spine in one of the affected individuals homozygous for a mutation in *MESP2*. The most striking segmentation abnormality is seen in the thoracic spine with
relative sparing of the lumber vertebrae. © 2004 by The University of Chicago Press.

mouse and human *MESP1*, *MESP2*, Thylacine 1 and 2 and chick mesogenin. In addition, both
MESP1 and *MESP2* contain a unique CPXCP motif immediately carboxy-terminal to the bHLH
domain (Fig. 8). The amino- and carboxy-terminal domains are separated in human *MESP2* by a
GQ repeat region also found in human *MESP1* (2 repeats) but expanded in human *MESP2* (13
repeats). Mouse MesP1 and MesP2 do not contain GQ repeats but they do contain two QX repeats
in the same region: mouse MesP1 QSQS; mouse MesP2 QAQM. Hydrophobicity plots indicate

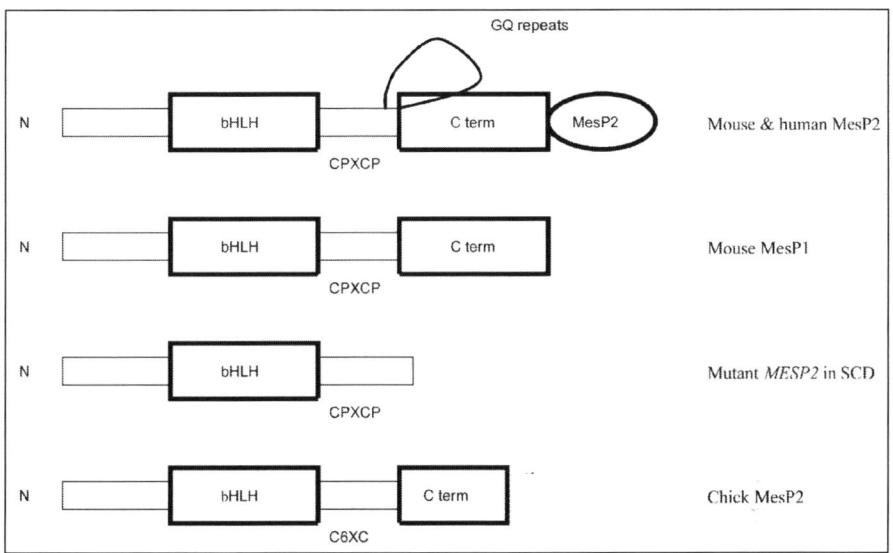

Figure 8. Schematic showing the MesP2 and MesP1homologues from human, mouse and chick. The carboxy-terminus is highly divergent between species.

that MesP1 and MesP2 share a carboxy-terminal region that is predicted to adopt a similar fold, although MesP2 sequences do contain a unique region at the carboxy-terminus.

Sequence analysis of 20 ethnically matched and 10 nonmatched individuals revealed the presence of a variable length polymorphism in the GQ region of human *MESP2*, beginning at nucleotide 535. This region contains a series of 12 bp repeat units. The smallest GQ region detected contains two type A units (GGG CAG GGG CAA, encoding the amino acids GQGQ), followed by two type B units (GGA CAG GGG CAA, encoding GQGQ) and one type C unit (GGG CAG GGG CGC, encoding GQGR). Analysis of this polymorphism in the matched and nonmatched controls revealed allele frequencies that were not significantly different statistically between the two groups.

We showed that a frameshift mutation in the human homologue of *MESP2* results in AR SCD in man and designated this as SCD 'type 2', or *MESP2*-associated SCD. *MESP2* is a member of the bHLH family of transcriptional regulatory proteins essential to a vast array of developmental processes.[101] Murine *Mesp1* and *Mesp2*, located on chromosome 7, are separated by approximately 23 kb. They are positioned head to head and transcribed from the interlocus region.[102] At least two enhancers are involved in the expression of these genes in mouse,[103] one in early mesoderm expression and the other in presomitic mesoderm (PSM) expression. In addition, a suppressor responsible for the rostrally-restricted expression in the PSM has been identified.[103] These enhancers are essential to the specific and coordinated expression of the MesP proteins. The expression of *MesP1* and *Mesp2* is first detected in the mouse embryo at the onset of gastrulation (~6.5 days *post coitum* [dpc]) and is restricted to early nascent mesoderm.[102,104] At this stage the expression domain of *Mesp1* is broader than that of *Mesp2* and lineage analysis of *Mesp2* expressing cells shows that they contribute to cranial, cardiac and extra-embryonic mesoderm.[105] Expression is then downregulated as *Mesp* transcripts are not detected later in development.[105] A second site of *Mesp* expression is detected at 8.0 dpc immediately prior to somitogenesis. A pair of MesP-expressing bands appear on each side of the embryonic midline, at the anterior part of the PSM where the somites are anticipated to form.[99,105] During somitogenesis *MesP1* and *MESP2* continue to be expressed in single bilateral bands in the anterior PSM. MesP2 expression within the PSM continues until about the time when somite formation ceases (~13.5 dpc)[105] after which *Mesp* expression is rapidly downregulated as transcripts are not detected in newly formed somites.

Alignment of MesP2 homologues from human, mouse, *Xenopus* and chick demonstrates a highly divergent carboxy-terminus between species (Fig. 8) and it is unknown whether functional domains are similarly arranged in all Mesp2 orthologues, though this is well characterised for the *Xenopus* Mesp2 orthologue, Thylacine.[106] If the carboxy-terminus in human MesP2 is required for transcriptional activation, then the mutant form of the protein described here, lacking the carboxy-terminus, would lose this function. Mouse pups lacking MesP2 die within 20 minutes of birth, presenting with short tapered trunks and abnormal segmentation affecting all but a few caudal (tail) vertebrae.[99] This resembles *Dll3*-null mice, where the rib and vertebral architecture is disturbed along the entire axis.

Unlike the mouse null MesP2 allele, the human mutant protein retains its bHLH region and, although truncated, could still dimerize, bind DNA and act in a dominant negative manner. However, in this family the heterozygous parents demonstrated no axial skeletal defects, from which we deduce that this *MESP2* mutation is recessive. Similarly, heterozygous mice are normal and fertile.[99] It is possible that a single functional carboxy-terminus is sufficient to activate transcription but alternatively the MESP1 protein may compensate for the absence of MESP2 function, just as MesP1 can rescue the axial skeletal defects in MesP2 deficient mice in a dosage dependent manner.[107] Another possibility is that the mutant *MESP2* transcript, containing a premature stop codon at 1099-1101 bp, is degraded by nonsense-mediated RNA decay with no truncated protein being produced.[108]

In murine somitogenesis MesP2 has a key role in establishing rostro-caudal polarity by participating in distinct Notch-signaling pathways.[109,110] *Mesp2* expression is induced by Dll1-mediated Notch signaling (presenilin1-independent) and Dll3-mediated Notch signaling (presenilin1-dependent), while inhibition of *Mesp2* expression is achieved through presenilin1-independent Dll3-Notch signaling. Since Mesp2 can inhibit *Dll1* expression this complex signaling network results in stripes of *Dll1*, *Dll3* and *Mesp2* gene expression in the anterior PSM.

The extent to which this directly correlates with somitogenesis in man is unknown, except that the phenotypes of the *Mesp2* and *Dll3* mutant mice closely resemble human SCD.[86,91,99] The findings in this one family provided the first evidence that *MESP2* is critical for normal somitogenesis in man. A second affected family with the identical mutation was presented at the 2005 meeting of the International Skeletal Dysplasia Society (L. Bonafé, personal communication), Martigny, Switzerland and manifests a very similar phenotype. Whether there is common ancestry between the two families is not known.

Lunatic Fringe (LNFG) and Spondylocostal Dysostosis

The identification of Notch pathway genes as causes of human SCD inevitably led to the hypothesis that other genes of the pathway may be implicated in other forms of AVS/MVSD. Among these the Lunatic Fringe (*LNFG*) gene was considered a strong candidate because the *Lnfg*-null mouse has a nonlethal phenotype that includes costo-vertebral abnormalities. Initially, however, mutation screening in a series of affected subjects with diverse radiological phenotypes initially failed to identify any positive cases. *LFNG* encodes a glycosyltransferase that posttranslationally modifies the Notch family of cell surface receptors, a key step in the regulation of this signaling pathway.[111] The LFNG protein is a fucose-specific ß 1,3 N-acetylglucosaminyltransferase[112,113] that functions in the Golgi to posttranslationally modify the Notch receptors, altering their signaling properties.[111] Earlier studies have shown that *Lfng* gene expression is severely disregulated in *Dll3*-null mice, suggesting that Lfng expression is dependent on Dll3 function.[91,114]

The proband under study was an adolescent boy originating from northern Lebanon, the second of five children born to consanguineous parents. He had a short neck and a short trunk at birth and at 15 years had marked shortening of the thorax with a pectus carinatum and kyphoscoliosis. A spinal MRI confirmed MSDS in the cervical and the thoracic spine with a serpentine curve in the cervical spine and concavity to the right in the upper spine and concavity to the left inferiorly.

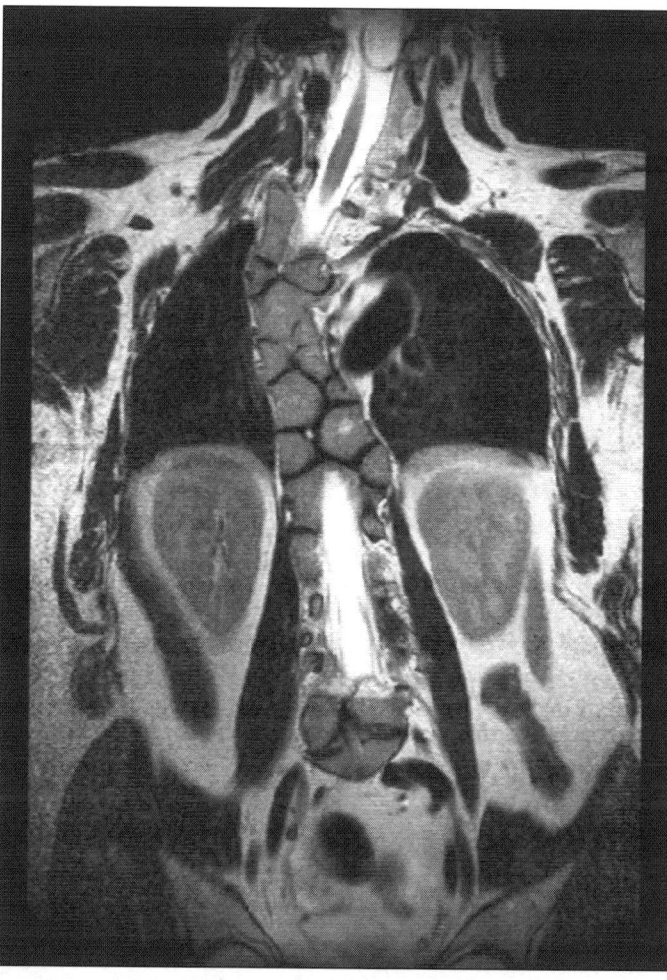

Figure 9. MR image of the spine in the affected individuals homozygous for a mutation in *LNFG*. Severe segmentation abnormalities throughout the spine has given rise to marked shortening of the trunk. © 2006 by The University of Chicago Press.

The thoraco-lumbar spine showed multiple hemivertebrae (Fig. 9) and there were also multiple rib anomalies. The spinal cord was normal with no evidence of a syrinx. At 15 he had a markedly short trunk with a height of 155 cm (5th percentile), lower segment of 92.5 cm and a span of 186.5 cm. As span:height ratio is close to unity in the normal situation, his stature was reduced by approximately 30 cm. At birth he had been noted to have a contracture of the left index finger and at age 15 he had hypoplasia of all the distal interphalangeal joints of the fingers, which were long and slender. Radiographs of the wrists and ankles were normal and the link between the spinal malformation and mild digital contractures remains speculative.

On sequencing the entire coding region and splice sites of the *LFNG* gene a homozygous missense mutation (c.564C > A) in exon 3 was detected,[115] resulting in substitution of leucine for phenylalanine (F188L). The proband's parents, with normal spinal anatomy, were both heterozygous for the mutant allele. The phenylalanine residue substituted is highly conserved[116] and close to the active site of the enzyme. The mutation created a novel MseI restriction enzyme site, which was used to confirm the sequencing results in the pedigree. This variant was not found in 34 ethnically matched control subjects (68 chromosomes) and the underlying base substitution was not present in the NCBI SNP database (www.ncbi.nlm.nih.gov). Examination of the mutation within a fringe model based on solved glycosyltransferase structures showed the conserved phenylalanine residue (F188) to be located in a helix that packs against the strand containing Mn^{2+}-ligating residues, rather than being directly involved in UDP-N-acetylglucosamine or protein binding.

Further evidence of causality was provided by functional assays of the *LFNG* mutant. Two F187L mutations in mouse Lfng that correspond to F188L in human *LFNG* were generated: a c.564C > A mutation that encodes the rare leucine codon (TTA) observed in the proband and the [c.562T > C + c.567C > G] mutation encoding the most common human leucine codon (CTG). In addition, a previously characterised enzymatically inactive form of Lfng (D202A) was created that disrupts the conserved DDD Mn^{2+}-binding active site.[117] Protein expression studies showed that both F187L mutant Lfng proteins were expressed at higher levels than wild type or D202A Lfng forms, indicating that both translation efficiency and protein stability were not adversely affected by the F187L amino acid change. As Lfng protein is normally present in the Golgi apparatus, intracellular protein localization was examined using immunofluoresence. Wild type and D202A mutant Lfng were localized predominantly to the Golgi whilst the F187L mutant Lfng did not colocalize with a marker. It was concluded that the F187L mutant form of Lfng is expressed but mislocalized within the cell.

This single case of SCD due to mutated *LNFG* has provided further evidence that proper regulation of the Notch signaling pathway is an absolute requirement for correct patterning of the axial skeleton, at the same time defining SCD type 3, or *LNFG*-associated SCD.

Jagged1 (JAG1), NOTCH2 and Alagille Syndrome (Arteriohepatic Dysplasia)

Alagille syndrome (AGS) is a multi-system disorder characterised primarily by paucity of bile ducts, giving rise to progressive cholestatic liver disease (sometimes manifesting as prolonged neonatal jaundice) and congenital heart disease. It follows autosomal dominant inheritance, in contrast to recessively inherited SCD due to mutated *DLL3* and *MESP2* and in common with most dominantly inherited conditions it demonstrates marked variability. First described in 1969,[118] it was delineated as an entity in 1975.[119] and diagnostic criteria established.[120] A clinical diagnosis has traditionally depended on the finding of interlobular bile duct paucity from a liver biopsy in combination with involvement of at least three of the five major features. These features are cholestasis, congenital heart disease, skeletal abnormalities, ocular anomalies and characteristic facial features. Additional abnormalities have been described in some patients, particularly minor distal limb defects and structural anomalies of the kidney and pancreas. A pigmentary retinopathy occurs in some patients. The diagnostic criteria and range of abnormalities are listed in Table 5. Clearly, the features of this condition are extensive and very different to the pure axial skeletal malformation seen in SCD, as previously described. However, the axial skeleton is frequently involved, though to a far lesser extent than in SCD, with the variable presence of butterfly vertebrae, hemivertebrae and absence of the sacrum in occasional cases. These skeletal anomalies are rarely of any clinical significance in AGS patients.

In classic cases the experienced clinician can diagnose AGS from the mildly dysmorphic facies. The forehead is often high and sometimes bossed and the eyes may be deep set and sometimes the palpebral fissures are upslanting and slightly narrow (Fig. 10). Whether the facies are secondary to other features of AGS, such as long term consequences of cholestasis, is a matter of conjecture.[121] In many subjects the facies are extremely subtle with respect to the dysmorphic features. Involvement of body systems, as with the facies, is extremely variable, as extended family studies following molecular diagnosis of index cases has shown.[122] In this study 53 mutation positive relatives of 34 AGS probands were assessed. Only 21% (11 subjects) of the mutation positive relatives had sufficient features that would have led to a clinical diagnosis of AGS. A further 32% (17 subjects) had mild features of AGS on detailed evaluation, whilst 47% (25 subjects) did not meet the clinical criteria

Table 5. *Diagnostic criteria (in italics) and additional features of alagille syndrome*

Criteria	Body System	Most Specific Abnormality	Additional Features
Mandatory feature	Paucity of intrahepatic bile ducts on liver biopsy		
Involvement of three or more of these systems	Liver	*Chronic cholestasis*	
	Congenital heart disease	*Pulmonary/peripheral pulmonary stenosis*	Tetralogy of Fallot, patent ductus arteriosus, aSD/VSD
	Skeleton	*Butterfly vertebrae*	Hemivertebrae, failure of sacrum formation, radio-ulnar synostosis, digital anomalies with short phalanges
	Ophthalmic	*Posterior embryotoxon*	Keratoconus, cloudy cornea, pigmentary retinopathy
	Craniofacial facies	*High forehead, deep set eyes +/– upslanting palpebral fissures, long nose with flattened tip, prominent chin*	Craniosynostosis, orofacial clefts, oligodontia
Additional features	Endocrine/metabolic		Hypothyroidism, xanthomas, diabetes mellitus, short stature, delayed puberty
	Abdomen		Renal dysplasia, renal cystic disease, unilateral renal agenesis, jejunal/ileal atresia/stenosis, intestinal malrotation, pancreatic atrophy
	Vasculature		Intracranial haemorrhage

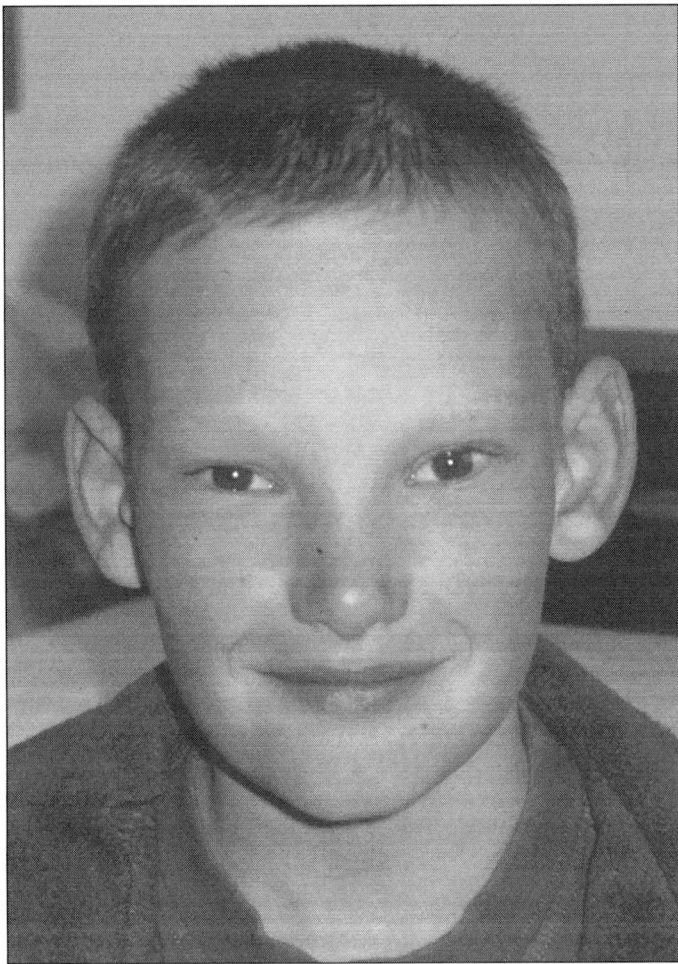

Figure 10. The face in Alagille syndrome. Broad forehead with deep set eyes, narrow upslanting palpebral fissures, a prominent columella to the nose and prominent chin.

for the condition, including 2 subjects with no features at all. The authors highlighted the fact that the characteristic facies demonstrated the highest penetrance of all the features.

At the other end of the disease spectrum, severely affected subjects may have life threatening complications from liver disease and/or congenital heart disease. The liver disease typically shows paucity of the intrahepatic bile ducts on biopsy but progression may occur and eventually lead to cirrhosis and hepatic failure, with perhaps one-quarter of patients presenting in infancy going on to require organ transplantation. Incipient liver failure can clearly give rise to a large number of other features of AGS, such as failure to thrive, short stature, delayed puberty and xanthomata. Congenital heart disease is present in a large proportion of patients and the pulmonary vasculature is most commonly involved. Thus, pulmonary valve or artery stenosis and peripheral pulmonary artery stenosis are the common lesions, occurring in two-thirds of AGS patients.[123] However, tetralogy of Fallot (TOF) is a complex malformation that occurs in up to 10% of patients and ventricular and atrial septal defects are also seen.[119,122] We have personally seen one patient with hypoplastic left heart syndrome,[124] a form of congenital heart disease previously unreported. It

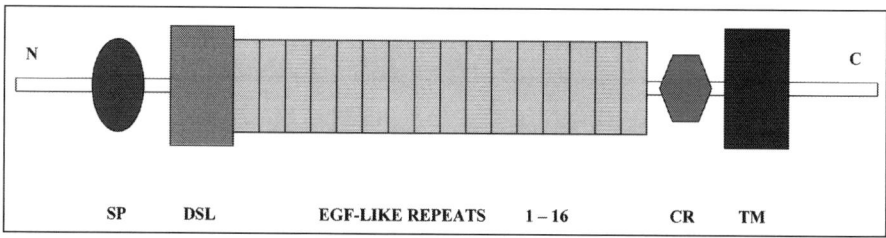

Figure 11. Schematic of the *JAG1* gene in man. SP—signal peptide; DSL—delta-serrate-lag; EGF—epidermal growth factor; TM—transmembrane.

is of interest that one family has been reported in which congenital heart disease segregates with a *JAG1* mutation in the absence of hepatic or other features of AGS.[125] Many cardiac lesions are either clinically insignificant or manageable, but those with complex congenital heart disease have a higher mortality rate than similar, sporadically occurring cases.[126] The excess mortality—33% with TOF in AGS patients vs 11% in sporadically occurring TOF—may be due to associated hepatic involvement but this is not fully resolved.

The *JAG1* gene is located on chromosome 20p12, comprises 26 exons and, like *DLL3*, is a ligand for Notch receptors with similar domain structure (Fig. 11). It was cloned in 1997[126,127] and numerous mutations have been found in affected patients. All types of mutation have been described—nonsense, missense, splice site, frameshift and deletion. It was reported early on that mutations were identified in up to 70% of patients,[128,129] though this figure increases to greater than 90% when strict clinical criteria are used (B. Kamath, personal communication). Up to 5% of AGS patients may have a microdeletion involving chromosome 20p12[130] and these patients are not readily distinguished from patients with *JAG1* mutations. For those heterozygous for mutations, which map to the extracellular and transmembrane domains of *JAG1*, there is no clear genotype-phenotype correlation,[122] and mutations give rise in 70% of cases to a premature termination codon. Haploinsufficiency has therefore been proposed as the main mechanism of AGS. However, there has been interest in the possibility of a dominant negative effect of truncated forms of Serrate/Jagged, though few studies of the mutant mRNAs and proteins from AGS patients have been performed in efforts to elucidate the molecular mechanisms. Recently published work[131] on the stability of mutant mRNA transcripts leaves open the possibility that a dominant negative effect may be the molecular mechanism in some patients. Transcripts from the livers of 5 patients and 24 lymphoblastoid cell lines of AGS patients were studied. Mutant *JAG1* transcripts were recovered from RNAs with 5 missense mutations, 2 in-frame deletions and from 19 out of 21 with premature termination codons. Mutant transcripts were also recovered from the tissues of a 23-week-old AGS foetus with a mutation giving rise to a premature termination codon. The results suggested that mutant transcripts with premature termination codons generally escape nonsense-mediated mRNA decay, which might then lead to the synthesis of soluble forms of JAG1. Similar truncated protein transcripts were identified in other cell lines transfected with a mutant *JAG1* cDNA and the presence of mutant proteins, regardless of mutation type, suggested the possibility of a dominant negative effect. The precise mechanism in *JAG1* mutated cases of AGS therefore remains to be elucidated.

The contribution of the Notch signaling pathway and *JAG1* to intrahepatic bile duct formation (IHBD) remains unknown. Studies of the expression patterns of *Jagged1*, *Notch2* and *Hes1* in *Hes1* null mice, in comparison with wild-type mice, suggest that Notch signaling is very important in the differentiation of biliary epithelial cells and is essential for their tubular formation during IHBD. *Jagged1* was found to be expressed in portal mesenchyme during the neonatal period and during the same period *Notch2* and *Hes1* expression was observed in the biliary epithelial cells adjacent to the *Jagged1*-positive cells. During ductal plate remodeling, *Notch2* and *Hes1* were up-regulated exclusively in the biliary epithelial cells that form tubular structures but tubular formation of IHBD was absent in *Hes1* null mice.[132]

The observation that a small number of AGS patients were negative on JAG1 mutation screening, despite meeting the diagnostic criteria, led to consideration that a second locus may be implicated in AGS. A Jagged1/Notch2 double heterozygote mouse was shown to have hepatic, cardiac, ocular and renal abnormalities not dissimilar to those seen in AGS patients.[133] Furthermore, the expression pattern of Notch2 suggested it might have a close developmental pathway relationship with Jagged1.[134] Sequencing of *NOTCH2* in 11 probands identified mutations in two cases, whose families together contained five affected individuals.[135] In both probands renal disease was severe, as well as being present in all three mildly affected relatives, raising the possibility of a slightly different spectrum of clinical severity compared to *JAG1*-positive AGS cases.

One of the mutations, c.5930– 1G→A (exon 33), predicts a premature termination codon in exon 34 such that 3 of the 7 ankyrin repeats and ensuing sequence, would be lost. Ankyrin repeats are localized intracellularly and interact with nuclear cofactors that modulate Notch signaling. The second mutation, c.1331G→A (exon 8), results in substitution of a tyrosine residue for a cysteine in the 11th EGF-like repeat (C444Y). As with missense mutations affecting conserved cysteine residues in EGF repeats in *DLL3* and other genes, this can be expected to be pathogenic and, indeed, was not found in 220 control chromosomes. A mutated Notch2 receptor is therefore confirmed as a rare cause of AGS. The *NOTCH2* gene is large, comprising a coding region of 34 exons and incorporating 36 EGF-like repeats besides the intracellular ankyrin repeat domain.

In terms of patho-physiology, considerable attention has been given to the possibility that AGS is essentially a defect of vasculogenesis.[136] Apart from the frequency of congenital heart disease, it has become clear that other forms of vascular anomaly occur frequently, in up to 9% of AGS patients. In the large review reported by Kamath et al[136] there were patients with aneurysms of the basilar arteries, middle cerebral arteries and aorta and some patients had structural anomalies of the internal carotid arteries. From a total of 268 patients reviewed 25 (9%) had significant noncardiac vascular anomalies and a further 9 subjects had intracranial events in the absence of documented intracranial vascular anomalies. Of the 29 patients in this group who died, 10 (34%) succumbed to noncardiac vascular complications. There was no genotype-phenotype correlation in those with noncardiac vascular pathology; indeed, most types of mutation occurred in *JAG1*—nonsense, missense, splice site and frameshift.

Defective vasculogenesis in AGS is consistent with a pattern observed in defective Notch pathway signaling. Most familiar to the clinician is the autosomal dominant condition CADASIL (Cerebral Autosomal Dominant Arteriopathy with Sub-cortical Infarcts and Leukoencephalopathy) due to heterozygous mutation in the Notch 3 receptor (not discussed in this chapter). Cerebrovascular accidents and dementia typically occur from late middle-age in this condition, preceded often by a history of migraine attacks at a younger age. Mutations in the Notch 3 receptor appear to result in alteration to cells of the smooth muscle in blood vessels.[137] In *Notch1* knockout mice and *Notch1/Notch4* double mutants, as well as mice homozygous for *Jag1* mutations, nonviability is due to failure of normal vasculogenesis.[138,139] In the developing liver the formation of mature tubular bile ducts is preceded by the formation of intrahepatic arterial branches.[140] Defective Notch signaling in AGS, therefore, may be the mechanism that leads to paucity of bile ducts, mediated by poor vascular development. One could also postulate that butterfly vertebrae arise because of partial failure of vascular supply in early development. However, such a mechanism is unlikely in the axial skeletal malformation of SCD due to mutated *DLL3*, *MESP2* and *LNFG*, whose effects appear to be restricted to spatial patterning in somitogenesis.

Concluding Remarks

Abnormal segmentation of the spine is an important 'handle' in clinical dysmorphology. However, the number and range of conditions and syndromes that include this malformation is feature is very large. Slowly, the techniques of molecular genetics are enabling us to identify the cause of some of those conditions following Mendelian inheritance, usually through studies undertaken on affected families suitable for DNA linkage analysis. The insights and clues gained from animal studies and model systems have also been crucial to identifying candidate genes and are likely to continue to be so. Somitogenesis genes have justifiably been the focus of huge attention but many other genes and mechanisms have been implicated in normal axial skeletal development and these need to be explored in large cohorts of affected patients whose clinical features are stratified systematically. There are even substantial challenges in identifying the genetic basis of AVS in families demonstrating variable autosomal dominant inheritance and here a candidate gene approach seems inevitable given the small size of most families.

There is undoubtedly a great deal more to learn and the unraveling of developmental pathways in man remains one of the challenges of modern genetic medicine. It seems likely that the monogenic causes of axial skeletal defects in man will, in due course, yield their secrets. However, it is likely that a large proportion of the phenotypes seen in clinical practice, if not the majority, are not monogenic in causation but multifactorial. Understanding this large and diverse group, which often present the most difficult medical and surgical problems, is likely to be the biggest challenge of all.

References

1. Jarcho S, Levin PM. Hereditary malformation of the vertebral bodies. Bull Johns Hopkins Hosp 1938; 62:216-226.
2. Norum RA, McKusick VA. Costovertebral anomalies with apparent recessive inheritance. Birth Defects OAS 1969; 18:326-329.
3. Cantú JM, Urrusti J, Rosales G et al. Evidence for autosomal recessive inheritance of costovertebral dysplasia. Clin Genet 1971; 2:149-154.
4. David TJ, Glass A. Hereditary costovertebral dysplasia with malignant cerebral tumour. J Med Genet 1983; 20:441-444.
5. Takikawa K, Haga N, Maruyama T et al. Spine and Rib Abnormalities and Stature in Spondylocostal Dysostosis. Spine 2006; 31:E192-E197.
6. Moseley JE, Bonforte RJ. Spondylothoracic dysplasia—a syndrome of congenital anomalies. Am J Roentgenol 1969; 106:166-169.
7. Cornier AS, Ramrez-Lluch N, Arroyo S et al. Natural history of Jarcho-Levin syndrome (Abstract). Am J Hum Genet 2000; 67(4)(Suppl.2):56(A238).
8. Cornier AS, Ramírez N, Arroyo S et al. Phenotype characterisation and natural history of spondylothoracic dysplasia syndrome: a series of 27 new cases. Am J Med Genet 2004; 128A:120-126.
9. Lavy NW, Palmer CG, Merritt AD. A syndrome of bizarre vertebral anomalies. J Pediatr 1966; 69:1121-1125.
10. Pochaczevsky R, Ratner H, Perles D et al. Spondylothoracic dysplasia. Radiology 1971; 98:53-58.
11. Pérez-Comas A, García-Castro JM. Occipito-facial-cervico-thoracic-abdomino-digital dysplasia: Jarcho-Levin syndrome of vertebral anomalies. J Pediatr 1974; 85:388-391.
12. Gellis SS, Feingold M. Picture of the month: spondylothoracic dysplasia. Am J Dis Child 1976; 130:513-514.
13. Trindade CEP, de Nóbrega FJ. Spondylothoracic dysplasia in two siblings. Clin Pediatr 1977; 16:1097-1099.
14. Solomon L, Jimenez B, Reiner L. Spondylothoracic dysostosis. Arch Pathol Lab Med 1978; 102:201-205.
15. Tolmie JL, Whittle MJ, McNay MB et al. Second trimester prenatal diagnosis of the Jarcho-Levin syndrome. Prenat Diagn 1987; 7:129-134.
16. Schulman M, Gonzalez MT, Bye MR. Airway abnormalities in Jarcho-Levin syndrome: a report of two cases. J Med Genet 1993; 30:875-876.
17. McCall CP, Hudgins L, Cloutier M et al. Jarcho-Levin syndrome: unusual survival in a classical case. Am J Med Genet 1994; 49:328-332.
18. Mortier GR, Lachman RS, Bocian M et al. Multiple vertebral segmentation defects: analysis of 26 new patients and review of the literature. Am J Med Genet 1996; 61:310-319.
19. Poor MA, Alberti A, Griscom T et al. Nonskeletal malformations in one of three siblings with Jarcho-Levin syndrome of vertebral anomalies. J Pediatr 1983; 103:270-272.
20. Karnes PS, Day D, Barry SA et al. Jarcho-Levin syndrome: four new cases and classification of subtypes. Am J Med Genet 1991; 40:264-270.
21. Simpson JM, Cook A, Fagg NLK et al. Congenital heart disease in spondylothoracic dysostosis: two familial cases. J Med Genet 1995; 32:633-635.

22. Aurora P, Wallis CE, Winter RM. The Jarcho-Levin syndrome (spondylocostal dysplasia) and complex congenital heart disease: a case report. Clin Dysmorphol 1996; 5:165-169.
23. Eliyahu S, Weiner E, Lahav D et al. Early sonographic diagnosis of Jarcho-Levin syndrome: a prospective screening program in one family. Ultrasound Obstet Gynecol 1997; 9:314-318.
24. Rastogi D, Rosenzweig EB, Koumbourlis A. Pulmonary hypertension in Jarcho-Levin syndrome. Am J Med Genet 2002; 107:250-252.
25. Shehata SMK, El-Banna IA, Gaber AA et al. Spondylothoracic dysplasia with diaphragmatic defect: a case report with literature review. Eur J Pediatr Surg 2000; 10:337-339.
26. Vásquez-López ME, López-Conde MI, Somoza-Rubio C et al. Anomalies of vertebrae and ribs: Jarcho Levin syndrome. Description of a case and literature review. Joint Bone Spine 2005; 72:275-277.
27. Kulkarni ML, Sarfaraz Navaz R, Vani HN et al. Jarcho-Levin Syndrome. Indian J Pediatr 2006; 73:245-247.
28. Phadke SR, Patil SJ, Kumari N et al. Spondylothoracic Dysplasia: Prenatal Diagnosis and the Problems of Nosologic Overlap. Am J Med Genet 2007; 143A:899-902.
29. Schambony A, Wedlich D. Wnt-5A/Ror2 regulate expression of XPAPC through an alternative non-canonical signaling pathway. Dev Cell 2007; 12:779-92.
30. Gao X, Gordon D, Zhang D et al. CHD7 gene polymorphisms are associated with susceptibility to idiopathic scoliosis. Am J Hum Genet 2007; 80:957-65.
31. Castroviejo P, Rodriguez-Costa T, Castillo F. Spondylo-thoracic dysplasia in three sisters. Dev Med Child Neurol 1973; 15:348-354.
32. Franceschini P, Grassi E, Fabris C et al. The autosomal recessive form of spondylocostal dysostosis. Radiol 1974; 112:673-675.
33. Silengo MC, Cavallaro S, Francheschini P. Recessive spondylocostal dysostosis: two new cases. Clin Genet 1978; 13:289-294.
34. Bartsocas CS, Kiossoglou KA, Papas CV et al. Spondylocostal dysostosis in South African sisters. Clin Genet 1981; 19:23-25. 110.
35. Beighton P, Horan FT. Spondylocostal dysostosis in South African sisters. Clin Genet 1981; 19:23-25.
36. Satar M, Kozanoğlu MN, Atilla E. Identical twins with an autosomal recessive form of spondylocostal dysostosis. Clin Genet 1992; 41:290-292.
37. Van der Sar A. Hereditary multiple hemivertebrae. Docum Med Geographica et Tropica 1952; 4:23-28.
38. Rütt A, Degenhardt KH. Beitrag zur Ätiologie und Pathogenese von Wirbelsäulenmißbildungen. Arch Orthop Unfallchir 1959; 51:120-139.
39. Peralta A, Lopez C, Gracia R et al. Polidispondilia familiar. Rev Pediatr Obstet Gynecol [Pediatr] 1967; VII:93-96.
40. Rimoin DL, Fletcher BD, McKusick VA. Spondylocostal dysplasia. Am J Med Genet 1968; 45:948-953.
41. Kubryk N, Borde M. La dysostose spondylocostale. Pédiatrie 1981; 17:137-146.
42. Temple IK, Thomas TG, Baraitser M. Congenital spinal deformity in a three generation family. J Med Genet 1988; 25:831-834.
43. Lorenz P, Rupprecht E. Spondylocostal dysostosis: dominant type. Am J Med Genet 1990; 35:219-221.
44. Casamassima AC, Morton CC, Nance WE et al. Spondylocostal dysostosis associated with anal and urogenital anomalies in a Mennonite sibship. Am J Med Genet 1981; 8:117-127.
45. Delgoffe C, Hoeffel JC, Worms AM et al. Dysostoses spondylocostales et cardiopathies congénitales. Ann Pédiat 1982; 29:135-139.
46. Hatakayama K, Fuse S, Tomita H et al. Jarcho-Levin Syndrome Associated with a Complex Congenital Heart Anomaly. Pediatr Cardiol 2003; 24:86-88.
47. Bonaime JL, Bonne B, Joannard A et al. Le syndrome de dysostose spondylothoracique ou spondylo-costale. Pédiatrie 1978; 33:173-188.
48. Wadia RS, Shirole DB, Dikshit MS. Recessively inherited costovertebral segmentation defect with mesomelia and peculiar facies (Covesdem syndrome). J Med Genet 1978; 15:123-127.
49. Young ID, Moore JR. Spondylocostal dysostosis. J Med Genet 1984; 21:68-69.
50. Van Thienen M-N, Van der Auwera BJ. Monozygotic twins discordant for spondylocostal dysostosis. Am J Med Genet 1994; 52:483-486.
51. Martínez-Frías ML, Bermejo E, Paisán L et al. Severe spondylocostal dysostosis associated with other congenital anomalies: a clinical/epidemiological analysis and description of ten cases from the Spanish Registry. Am J Med Genet 1994; 51:203-212.
52. Eller JL, Morton JM. Bizarre deformities in offspring of user of lysergic acid diethylamide. New Eng J Med 1976; 283:395-397.

53. Devos EA, Leroy JG, Braeckman JJ et al. Spondylocostal dysostosis and urinary tract anomaly: definition and review of an entity. Eur J Paed 1978; 128:7-15.
54. Kozlowski K. Spondylo-costal dysplasia. Fortschr Röntgenstr 1984; 140:204-209.
55. Roberts AP, Conner AN, Tolmie JL et al. Spondylothoracic and spondylocostal dysostosis. J Bone Jt Surg 1988; 70B:123-126.
56. Giacoia GP, Say B. Spondylocostal dysplasia and neural tube defects. J Med Genet 1991; 28:51-53.
57. Murr MM, Waziri MH, Schelper RL et al. Case of multiple vertebral anomalies, cloacal dysgenesis and other anomalies presenting prenatally as cystic kidneys. Am J Med Genet 1992; 42:761-765.
58. Lin AE, Harster GA. Another case of spondylocostal dyplasia and severe anomalies (letter). Am J Med Genet 1993; 46:476-477.
59. Ohzeki T, Shiraishi M, Matsumoto Y et al. Sporadic occurrence of spondylocostal dysplasia and meso-cardia in a Japanese girl. Am J Med Genet 1990; 37:427-428.
60. Wynne-Davies R. Congenital vertebral anomalies: aetiology and relationship to spina bifida cystica. J Med Genet 1975; 12:280-288.
61. McLennan JE. Rib anomalies in myelodysplasia. Biol Neonate 1976; 29:129-141.
62. Naik PR, Lendon RG, Barson AJ. A radiological study of vertebral and rib malformations in children with myelomeningocele. Clin Radiol 1978; 29:427-430.
63. Lendon RG, Wynne-Davies R, Lendon M. Are congenital vertebral anomalies and spina bifida cystica aetiologically related? J Med Genet 1981; 18:424-427.
64. Sharma AK, Phadke SR. Another case of spondylocostal dysplasia and severe anomalies: a diagnostic and counseling dilemma. Am J Med Genet 1994; 50:383-384.
65. Martínez-Frías ML. Multiple vertebral segmentation defects and rib anomalies. Am J Med Genet 1996; 66:91.
66. Aymé S, Preus M. Spondylocostal/spondylothoracic dysostosis: the clinical basis for prognosticating and genetic counselling. Am J Med Genet 1986; 24:599-606.
67. Herold HZ, Edlitz M, Barochin A. Spondylothoracic dysplasia. Spine 1988; 13:478-481.
68. Reyes MC, Morales A, Harris V et al. Neural defects in Jarcho-Levin syndrome. J Child Neurol 1989; 4:51-54.
69. Bohring A, Lewin SO, Reynolds JF et al. Polytopic anomalies with agenesis of the lower vertebral column. Am J Med Genet 1999; 87:99-114.
70. Novak RW, Robinson HB. Coincident DiGeorge anomaly and renal agenesis and its relation to maternal diabetes. Am J Med Genet 1994; 50:311-312.
71. Riccardi VM. Trisomy 8: an international study of 70 patients. Birth Defects OAS 1977; XIII (3C):171-184.
72. Dowton SB, Hing AV, Sheen-Kaniecki V et al. Chromosome 18q22.2-qter deletion and a congenital anomaly syndrome with multiple vertebral segmentation defects. J Med Genet 1997; 34:414-417.
73. Nakano S, Okuno T, Hojo H et al. 18p- syndrome associated with hemivertebrae, fused ribs and micropenis. Jpn J Hum Genet 1977; 22:27-32.
74. Crow YJ, Tolmie JL, Rippard K et al. Spondylocostal dysostosis associated with a 46,XX,+15,dic(6;15) (q25;q11.2) translocation. Clin Dysmorphol 1997; 6:347-350.
75. De Grouchy J, Mlynarski JC, Maroteaux P et al. Syndrome polydysspondylique par translocation 14-15 et dyschondrostéose chez un même sujet. Ségrégation familiale. C R Acad Sci [D] Paris 1963; 256:1614-1616.
76. Joutel A, Tournier-Lasserve E. Notch signalling and human disease. Cell Dev Biol 1998; 9:619-625.
77. Gridley T. Notch Signaling in Vertebrate Development and Disease. Mol Cell Neurosci 1997; 9:103-108.
78. Gridley T. Notch signaling and inherited disease syndromes. Hum Mol Genet 2003; 12 Rev 1: R9-R13.
79. Pourquié O, Kusumi K. When body segmentation goes wrong. Clin Genet 2001; 60:409-416.
80. Harper JA, Yuan JS, Tan JB et al. Notch signaling in development and disease. Clin Genet 2003; 64:461-472.
81. Turnpenny PD, Thwaites RJ, Boulos FN. Evidence for variable gene expression in a large inbred kindred with autosomal recessive spondylocostal dysostosis. J Med Genet 1991; 28:27-33.
82. Turnpenny PD, Bulman MP, Frayling TM et al. A gene for autosomal recessive spondylocostal dysostosis maps to 19q13.1-q13.3. Am J Hum Genet 1999; 65:175-182.
83. Giampietro PF, Raggio CL, Blank RD. Synteny-defined candidate genes for congenital and idiopathic scoliosis. Am J Med Genet 1999; 83:164-177.
84. Grüneberg H. Genetical studies on the skeleton of the mouse. Genet Res Camb 1961; 2:384-393.
85. Dunwoodie SL, Henrique D, Harrison SM et al. Mouse Dll3: a novel divergent Delta gene which may complement the function of other Delta homologues during early pattern formation in the mouse embryo. Development 1997; 124:3065-3076.

86. Kusumi K, Sun ES, Kerrebrock AW et al. The mouse pudgy mutation disrupts Delta homologue Dll3 and initiation of early somite boundaries. Nature Genet 1998; 19:274-278.

87. Bulman MP, Kusumi K, Frayling TM et al. Mutations in the human Delta homologue, DLL3, cause axial skeletal defects in spondylocostal dysostosis. Nature Genet 2000; 24:438-441.

88. McGrew MJ, Pourquié O. Somitogenesis: segmenting a vertebrate. Curr Opinion Genet Dev 1998; 8:487-493.

89. Pourquié O. Notch around the clock. Curr Opinion Genet Dev 1999; 9:559-565.

90. Martínez-Frías ML. Segmentation Anomalies of the Vertebrae and Ribs: One Expression of the Primary Developmental Field. Am J Med Genet 2004; 128A:127-131.

91. Kusumi K, Dunwoodie SL, Krumlauf R. Dynamic expression patterns of the pudgy/spondylocostal dysostosis gene Dll3 in the developing nervous system. Mech Dev 2001; 100:141-144.

92. Dunwoodie SL, Clements M, Sparrow DB et al. Axial skeletal defects caused by mutation in the spondylocostal dysplasia/pudgy gene Dll3 are associated with disruption of the segmentation clock within the presomitic mesoderm. Development 2002; 129:1795-1806.

93. Floor E, De Jong RO, Fryns JP et al. Spondylocostal dysostosis: an example of autosomal dominant inheritance in a large family. Clin Genet 1989; 36:236-241.

94. Whittock NV, Ellard S, Duncan J et al. Pseudo-dominant inheritance of spondylocostal dysostosis type 1 caused by two familial delta-like 3 mutations. Clin Genet 2004; 66:67-72.

95. Kusumi K, Stevens SA, Mimoto MS et al. Dll3-Notch1 double mutant mice are a model for congenital scoliosis and craniofacial disorders. Am J Hum Genet 2003; 73 (Suppl 5):A51:172.

96. Turnpenny PD, Whittock NV, Duncan J et al. Novel mutations in DLL3, a somitogenesis gene encoding a ligand for the Notch signalling pathway, cause a consistent pattern of abnormal vertebral segmentation in spondylocostal dysostosis. J Med Genet 2003; 40: 333-339.

97. Maisenbacher MK, Han JS, O'Brien ML et al. Molecular analysis of congenital scoliosis: a candidate gene approach. Hum Genet 2005; 116(5):416-419.

98. Giampietro PF, Raggio CL, Reynolds C et al. DLL3 as a candidate gene for vertebral malformations. Am J Med Genet A 2006; 140(22):2447-2453.

99. Saga Y, Hata N, Koseki H et al. Mesp2: a novel mouse gene expressed in the presegmented mesoderm and essential for segmentation initiation. Genes Dev 1997; 11:1827-1839.

100. Whittock NV, Sparrow DB, Wouters MA et al. Mutated MESP2 causes spondylocostal dysostosis in humans. Am J Hum Genet 2004; 74:1249-1254.

101. Massari ME, Murre C. Helix-loop-helix proteins: regulators of transcription in eucaryotic organisms. Mol Cell Biol 2000; 20:429-440.

102. Saga Y, Hata N, Kobayashi S et al. MesP1: a novel basic helix-loop-helix protein expressed in the nascent mesodermal cells during mouse gastrulation. Development 1996; 122:2769-2778.

103. Haraguchi S, Kitajima S, Takagi A et al. Transcriptional regulation of Mesp1 and Mesp2 genes: differential usage of enhancers during development. Mech Dev 2001; 108:59-69.

104. Kitajima S, Takagi A, Inoue T et al. MesP1 and MesP2 are essential for the development of cardiac mesoderm. Development 2000; 127:3215-3226.

105. Saga Y, Miyagawa-Tomita S, Takagi A et al. MesP1 is expressed in the heart precursor cells and required for the formation of a single heart tube. Development 1999; 126:3437-3447.

106. Sparrow DB, Jen WC, Kotecha S et al. Thylacine 1 is expressed segmentally within the paraxial mesoderm of the Xenopus embryo and interacts with the Notch pathway. Development 1998; 125:2041-2051.

107. Saga Y. Genetic rescue of segmentation defect in MesP2-deficient mice by MesP1 gene replacement. Mech Dev 1998; 75:53-66.

108. Culbertson MR. RNA surveillance. Unforeseen consequences for gene expression, inherited genetic disorders and cancer. Trends in Genetics 1999; 15:74-80.

109. Takahashi Y, Koizumi K, Takagi A et al. Mesp2 initiates somite segmentation through the Notch signalling pathway. Nat Genet 2000; 25:390-396.

110. Takahashi Y, Inoue T, Gossler A et al. Feedback loops comprising Dll1, Dll3 and Mesp2 and differential involvement of Psen1 are essential for rostrocaudal patterning of somites. Development 2003; 130:4259-4268.

111. Haines N, Irvine KD. Glycosylation regulates Notch signalling. Nat Rev Mol Cell Biol 2003; 4:786-97.

112. Bruckner K, Perez L, Clausen H et al. Glycosyltransferase activity of Fringe modulates Notch-Delta interactions. Nature 2000; 406:411-5.

113. Moloney DJ et al. Fringe is a glycosyltransferase that modifies Notch. Nature 2000; 406:369-75.

114. Kusumi K et al. Dll3 pudgy mutation differentially disrupts dynamic expression of somite genes. Genesis 2004; 39:115-21.

115. Sparrow DB, Chapman G, Wouters MA et al. Mutation of the LUNATIC FRINGE gene in humans causes spondylocostal dysostosis with a severe vertebral phenotype. Am J Hum Genet 2006; 78:28-37.

116. Correia T et al. Molecular genetic analysis of the glycosyltransferase Fringe in Drosophila. Proc Natl Acad Sci USA 2003; 100:6404-9.
117. Chen J, Moloney DJ, Stanley P. Fringe modulation of Jagged1-induced Notch signaling requires the action of beta 4galactosyltransferase-1. Proc Natl Acad Sci USA 2001; 98:13716-21.
118. Alagille D, Habib EC, Thomassin N. L'atresie des voies biliaires intrahepatiques avec voies biliaires extrahepatiques permeables chez l'enfant. J Par Pediatrie 1969; 301-318.
119. Alagille D, Odievre M, Gautier M et al. Hepatic ductular hypoplasia associated with characteristic facies, vertebral malformations, retarded physical, mental and sexual development and cardiac murmur. J Pediatr 1975; 86:63-71.
120. Alagille D, Estrada A, Hadchouel M et al. Syndromic paucity of interlobular bile ducts (Alagille syndrome or arteriohepatic dysplasia): review of 80 cases. J Pediatr 1987; 110:195-200.
121. Kamath BM, Loomes KM, Oakey RJ et al. Facial features in Alagille syndrome: Specific or cholestasis facies? Am J Med Genet 2002; 112:163-170.
122. Kamath BM, Bason L, Piccoli DA et al. Consequences of JAG1 mutations. J Med Genet 2003; 40:891-895.
123. Emerick KM, Rand EB, Goldmuntz E et al. Features of Alagille syndrome in 92 patients: frequency and relation to prognosis. Hepatology 1999; 29:822-829.
124. Robert MLP, Lopez T, Crolla J et al. Alagille syndrome with deletion 20p12.2-p12.3 and hypoplastic left heart. Clin Dysmorphol (in press).
125. Eldadah ZA, Hamosh A, Biery NJ et al. Familial tetralogy of Fallot caused by mutation in jagged1 gene. Hum Mol Genet 2001; 10:163-169.
126. Li L, Krantz ID, Deng Y et al. Alagille syndrome is caused by mutations in human Jagged1, which encodes a ligand for Notch1. Nature Genet 1997; 16:243-251.
127. Oda T, Elkahloun AG, Pike BL et al. Mutations in the human Jagged1 gene are responsible for Alagille syndrome. Nature Genet 1997; 16:235-242.
128. Krantz ID, Colliton RP, Genin A et al. Spectrum and frequency of Jagged1 (JAG1) mutations in Alagille syndrome patients and their families. Am J Hum Genet 1998; 62:1361-1369.
129. Crosnier C, Driancourt C, Raynaud N et al. Mutations in JAGGED1 gene are predominantly sporadic in Alagille syndrome. Gastroenterology 1999; 116:1141-1148.
130. Krantz ID, Rand EB, Genin A et al. Deletions of 20p12 in Alagille syndrome:frequency and molecular characterization. Am J Med Genet 1997; 70:80-86.
131. Boyer J, Crosnier C, Driancourt C et al. Expression of mutant JAGGED1 alleles in patients with Alagille syndrome. Hum Genet 2005; 116:445-53.
132. Kodama Y, Hijikata M, Kageyama R et al. The role of notch signaling in the development of intrahepatic bile ducts. Gastroenterology 2004; 127:1775-86.
133. McCright B, Lozier J, Gridley T. A mouse model of Alagille syndrome: Notch2 as a genetic modifier of Jag1 haploinsufficiency. Development 2002; 129:1075-82.
134. Loomes KM, Taichman DB, Glover CL et al. Characterization of Notch receptor expression in the developing mammalian heart and liver. Am J Med Genet 2002; 112(2):181-9.
135. McDaniell R, Warthen DM, Sanchez-Lara PA et al. NOTCH2 mutation cause Alagille syndrome, a heterogeneous disorder of the notch signaling pathway. Am J Hum Genet 2006; 79:169-173.
136. Kamath BM, Spinner NB, Emerick KM et al. Vascular Anomalies in Alagille Syndrome. A Significant Cause of Mortality and Morbidity. Circulation 2004; 109:1354-1358.
137. Joutel A, Corpechot C, Ducros A et al. Notch3 mutations in CADASIL, a hereditary adult-onset condition causing stroke and dementia. Nature 1996; 383:707-710.
138. Xue Y, Gao X, Lindsell CE et al. Embryonic vascular lethality and vascular defects in mice lacking the Notch ligand Jagged1. Hum Mol Genet 1999; 8:723-730.
139. Krebs LT, Xue Y, Norton CR et al. Notch signaling is essential for vascular morphogenesis in mice. Genes Dev 2000; 14:1343-1352.
140. Libbrecht L, Cassiman D, Desmet V et al. The correlation between portal myofibroblasts and development of intrahepatic bile ducts and arterial branches in human liver. Liver 2002; 22:252-258.
141. Sparrow DB, Clements M, Withington SL et al. Diverse requirements for Notch signalling in mammals. Int J Dev Biol 2002; 46:365-374.
142. Bonafé L, Giunta C, Gassner M et al. A cluster of autosomal recessive spondylocostal dysostosis caused by three newly identified DLL3 mutations segregating in a small village. Clin Genet 2003; 64:28-35.

INDEX